INTRODUCTION TO SYSTEMS PHILOSOPHY

ERVIN LASZLO

INTRODUCTION TO SYSTEMS PHILOSOPHY

Toward a New Paradigm of Contemporary Thought

With a Foreword by

Ludwig von Bertalanffy

HARPER TORCHBOOKS
Harper & Row, Publishers
New York, Evanston, San Francisco, London

For Christopher and Alexander

A hardcover edition of this book was originally published by Gordon and Breach, Science Publishers in 1972. It is reprinted here by arrangement.

First HARPER TORCHBOOK edition published 1973.

STANDARD BOOK NUMBER: 06–131762–4

PREFACE TO THE TORCHBOOK EDITION

In the year and a half that elapsed since the manuscript for the original hard-cover edition of this book was completed, a number of relevant works have been published in systems theory, science, and philosophy. I have also received a number of comments, reviews, and discussion articles which raised interesting and important points. In the present edition reference has been incorporated to the most immediately pertinent recent publications and revisions have been made where I judged them to be necessary, in the light of my own thinking and the ideas of my commentators. I am grateful to Mr. Hugh Van Dusen, my editor at Harper's, for his help in preparing this edition.

Switzerland
July 1972

PREFACE

I came to philosophy in my mid-twenties because I had questions to which I wanted answers, rather than in the late teens because it seemed like a good profession to enter upon. Thus I avoided the overkill typical of the majority of academic philosophy programs: brainwashing with complex theories from the history of thought which largely fail to relate to anything meaningful in the student's experience. I was old enough, and my questions were pressing enough, to make me choose my philosophers with care. I paged through the history of philosophy, and adopted Aristotle as my first tutor; later I went back to Plato. At the same time I could not help feeling that many of the questions they were grappling with might benefit from the findings of the contemporary sciences, and I began to read Jeans, Eddington, Einstein, and also Driesch, Mach and Pavlov. I was struck by the tremendous refinement of the answers proposed by great philosophers, and the equally tremendous wealth of information offered by great scientists. However, I was still not getting satisfactory answers to my questions, because the philosophical answers lacked an adequate factual basis, and the scientific answers tended to be either restrictive or naïvely generalizing from the viewpoint of a specialty. There must be someone on the contemporary scene who is capable of combining philosophic astuteness with scientific informedness, I thought; and eventually I came across Whitehead. In his "philosophy of organism" I believed I had found an answer meriting sustained consideration. Here was a wealth of data from the sciences used as the foundation for a many-sided philosophical synthesis, telling me what I wanted to know: what the nature of the world is into which I was born, and what am I, if more than an ephemeral burst of unexplored consciousness. That life is a "tale told by an idiot" I could not accept, nor did it seem reasonable to me that I can find out about life and the world by introspecting on my own experience. Hence I asked for a considered philosophical synthesis of the best knowledge available in our day, and I embarked on the quest of finding a synthesis.

Whitehead was illuminating, but not the final answer. For one thing, his ultimate principles were debatable—God, pure possibility, conceptual prehension and related concepts, permit of alternative solutions. For another, the gathering of scientific information did not come to a stop in the first half of this century, but led to the accumulation of incomparably richer storehouses of tested knowledge. Hence I concluded that the Whiteheadian synthesis has to be done over again, in the light of contemporary findings, and perhaps without the superstructure of metaphysical principles gained by personal insight alone. Somewhat foolhardily (I doubt if I would start such an undertaking again if I could relive those days), I set about putting together notes for my own philosophical synthesis of contemporary scientific theories. First called "A Theory of Organic Relations," my notes eventually took on the character of a Whiteheadian process philosophy, centered on man and society, but founded on the natural scientific world picture. Earlier, they went through a series of metamorphoses—Hegelian, Marxian, positivist. After three years of constant and rather lonely struggle, I showed the notes (which were already in a neo-Whiteheadian stage of transformation) to a philosophy publisher in The Hague, and, to my great surprise, he offered to publish them. I set about consolidating them in a more systematic fashion and, feeling for the first time in my life that my thoughts might have some function apart from answering my own questions and satisfying my personal thirst for meaning, I sent off my manuscript—which was eventually published as *Essential Society: An Ontological Reconstruction.*[1]

The first book was followed by several others, as I plunged wholeheartedly into the search for competent and meaningful synthesis. After working for some years in Switzerland, relatively isolated from thinkers with similar goals, I discovered, to my delight and amazement, that in America my modest efforts were paralleled by several eminent investigators in the still nebulous integrative and interdisciplinary areas. During the year of my fellowship at Yale I had the good fortune to meet and collaborate with F. S. C. Northrop and Henry Margenau, through whom I came to the Foundation (now Center) for Integrative Education and its pioneering journal, *Main Currents in Modern Thought.* It was on the pages of a back issue of that journal that I rediscovered Ludwig von Bertalanffy—known to me as a biologist—as an integrative philosopher. I found, to my further delight and amazement, that the organic synthesis of Whitehead can be updated by the synthesis of a general systems theory, replacing the notion of "organism" and its Platonic correlates with the concept of a dynamic, self-sustaining "system" discriminated against the background of a changing natural environment. Using this concept, the impressive findings of the contemporary sciences can become meaningful and relevant,

[1]The Hague: Martinus Nijhoff, 1963.

and answers can be forthcoming to questions which surely every thoughtful person asks, sometime in his life.

The key to a more meaningful philosophy may be at hand, but it remains to be put to use. Surveying the contemporary scene around me today, I still cannot conclude otherwise than by noting that the majority of philosophers continue to offer highly refined but factually anaemic theories; and that scientists, in their increasingly frequent excursions into general philosophical problems, continue to be shackled by the optic of their particular specialties. There are but a handful of people, like those mentioned above and their collaborators, who combine scientific information with philosophic sophistication. Yet questions of meaning continue to press us and become, with frightening rapidity, questions of survival.

The present situation in the field of theory is characterized by the fruits of the scientific "information explosion" on the one hand, and by highly sophisticated methodologies and conceptual analyses on the other. The obvious task is to bring the two together. That means making philosophers aware of relevant scientific findings, and scientists aware of the relevant methods and conceptual frameworks of philosophy. Not all of science is relevant to philosophy, and not all of philosophy is relevant to science. But there is a wealth of scientific data that can function as a foundation of an informed philosophy; and there is a similar wealth of philosophic methods and concepts which constitute the necessary condition of the synthesis of scientific findings. The general systems theory, pioneered by von Bertalanffy, Kenneth Boulding, Anatol Rapoport and their collaborators, gives us a theoretical instrument for assuring the mutual relevance of scientific information and philosophic meaning. Extended into a general *systems philosophy,* this instrument can polarize the contemporary theoretical scene as a magnet polarizes a field of charged particles: by ordering the formerly random segments into a meaningful pattern. If made good use of, this instrument could channel to us a stream of informed as well as sophisticated answers through the cross-fertilization of contemporary science and philosophy.

This, then, is my credo; the conviction which led to the formulation of the present *Introduction* to systems philosophy over a period of three years of research in alternating states of euphoria and dejection. My personal thanks go first of all to my wife who had to put up with what must surely have seemed like a permanent obsession even, I am afraid, in our brief periods of vacation. I have received encouragement, advice, and constructive criticism from many friends and colleagues, of whom I would especially like to thank Ludwig von Bertalanffy, Henry Margenau, Errol E. Harris, Stephen C. Pepper, Lee Thayer, Rubin Gotesky, Jere Clark, Fritz Kunz and Ralph Burhoe; my colleagues at the Philosophy Department at SUNY

Geneseo; and the graduate students and auditors of my systems philosophy seminar at Northwestern University. I am grateful to Stacey Edgar for reading the proofs and making helpful suggestions. Last but not least, my thanks and appreciation to Martin Gordon and Edmund Immergut, my publishers, for their faith and imaginative support of this venture.

ACKNOWLEDGMENTS

Portions of this book have appeared prior to, or are appearing concurrently with, the publication of the book itself. "The Case for Systems Philosophy" *(Metaphilosophy,* April, 1972), is based on Chapters 1 and 9; "Systems Philosophy" *(Main Currents in Modern Thought,* November–December, 1971) derives mainly from Chapter 1; "Basic Constructs of Systems Philosophy" *(Systematics,* September, 1972) incorporates sections of Chapter 4; "A Systems Philosophical Approach to Values" *(Behavioral Science,* forthcoming) makes use of ideas advanced in Chapter 13; "Reverence for Natural Systems" *(Fields Within Fields...Within Fields,* Vol. 3, No. 1 [1970], reprinted in *Emergent Man: His Chances, Problems and Potentials,* Julius Stulman and Ervin Laszlo, eds., [1973]) is excerpted from Chapter 14; and "Notes on the Poverty of Contemporary Philosophy" (*Zygon,* Vol. 6, No. 1 [March, 1971]) includes material here presented in the Conclusions (pp. 296f.). I would like to thank the editors and publishers for granting the necessary permissions and making the indicated arrangements.

E.L.

CONTENTS

FOREWORD

by

LUDWIG VON BERTALANFFY

It gives me great pleasure to introduce the present work by Ervin Laszlo. This pleasure is both of a personal nature, and in seeing an important task ably performed.

The notion of "system" has gained central importance in contemporary science, society and life. In many fields of endeavor, the necessity of a "systems approach" or "systems thinking" is emphasized, new professions called "systems engineering," "systems analysis" and the like have come into being, and there can be little doubt that this concept marks a genuine, necessary, and consequential development in science and world view.

The present writer may be known for his work in general systems theory and as a founder of the "Society for General Systems Research." Thus speaking personally, it is with satisfaction that I see our author following a path I had pursued over so many years. It was in the late Twenties that I emphasized, under the title of "organismic biology," the necessity of regarding the living organism as an "organized system," and defined "the fundamental task of biology as discovery of the laws of biological systems at all levels of organization." This trend arising in biology and other disciplines led me to conceive the idea of general systems theory (introduced in the 1930s and '40s), that is, an interdisciplinary doctrine "elaborating principles and models that apply to systems in general, irrespective of their particular kind, elements, and 'forces' involved."

Already in 1936, "organismic biology" was used as the basis of an inductive metaphysics by the Berlin philosopher, C. Fries.* It needed time, however, and the joint effort of workers in many fields to develop "systems science" as a new way of "seeing" in science and technology, a scientific field and a philosophy.

Classical science in its diverse disciplines, be it chemistry, biology, psychology, or the social sciences, tried to isolate the elements of the observed

---

*C. Fries, *Metaphysik als Naturwissenschaft. Betrachtungen zu Ludwig von Bertalanffy's Theoretischer Biologie*. Berlin, 1936.

universe—chemical compounds and enzymes, cells, elementary sensations, freely competing individuals, as the case may be—expecting that by putting them together again conceptually or experimentally, the whole or system —cell, mind, society—would result and be intelligible. Now we have learned that for an understanding not only the elements but their interrelations as well are required: say, the interplay of enzymes in a cell or of many mental processes conscious and unconscious, the structure and dynamics of social systems, and so forth. This requires exploration of the many systems in our observed universe in their own right and specificities. Moreover, it turns out that there are correspondences or isomorphisms in certain general aspects of "systems." This is the domain of general systems theory. Indeed, such parallelisms or isomorphisms appear—sometimes surprisingly—in otherwise totally different "systems." General systems theory, then, is the scientific exploration of "wholes" and "wholeness" which, not so long ago, were considered metaphysical notions transcending the boundaries of science. Hierarchic structure, stability, teleology, differentiation, approach to and maintenance of steady states, goal-directedness—these are a few of such general-system properties; and novel conceptions and mathematical fields have been developed to deal with them: dynamic system theory, automata theory, system analysis by set, net, graph theory, and others. This interdisciplinary endeavor, of elaborating principles and models applying to "systems" in general, at the same time provides a possible approach toward unification of science.

This was paralleled by developments in technology which, in cybernetics, control engineering, computer work, etc., similarly led to models and principles of an interdisciplinary nature, applying to generalized "systems" or classes of systems independently of their realization as mechanical, electrical, biological, social (etc.) systems. N. Wiener's "Cybernetics" as "study of communication and control" was pioneering, and here too a crop of new approaches and theories—information, circuit, decision, game theory and others—developed. Modern technology and society have become so complex that traditional ways and means are not sufficient any more—approaches of a "holistic" (or systems) and generalist nature have to be introduced. This is true in many ways. Innumerable problems in industry, commerce, politics, etc., ask for a "systems approach" in automation, computerization, and so on, handling interacting systems of a complexity not amenable to classical methods of mathematics. Systems of many levels ask for scientific control: ecosystems, the disturbance of which results in pressing problems like pollution; formal organizations like a bureaucracy or army; the grave problems appearing in socio-economic systems, in international relations, politics and deterrence. Regardless of the question of how far scientific understanding (contrasted to the admisssion of irrationality in cultural and historical events) is possible, and to what extent scientific control is desirable, there can be no dispute that these are essentially "system" problems, that is, problems of interrelations of a great number of "variables."

Finally, now there is "systems philosophy." The concept of "system" constitutes a new "paradigm," to use Th. Kuhn's expression in his brilliant exposition of the "structure of scientific revolutions"; or a "new philosophy of nature," as I myself put it, contrasting the blind laws of nature of the mechanistic world view and the world process as a Shakespearean tale told by an idiot, with an organismic outlook of the "world as a great organization."

I had the privilege of making Ervin Laszlo's acquaintance only after his present great opus was, in principle, finished. It therefore is an independent and original work expressing the current trend in its own terms. This gives us the benefit of our author's ideas, yet the meaning and bearing of his "Systems philosophy" is, in a way, determined by the movement briefly outlined.

Systems philosophy, then, first must find out the "nature of the beast." This is the question of what is meant by "system," and how systems are realized in reality at the various levels of observation. In Laszlo's terms, this is the methodology and theory of *natural* systems. Secondly, there is epistemology, i.e., the methodology and theory of *cognitive* systems. Having been educated in the philosophical tradition later called the Vienna Circle, I may perhaps hint at the change that has taken place in this regard. The epistemology (and metaphysics) of logical positivism was determined by the ideas of physicalism, atomism, and the "camera theory" of knowledge. These, in view of present-day knowledge, are quite simply obsolete. Instead of physicalism and reductionism, the problems and modes of thought occurring in the biological, behavioral and social sciences should equally be taken into account. Compared to the analytical procedure of classical science, with linear causality connecting two variables as the basic category, the investigation of organized wholes of many variables requires new categories of interaction, transaction, teleology, etc., with many problems arising in epistemology, mathematical models and techniques. Furthermore, knowledge is not a simple approximation to "truth" or "reality"; it is an interaction between knower and known, thus depending on a multiplicity of factors of a biological, cultural, linguistic, etc., nature. This leads to a "perspective philosophy" for which physics, fully acknowledging its achievements in its own and related fields, is not a monopolistic way of knowledge. Against reductionism and theories declaring that reality is "nothing but" a heap of physical particles, genes, reflexes, drives or whatever the case may be, we see science as one of the "perspectives" that man, with his biological, cultural and linguistic endowment and bondage, has created to deal with the universe he is "thrown into," or rather to which he is adapted owing to evolution and history.

Finally, systems philosophy as such is obviously concerned with the relations of man and world and the perennial problems of philosophy. If nature is a hierarchy of organized wholes, the image of man will be different from what it is in a world of physical particles as the ultimate and only "true" reality governed by chance events. Rather, the world of symbols, values, social entities and cultures is something very "real"; and its embeddedness in a cosmic order of hierarchies is apt to bridge the opposition of C. P. Snow's "Two Cultures" of science and the humanities, technology and history, natural and social sciences or in whatever way the antithesis is formulated.

Laszlo's pioneering work develops systems philosophy both in breadth and depth. As he argues convincingly, contemporary "analytic" philosophy is in danger of "analyzing itself out of existence"; to analyze the meaning of a term or, even better, to analyze the analysis of the meaning of a term as offered by a philosopher of current repute may make the wheels of academic philosophy go round and produce the required number of master's and doctoral theses; but it is a somewhat boring enterprise and rather frivolous in its unconcern with the issues that are at the root of our present crises, from the dangers of the "megamachine" of modern technology to the disbalance of the "ecosytem" of nature to the innumerable psychological, social, economic and political problems of the present.

What we need, says Laszlo, is rather a "synthetic" philosophy, that is, one which receives new inputs from the various developments in modern science and tries to follow the other way in philosophy, namely, endeavors to put together the precious pieces of specialized knowledge into a coherent picture—hazardous as such an enterprise may be. Rather than a sum whose components are exhaustively described by the specialized disciplines, our observed universe appears as "an interconnected system of nature"; and disregard of this basic fact leads to dire consequences, as well-known examples, from DDT to the discontent with technological-commercialist society, attest. The specialist's presupposition is a "pluralistic" universe whose parts are isolable enough to permit independent exploration in the conventional disciplines; but this needs as a complement the generalist's vision of systemic order and interdependence. This latter presupposition is neither void nor mere speculation; for the various system sciences have shown that there are concepts, models, and invariants revealing a general order that transcends the more or less special ones in the conventional sciences. Let us be aware that this view, too, is a "perspective" of reality, determined and limited by our human bondage; but so, after all, are the conventional and orthodox disciplines. Thus, new concepts, categories and structures emerge in the contemporary sciences, from physics to sociology, centered around general invariants like whole, system, and its various specifications. As Laszlo points out, they present the outlook of a synthetic philosophy whose "data come from the empirical sciences, problems from the history of philosophy, and concepts from modern systems research." And if a good part of our present trouble results from the encapsulation of modern man in different, water-tight compartments—man as an anthropological, psychological, economic unit while forgetting "man" as living and experiencing whole—such new conception, synthesizing different aspects and perspectives in terms common to all of them, may well contribute toward the resolution of contemporary problems whose "bracketing out" led to the sterility of conventional analytic philosophy.

Laszlo's work is the first comprehensive treatise of "systems philosophy." No one who looks beyond his own specialty and narrow interests will be able to deny the legitimacy of this quest. A bold enterprise, obviously, and one requiring broad scholarship and a mind at the same time synthetic and critical. But these are characteristics our author brings to his task. We sincerely wish that his work finds the broadest possible acceptance which, we do not doubt, will much contribute to rescue philosophy at our schools from isolation and scholasticism, and bring it back into the current of recent science and of the many urgent problems with which science and society are confronted.

Introduction

THE IDEA OF SYSTEMS PHILOSOPHY

Chapter 1

WHY SYSTEMS PHILOSOPHY?

Some reasons, for synthetic philosophy generally

The persistent theme of this study is the timeliness and the necessity of a return from analytic to synthetic philosophy. Analysis has performed its important task of doing away with unverified and unverifiable speculation, and insistence on its methods now yields but professional *self*-analysis: philosophizing with increasing logic but decreasing substance. Philosophy, like any other non-axiomatic field, needs a constant input of fresh empirical information. But, by and large, the channels of input into contemporary philosophy have been cut off—those that existed were largely analyzed out of existence and few new ones were allowed to develop. Information gained in fields other than philosophy was considered suspect—it was to be purified and analyzed for meaning. However, more needs to be done with such information than its logical analysis. Insistence on the latter brings about the paradoxical situation that extra-philosophical inquiries are often more philosophical than philosophy. A rose by any other name is still a rose, and philosophical inquiry by the name of science (or whatever else) is still philosophical. Lest philosophers analyze themselves out of philosophy, a return must be effected to synthesis: not as unwarranted speculation, but as a considered treatment of information of extra-philosophical origin. Synthesis can mean the conjoining of various sets of non-philosophically researched data, to furnish new avenues toward the constructive discussion of substantive philosophical issues.

No doubt, such endeavor is not really "safe." To analyze the meaning of a term, or even better, to analyze the analysis of the meaning of a term offered by some philosopher of current repute, is a less risky endeavor. It is grist down the mill of thousands of budding philosophers in the Anglo-Saxon world. It gives them something to comment upon and thus makes the wheels of professional philosophy go around. Unfortunately, these wheels have less and less of real substance to grind. It is not much of a service to the cause of philosophy, conceived as an inquiry into truth and reality, to add to the current technical literature. It is more of a service, if a much more risky one, to step outside the currently popular modes of philosophizing and help to call synthetic (yet carefully reasoned) philosophy back to life.

(i) *Extrinsic reasons*

I shall offer both extrinsic and intrinsic reasons for synthetic and well-informed philosophy. Consider some extrinsic reasons first.

Analysis demands specialization and the corresponding splitting of the philosophical camp into specialists of its various branches.[1] An analogous process of specialization obtains in the other cognate disciplines. As a result, the fields of knowledge are worked in patches, each man concerned with no more than his own territory, "cultivating his own garden." This may be the indicated and necessary method for exploring phenomena in depth, within limits imposed by the investigator himself. But, unfortunately, natural phenomena do not come in patches. The findings overlap and criss-cross. Cultivating one patch reacts upon another in quite unexpected ways. The world is not a series of gardens suitable for cultivation by friendly but independent neighboring gardeners. It is much more like a system or network in which the knowledge of one element presupposes familiarity with all others. And even if this presupposition diminishes in import beyond specifiable limits, the *effects* of interventions based on limited knowledge transcend it and embrace much wider boundaries. This is true of science as well as of philosophy. For example, it is sufficient for the practical-theoretical purpose of gaining further knowledge of chemical affinities between synthetic substances and organic reactions to study the effect of pesticides upon living organisms, but the effect of *using* the information thus gained to eliminate certain species of insects is felt in domains of inquiry as far removed as economics and politics—not to mention the more neighboring disciplines of ecology, both human and animal. If knowledge is to be gathered, and if it is to be put to use to reach predicted ends, the patchwork approach must be overcome. We appear to be part of an interconnected system of nature, and unless informed 'generalists' make it their business to develop systematic theories of the patterns of interconnection, our short-range projects and limited controlabilities may lead us to our own destruction. As Wiener warned us when he compared contemporary technology to the magic of the monkey's paw, which fulfills all desires but brings misery to its owner who did not know what to wish for, we may be asking things of our world which it can fulfill by appeasing an immediate want or need, but we may, in the long run,

---

[1]Cf. Hans Reichenbach's statement in *The Philosophy of John Dewey,* Paul Schilpp, ed. New York, 1939 and 1951.

frustrate other needs and upset our very conditions of existence. We may be on the brink of outprogramming ourselves.[2] And to whom should we appeal for relief from our predicament? The majority of contemporary Western philosophers are content to analyze its conceptual and linguistic problems and presuppositions. It is increasingly the theoretical scientist and concerned humanist and educator who take it upon themselves to attack the problem constructively and propose solutions. Yet, are these not problems which deserve the best efforts of *philosophers*?

Synthesis in the fields of knowledge has yet another crucial task to fulfill. It has to help us to overcome what has been described as the "existential vacuum."[3] An increasing number of people are forced into psychiatric care due to a failure to find meaning in life. They complain of an inner emptiness, a sense of total and ultimate meaninglessness. People are not guided by their instincts, as animals, and cannot find satisfaction in carrying out genetically programmed patterns of behavior. In earlier epochs they were guided by synthetic modes of thought which rested in part on faith and imagination; but the great myths of former ages and the religions of our immediate heritage have lost their cogency to millions. According to the "ideologues," they are capable of being replaced by action-oriented ideologies, such as Nazism and Communism, which present a total world-view with explicit directives for action. But Nazism was a perversion on the level of social morality and had to be suppressed, and Communism manifests its full power to give meaning to existence only when the revolution requested by its leaders is in full swing. Thus the "existential vacuum" envelops all but the actively rebelling and militant nations of the world: the Maoists do not complain of a sense of emptiness, nor do the Israelis, who likewise must fight and have a thoroughly understood reason to live. But the so-called "advanced" societies of the world, those, namely, where the struggle for existence and for social reform has yielded to technologically effected, relatively stable well-being, fail to offer their members meaningful reasons for their existence. As a result, strains and stresses are produced which manifest themselves in such phenomena as violence, anarchism, and political witch-hunts—directed against mostly imaginary scapegoats ("capitalists" or "communists" or merely "administrators"), and which generate intense interest

[2]I spoke of this danger in more detail in "Human Dignity and the Promise of Technology," *The Philosophy Forum,* Vol. IX, No. 3/4. (Reprinted in *Human Dignity: This Century and the Next,* Gotesky and Laszlo, eds., New York, London, Paris, 1971.)

[3]Viktor E. Frankl, "Reductionism and Nihilism," *Beyond Reductionism: New Perspectives in the Life Sciences,* Arthur Koestler and J. R. Smythies, eds., London and New York, 1969.

in traditional religion (typically in Communist countries) as well as in Oriental religions and mysticism (in the West). One American observer recently spoke of a "spiritual explosion" and a "search for the transcendental answers to fundamental questions of human life" and held that these have now intensified beyond measure. "Bookstores are crammed with Eastern sacred texts, studies of astrology, reincarnation, states of consciousness, and the like. Students across the country are demanding courses in Buddhism, Hinduism, and mysticism . . . Psychiatrists, psychologists, and clergymen of all faiths are joining the younger generation in this pursuit."[4]

The demand for "seeing things whole" and seeing the world as an interconnected, interdependent field or continuum, is in itself a healthy reaction to the loss of meaning entailed by overcompartmentalized research and piecemeal analysis, bringing in particularized "facts" but failing in relevance to anything of human concern. In the 19th Century, the existential vacuum was provoked by the then fashionable attitude of nihilism. Today, as Frankl points out, nihilism is out of fashion but it reappears under the guise of reductionism, the typical mark of specialization. ". . . I would say that reductionism today is a mask for nihilism. Contemporary nihilism no longer brandishes the word nothingness; today nihilism is camouflaged as *nothing-but-ness*."[5] Frankl is mainly concerned with the reduction of human phenomena to unconscious drives and aggressions and these in turn to physiological mechanisms. Yet reductionism is also the more widespread phenomenon of tracing observed or inferred processes to their smallest parts and explaining them by their causal interaction. Such practice, while contributing much noteworthy detailed knowledge of isolated events, leaves out of consideration larger interconnections which may be decisive for the understanding of the phenomena. It brings about specialization with all its attendant features: special languages, methods, constructs, foci of attention, and so on. In short, reductionism generates a multiplicity of limited-range theories, each of which applies to a small domain of highly specific events, but says nothing about the rest. (And if the specialist does use his particular theory to explain events outside its scope, he becomes a *"terrible generalisateur,"* which, replacing the legendary *"terrible simplificateur,"* is the mark of the generalizing specialist.[6]) In consequence the accumulation of highly specific bits of knowledge fails to give meaning to wider chunks of experience and does nothing to fill the present existential vacuum. Specialized science is simply irrelevant to the ques-

---

[4]Jacob Needleman, *The New Religions*, New York, 1970.
[5]Frankl, *op. cit.*, p. 398.
[6]*Ibid.*, p. 397.

tion of meaning in life. But the latter cannot be dismissed with a wave of the hand, as specialists tend to do; there are good indications that there is such a thing as a "will to meaning" in man as one of his most basic motivational forces, and that some 20% of contemporary neuroses are due to its frustration.[7]

There is a need today for bringing fresh and reliable empirical information into philosophy; for overcoming the patchwork approach in the use of knowledge as a means of safeguarding ourselves against disaster due to ignorance of systemic interconnections in nature; and for gaining insight into the general patterns of existence in this world as a means of providing meaning for the brute fact of being here. All this requires the resuscitation of a mode of rational and systematic thinking which has fallen into disrepute through overinsistence on detailed investigation and specialization.

Without disparaging the contribution to dependable science of carefully delimited specialized inquiry, let it be said, with Maslow, that of the two modes of thinking, the "atomistic" and the "holistic," it is the holistic which is the sign of the healthy, self-actualizing person. Insistence on the atomistic mode is in itself a form of mild psychoneurosis.[8] Among contemporary scientists and philosophers it is often a defense reaction—it is the most justifiable way of waving away questions for which their theories provide no answers. But there must be informed minds among us who take upon themselves the no doubt thankless task of attempting to see through the web of unclearly related special theories, and effect a *fusion* where today we only have *confusion*.

I have now made a brief plea for synthetic philosophy on extrinsic grounds, and I shall not press it further. Let the reader consider whether it is a needed corrective for the impoverishment of philosophy nurtured on an analytic diet, and whether it could help him and his fellow human beings to see the way out from the increasingly complex and darkening labyrinth leading toward the 21st Century—but let him then, having decided to give some of his no doubt precious time to it, weigh it on its own grounds, judge it for what it is worth in itself. Hence I will now argue that synthetic philosophy is as good as (regardless of whether it is extrinsically preferable to) limited-scope specialized inquiries, not on the grounds of what it does for philosophy or humanity, but in view of the consideration that there are no better reasons for the specialized inquiry than there are for the generalized one.

---

[7]Cf. the results of the "purpose-in-life (PIL) test," described by Frankl, *op. cit.*
[8]Abraham H. Maslow, *Motivation and Personality,* New York, 1970. Preface.

(ii) *Intrinsic reasons*

Coherent and systematic theories of the empirical world are based on two "primary presuppositions":

1. *The world exists;*

and

2. *The world is, at least in some respects, intelligibly ordered* (*open to rational inquiry*).

All empirical scientists share these presuppositions, as do most, but not all, empirical philosophers. Those that do not, believe in being able to demonstrate them as tenets grounded in some indubitable fact of experience, or as a necessary precondition of consciousness and knowledge (e.g., the Cartesian *cogito,* the Kantian *a priori* and the Husserlian *apodeicticity*). But a theory which would indubitably and consistently derive these tenets from the grounds of mind and experience still awaits its formulation. For the present, we are obliged to consider them as presuppositions, lying at the foundations of all coherent and systematic theories of the empirical world. (At this point I have very likely lost the more dogmatic among my Kantian, Cartesian, phenomenological and existentialist readers, but that cannot be helped.) The "epistemological turn" introduced into modern philosophy by Descartes and Kant provides effective checks for some philosophers, preventing them from systematically considering the laws and principles of the empirical world. Yet, as long as the existence and the rational knowability of that world are not assumed, but either questioned or sought to be derived from some prior datum of mind and experience, potentially valuable energies are wasted on problems which do not permit of solutions, only of sophistries. However, once these minimal primary presuppositions are permitted, the door is opened to the rational mapping of the empirical world: theory construction can begin. Such construction need not be a naïve proffering of various schemes for explaining the empirical world, but can be a critical enterprise, proceeding by examining each step along the way.

In this critical yet "bold" spirit we consider the next two presuppositions, one or the other of which is likewise the *sine qua non* of systematic empirical theories. These are the "secondary presuppositions":

(i) *The world is intelligibly ordered in special domains;*

or

(ii) *The world is intelligibly ordered as a whole.*[9]

These presuppositions are the watershed separating the "specialist" from the "generalist." One or the other presupposition is made by every investi-

[9]This is not to be read to suggest that *all* of the world is intelligibly ordered, only that whatever intelligible orders are the case for it extend over the entire world.

gator, but, as a rule, those that make the first one refuse to admit that it is a presupposition. They tend to look upon the intelligibility of the phenomena they investigate as a "fact of nature." On the other hand, the second presupposition is usually considered to be in need of demonstration. The specialist limits his inquiry to a strategically isolated genus of events and assumes that his domain is intelligible, but often refuses to assume that it is also intelligibly interwoven with all other domains and not just the neighboring ones. The generalist makes precisely this assumption and holds that the knowledge of events in any one domain becomes fully intelligible when brought into conjunction with the knowledge of events in the other domains. "The point is," said Whitehead, "that every proposition refers to a universe exhibiting some general systematic metaphysical character. . . Thus every proposition posing a fact must, in its complete analysis, propose the general character of the universe required for that fact."[10]

The burden of proof is unjustly distributed among these presuppositions. One is considered a "given" of the world which it would be foolish to deny; the other is accused of being metaphysical (meaning "untenable" or, at best, "based on unconfirmed speculation") and in need of empirical confirmation. Yet when we acknowledge that both presuppositions rest on the "primary presuppositions" of the existence and the intelligibility of the world, then demands for proof will be tempered and equalized. We may never know with absolute certainty whether or not the empirical world truly exists, and whether, even if it does, it is open to being grasped in its very "thusness" through rational inquiry. But we do know that rational theories can be built of many aspects of the world and that they serve to explain and predict many events. We also do know that in addition to a large number of special theories, applying to limited domains only, an increasing number of general theories are making their appearance today, and these apply to the empirical world as such, or at least to very large chunks of it. If all these theories rest on prior presuppositions (and if these presuppositions are either allowed or the inquiry bogs down in the epistemological tangle), then neither special nor general theories have a privileged apodeictic status.

It might be countered, however, that special theories grasp the empirical world with a relatively higher degree of precision and truth than general theories. The objection is based on the assumption that special orders are intelligible in themselves, or at least that they are not rendered significantly more intelligible by consideration of general, overarching orders. Yet the truth of this assumption is directly dependent on the prior choice between the "secondary presuppositions" : the assumption is true if the first presupposition is affirmed, and false in the light of the second.

[10] A. N. Whitehead, *Process and Reality,* New York, 1929, Part I, Ch. I, Sect. V.

For, if the world is intelligibly ordered as a whole, then the more regions of order are disclosed by a theory, the more that theory is divested of the personal bias of the investigator. We know, since the work of von Uexkuell, Whorff and their followers, that each investigator grasps with immediate apprehension that "slice" of the world which is relevant to his psycho-physical constitution, culture, language and training. In a generally ordered world these "spectacles" could be removed proportionately to the discovery of orders appearing in many regions. The great thinkers of the past have realized this and set about purposively to explore general order. Their efforts are summed up in Hegel's dictum: *das Wahre ist das Ganze* (the true is the whole).

I am not suggesting that the second of the "secondary presuppositions" alone is true, only that we regard them both as presuppositions of equal heuristic potential. Attempts at constructing general theories of the empirical world are intrinsically neither better, nor worse, than attempts at producing special theories. And since extrinsically there are excellent reasons for championing general theories today, the synthesizing enterprise of the general investigator achieves not only validity, but also urgency.

Further reasons, for systems philosophy specifically

General theory is emerging these days from the workshops of the systems scientists. Their disciplines—general systems theory, cybernetics, information and game theory, etc.—come up with theories applicable to a wide range of empirical phenomena. Their common denominator is the systems concept as the basic conceptual category. Such concepts are the tools *par excellence* of general theory: their advantage over other concepts is that they are capable of remaining invariant where others encounter limits of applicability. That is, the range of their transformations (more exactly, the number of operations in regard to which they are invariant) is greater. Hence they can exhibit general order where the classical concepts show only delimited special orders.

Systems concepts may thus be thought about in terms of a general meta-language of scientific discourse. Miller points out that general systems terms make it easier to recognize similarities that exist in systems of different types and levels, whereas the specialist languages limit the horizons of thought to the borders of the discipline: "They mask important intertype and interlevel generalities and make general theory as difficult as it is to think about snow in a language that has no word for it."[11]

The special languages of classical science describe systemic interconnections at unique levels of organization and are unable to bridge differences

[11]James G. Miller, "Living Systems: Basic Concepts," *Behavioral Science,* 10:193–237, reprinted in *General Systems Theory and Psychiatry,* Gray, Duhl, and Rizzo, eds., Boston, 1969.

in levels. The concepts and laws of the more complex phenomena seem to "emerge" at that level, without prior connection with concepts and laws applicable to lower levels. For example, the state of a small systems of particles is fully described in terms of the positions and velocities of the particles and the forces that act between them. But at greater complexities of organization new laws need to be postulated; for example, those of the Pauli exclusion principle, Sommerhof's "directive correlation," Cannon's "homeostasis," von Bertalanffy's "equifinality," etc. Likewise, new concepts are required to denote the phenomenon and describe its states: temperature, pressure, entropy, feedback, and so on. These appear "emergent" when viewed in the framework of classical mechanics of particles: a single particle, or a single molecule in a gas, has neither temperature, nor pressure, and of course does not exhibit homeostatic or similar complex forms of behavior. If the special languages of classical disciplines are adhered to exclusively, nature becomes compartmented into distinct segments, each characterized by its own set of entities, properties and laws. But if the general metalanguage of systems theory is adopted, then the dictum "special concepts for special phenomena" loses validity. We can pass from one field of investigation to others without losing one system of reference and having to adopt another. Instead, we can "translate" the special languages by means of a general systems language, which offers invariant meanings of which the local values are isomorphic transformations.

Without falling into the dogmatic trap of saying that such high-level invariant concepts demonstrate general order with certainty, I shall merely claim that the wide empirical applicability of systems concepts argues for the justification of assuming general order as a cogent hypothesis. If the presupposition of general order has its own contribution to make to the progress of knowledge, theories based on general systems concepts will be justified. And the unique value of the exploration of general order is the possibility of elucidating connections between disparate types of phenomena, disjunctively treated by lower-level specialized theories. Such elucidation has extrinsic merit in this day and age (as I argued above), and it also has intrinsic justification. The latter lies in finding order and interconnection where none has been found before: the very aim of cognitive theoretical science. Simpson said that "scientists tolerate uncertainty and frustration, because they must. The one thing that they do not and must not tolerate is disorder. The whole aim of theoretical science is to carry to the highest possible and conscious degree the perceptual reduction of chaos that began in so lowly and (in all probability) unconscious way with the origin of life."[12] Even the myths of so-called primitive societies are now found to be

---

[12]G. G. Simpson, *Principles of Animal Taxonomy,* New York, 1961, p. 5.

high-level order-seeking enterprises—"sciences of the concrete."[13] Ethical norms, aesthetic sensibilities, and, in general, the organization of behavior, both civilized and primitive, appear to be based on a search for order and coherence.[14] Seeking such order is in accord with those very processes upon which our entire way of life is built. The generalist, who seeks order and interconnection at levels of greater generality than the specialist, is not doing anything contrary to the principles which underlie and guide philosophy and science; he merely recognizes and attempts to satisfy them. The presupposition of general order in nature is not any the less valid than special order, and this presupposition is lent significant weight today by the success of systems theorists in building integrative models of the empirical world.

The promise of systems philosophy

The present study argues for a systematic and constructive inquiry into natural phenomena on the assumption of general order in nature. Building on this presupposition (clearly labelled as such; and also clearly understood that it is no worse, and possibly better, than the supposition of special orders) proceeds by assimilating information gathered by the empirical sciences to basic philosophical problems. Thus the empirical sciences are viewed as active storehouses of information forming the substance of potential solutions to typically philosophical issues. By this method scientific findings are used instrumentally, to construct a conceptual framework adequate to the understanding of nature as an integral network of ordered interdependency of which man is a part. The world becomes humanly relevant without becoming anthropomorphized. Man is not the center of the universe, nor is the universe built in his image, but he is a part of the overriding order which constitutes the universe.

The philosophy capable of performing this task evolved in slow progression from the philosophy of real universals of Plato and the categorial scheme of Aristotle, through scholastic metaphysics in the Middle Ages, to the modern process-philosophies of Bergson, Lloyd Morgan, Samuel Alexander and Alfred North Whitehead. Systems philosophy is a logical next step in this progression. It reintegrates the concept of enduring universals with transient processes within a non-bifurcated, hierarchically differentiated realm of invariant *systems*, as the ultimate actualities of self-structuring nature. Its data come from the empirical sciences; its problems from the history of philosophy; and its concepts from modern systems research.

[13]Claude Lévi-Strauss, *The Savage Mind,* Chicago, 1966.

[14]Cf. Ervin Laszlo, "Integrative Principles of Art and Science," in *Integrative Principles of Modern Thought,* Henry Margenau, ed., New York, 1972.

The need for such a philosophy is becoming recognized today by progressive thinkers in many parts of the world. From the Soviet side, Blauberg, Sadovsky and Yudin recently requested a "philosophy of systems" to "explicate the characteristics of the systems world view" and tackle its methodological and epistemological problems.[15] In the Western world, many such demands are made. Von Bertalanffy, for example, points out that "Any theory of wider scope implies a world view . . . any major development in science changes the world outlook and is 'natural philosophy' or 'metascience,' to use a modern expression." Today, he tells us, we are looking for a new basic outlook—*the world as organization.* We need an expansion of the system of traditional physics; we need concepts and models adequate to deal with the biological, behavioral and social universes; and we need abstract models which, by means of isomorphisms in their formal structure when applied to diverse realms of phenomena, can function in an interdisciplinary and integrative manner. And I am in agreement with von Bertalanffy's optimistic assessment that "there are recent developments, loosely circumscribed by the concept of system, which try to answer the demands mentioned. In contrast to the progressive and necessary specialization of modern science, they let us hope for a new integration and conceptual organization. Speaking in terms of natural philosophy, as against the world as chaos, a new conception of the world as organization seems to emerge."[16]

Von Bertalanffy lists, among the new disciplines which carry the nucleus of this new natural philosophy: *general systems theory; cybernetics; information, decision,* and *game theories,* and others. I would now like to add to this list *systems philosophy,* as the field wherein the "new natural philosophy" of the "systems world view" is to find detailed and self-critical formulation, as the paradigm of general theory in contemporary thought.

---

[15]I. V. Blauberg. V. N. Sadovsky and E. G. Yudin, "Some Problems of General Systems Development," *Unity Through Diversity,* W. Gray and N. Rizzo, eds., New York (in press).

[16]Ludwig von Bertalanffy, *Robots, Men and Minds,* New York, 1967. Part Two.

Chapter 2

METHOD

Systems philosophy, as any complex theoretical enterprise, requires explicit methodological and conceptual foundations. I submit that it must be based on (i) a *general theory of systems,* which is to offer the principles and categories for confronting the typically philosophical problems which constitute (ii) *systems philosophy* itself. Hence Part One shall be concerned to develop the methodology and basic principles and categories of a general theory of systems, to be followed in Part Two by a constructive systems theoretical exploration of perennial philosophical problems.

Perspectivism

I do not suggest that the general theory of systems I shall sketch here (as well as other, more accomplished, general systems theories emerging today) represents the sole valid, and hence necessary, approach to sound empirical theory. I only suggest that such general systems theories grasp some forms of order in the world which elude other types of theories. Thus general systems theories are one species of "world hypothesis": they find their use and justification in the elucidation of otherwise chaotic patterns of events in the natural world. That currently general systems theories represent the most intensely worked field of general theory is not accidental: the systems concepts used in this theory permit a more adequate interpretation of a greater range of phenomena in a more consistent and unitary manner than any other.[1] Nevertheless, hypotheses of this (and any other) kind are best thought of as representations of certain perspectival

---

[1]It is noteworthy that Pepper, who originally advanced the concept of world-hypotheses, held, in 1942 (*World Hypotheses,* University of California Press), that there are four equally adequate world hypotheses: Mechanism, Contextualism, Organism, and Formism. In 1966, then, Pepper added a *fifth* world hypothesis and adopted it as his own (*Concept and Quality,* La Salle, Ill.); this is selectivism based on the "root metaphor" of a purposive, self-regulating *system.* In 1970 he said of it that "It seems to me to have promise of greater adequacy than any of the four picked out previously." ("Autobiography of an Aesthetics," *Journal of Aesthetics and Art Criticism,* Spring, 1970.) In reviewing the original edition of the present book, Pepper reaffirmed this choice in glowing terms. ("Systems Philosophy As a World Hypothesis," *Phil. & Phenom. Research,* June, 1972.)

features of nature. This applies to all theories of the empirical world, including that paradigm of modern theoretical science, physics. Even the most rigorous of scientific theories represents but one type of model for mapping some of the features of the observed world in a manner that is neither exhaustive, unique, nor certain.[2] Thus dogmatism is misplaced in *all* empirical inquiry, and in a general philosophical one especially. Realizing this, the present work, unlike many traditional philosophies, is neither dogmatic nor exclusive of other theories, mapping perhaps different aspects of what is presumably the same universe. Certain aspects of this universe may be best amenable to treatment by constructs of the kind I have chosen; these appear to provide the greatest range of invariance and are hence best suited to grasping a general order, recurring in locally differentiated transformations. But this appropriateness of the chosen constructs suggests neither that they are absolutely true and exclusively valid in empirical theory-construction, nor that a conjunction with other constructs, mapping different features, would not be feasible. My choice of systems constructs has been dictated by the prior realization that a general order is, in all intrinsic respects, as good a presupposition as any special order and that, in extrinsic respects, it is a considerably better one.

Creative deduction

Consider, then, the here advanced hypothesis as a conceptual model, of the systems theoretical species, mapping into potentially quantifiable constructs certain recurrent general features of the scientifically observable universe. Such models can be built by using at least two basic methods. One takes the world as we find it and examines the various systems that occur in it (physical, biological, sociological, etc.) and then draws up statements about the observed regularities. The second method goes to the other extreme, and considers the set of all conceivable systems and then reduces the set to a more reasonable size. Ashby identifies the former as the empirical method of von Bertalanffy and his collaborators, and the latter as the axiomatic method which he himself has recently followed.[3] These methods represent the logical alternatives of all theory construction. But they are seldom if ever found in such pure forms as Ashby appears to suggest. Empirical observation is meaningless without the imaginative envisagement of various abstract possibilities which are then seen either

---

[2]Cf. Albert Einstein, *Sidelights on Relativity* (1922); also Ludwig von Bertalanffy, *General System Theory,* New York, 1968.

[3]W. Ross Ashby, "Principles of the Self-Organizing System," *Principles of Self-Organization,* Foerster and Zopf, eds., New York, 1962.

to be, or to fail to be, exemplified in the content of observation. And axiomatic construction, apart from the non-empirical mathematical sciences, rarely considers "the set of all conceivable" systems or models, but is guided in the selection of the hypothesis by an assessment, based on previous observation, of the kind of constructs which are likely to find empirical exemplification. Of these two "modified" alternatives, emphasis —even within the special sciences themselves—has shifted in this century from the empirical-imaginative to the deductively applicable axiomatic method. The shift has been mainly prompted by the fate of Newtonian physics. Newton created the impression that there were no assumptions in his theory which were not necessitated by observational and experimental data. Had this notion been correct, his theory would never have required modification: observation and experiment could not have contradicted it. But such contradiction occurred in 1885 in the famed Michelson-Morley experiments. Some ten years later, experiments on black body radiation enforced the fallibility of Newtonian concepts. Northrop expressed the conclusion concerning the method of the new physics positively, in saying that "the theory of physics is neither a mere description of experimental facts nor something deducible from such a description; instead, as Einstein has emphasized, the physical scientist only arrives at his theory by speculative means. The deduction in his method runs not from facts to the assumptions of the theory but from the assumed theory to the facts and the experimental data. Consequently theories have to be proposed speculatively and pursued deductively with respect to their many consequences so that they can be put to indirect experimental tests."[4] No datum of observation and experimentation leads with any certainty to a theoretical construct; the latter is elicited by the creative imagination of the observer, tested and modified by his methodological restraints. The method in question consists, according to Rosenblueth, "of hypothesizing an abstract general law or theory which will fit a specific set of observational or experimental data." He adds that this method is the one that has been used in the formulation of all the important scientific theories, since such theories cannot be reached by inductive, deductive or analogical inferences. The theories were *created* by their propounders, rather than *induced* on the basis of large numbers of consistent observations and experiments.[5] Yet, Rosenblueth complains, the majority of philosophers view scientific theory formulation as a process of induction followed by deduction, according to the norms suggested by

---

[4]F. S. C. Northrop, "Introduction," in Werner Heisenberg, *Physics and Philosophy*, New York, 1958.

[5]Arturo Rosenblueth, *Mind and Brain; a Philosophy of Science*. Cambridge, Mass., 1970, pp. 80–81.

traditional logic. These arguments are quite naïve and inadequate.[6] In fairness to philosophers it should be pointed out that there are notable exceptions among them, as the statement by Northrop, quoted above, and that from Whitehead, which follows, testify. "The true method of discovery is like the flight of an aeroplane. It starts from the ground of particular observation; it makes a flight in the thin air of imaginative generalization; and it again lands for renewed observation rendered acute by rational interpretation."[7] This method is appropriate to propose theories which are neither purely induced nor deduced, but are so formulated that empirically applicable laws can be derived *from* them.

"Creative deduction" is a method which forms a common bond between contemporary science and general systems theory. The latter will be exemplified by a hypothesized theory so formulated that statements defining invariances ("laws") may be derived from it. Producing such theories represents a creative "flight of imagination" exemplified, among others, in Einstein's theory of relativity, Morgan's theory of genes, and Yukawa's theory of mesons. The task of this attempt at the formulation of such a theory is to perform this "flight" and then to test its deduced consequences against data of observation and experimentation.

The theory here to be presented is a mapping of certain recurring features of the scientifically constructed empirical world. Now, as discussed in Chapter 1, the presupposition of limited order in nature is the minimum condition for the applicability of any special scientific theory, and the presupposition of general order is the precondition of the applicability of a *general* theory. Aside from the unanswerable question, whether or not nature *is* in fact ordered, in one way or another, the practical question, whether nature (or the information we have of it) *can* thus be ordered, is meaningfully answered by the recently demonstrated success of systems constructs in applying to phenomena in diverse fields of investigation. Of the many philosophically and scientifically acceptable statements concerning the empirical world, statements from the field of general systems theory appear repeatedly in a wide variety of cases. Their relevance for the presupposition of general *orderability* (if not necessarily *order*) in nature expresses itself in the isomorphism of the laws and principles formulated in regard to systems of different kinds. Such isomorphism (or parallellism) of the applicable constructs suggests a fundamental unity of the observables—as von Bertalanffy[8] and other general systems theorists point out. "Speaking in the 'formal' mode, i.e., looking at the conceptual constructs of science, this [the

---

[6] *Ibid.*, Preface.

[7] A. N. Whitehead, *Process and Reality, op. cit.* Part I, Ch. I. Sect. II.

[8] *General System Theory, op. cit.*, pp. 48–49.

isomorphy of laws in different fields] means structural uniformities of the schemes we are applying. Speaking in the 'material' language, it means that the world, i.e. the total of observable events, shows structural uniformities, manifesting themselves by isomorphic traces of order in the different levels or realms."

The creatively postulated and deductively applied systems constructs include wholeness, feedback, steady and stationary states, entropy, and many others. In virtue of the inclusive applicability of such constructs, much of the universe available to our scrutiny can be conceptually "mapped" as hierarchies, i.e., as a realm of systems-in-environments, constituting higher order systems which, within their particular environments, constitute systems of still more inclusive order.

Second-order models

Systems constructs are possibly the best contemporary instruments for building models of other (empirical) models. The many sciences, and other disciplines dealing with aspects of the experienced world, build "models" of their own particular area of experience. For example, in economics, man appears as a rational, profit-seeking, consuming and producing entity; in existential psychology he appears as an irrational or supra-rational, spontaneously cognizing appreciator and actor in life. Man, of course, is both. He also incorporates the traits ascribed to him by other empirical models, e.g., in biology, sociology, ecology, and so on. And the same applies, *mutatis mutandis*, to the other entities and processes populating the empirical world (although the wealth of models is greatest in regard to man). If we assume that reality is merely mapped by the models and not determined by them, it follows that the models give so many perspectives of what may be a common underlying core of events. General systems theory's task is to uncover that core. Failing some direct communion with objective reality, it can only do this through the existing models. A mere analysis of the models yields an undoubtedly clarified but still one-sided and uncoordinated view of the world; thus analysis must be supplemented by synthesis. And synthesis need not be the arbitrary procedure of declaring one concept to be "the same as" another. Rather, synthesis can follow essentially the same procedure as model-building in the empirical sciences, taking place however on a "second-order" meta-level.

General systems synthesis, I suggest, is the building of models of models. Its procedures can be controlled and explicit, and need not be less rigorous in aim and expectancy than those of any empirical science. Its main area of difference is given by the fact that the data of systems synthesis are theories —"first-order" models of the experienced world—and not experiences themselves. *Its basic conceptual assumption is that the first-order models refer to some common underlying core termed "reality," and that this core is generally ordered.* Thus the special orders elucidated by the many empirical-level models serving as its data can be integrated into a second-order model exhibiting a species of general order. In view of the fitness of systems concepts to remain invariant when passing from one first-order model to another, thereby permitting the translation of terms and concepts as particular transformations of the invariance stated in the systems language, the second-order model constitutes a *general systems theory*. It integrates the findings of the many specialized first-order sciences in an optimally consistent framework, and thus serves as the foundation upon which we can build the here advocated structure of *systems philosophy*. (Figure 1)

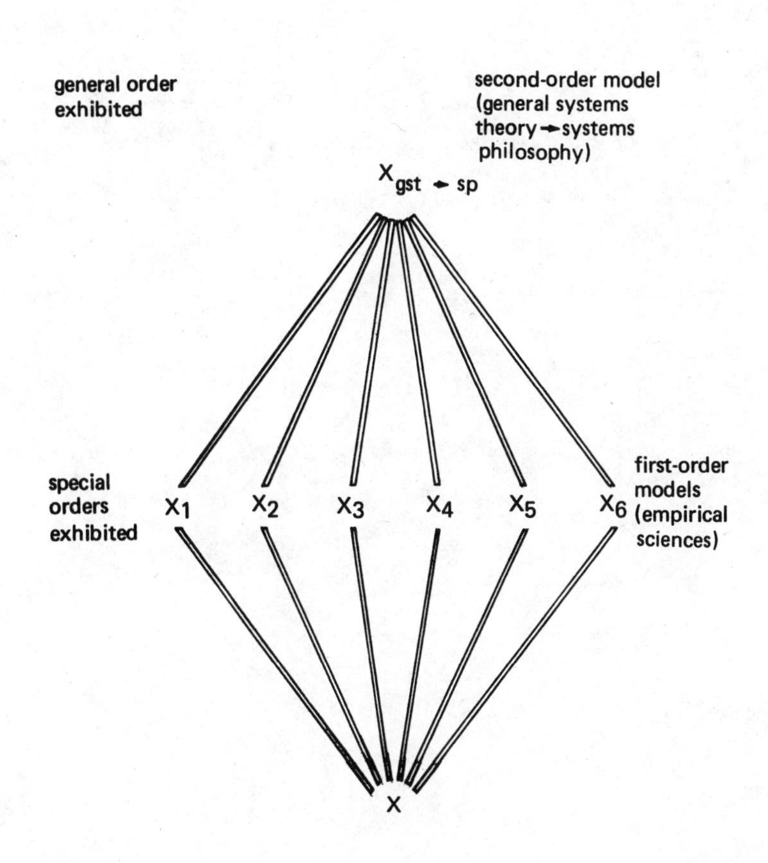
general order
exhibited
second-order model
(general systems
theory → systems
philosophy)
$X_{gst \rightarrow sp}$
special
orders
exhibited
X_1
X_2
X_3
X_4
X_5
X_6
first-order
models
(empirical
sciences)
X
order?
nature

Figure 1

Chapter 3

SPECIFICATIONS

From substantial to relational entities

Attention in the contemporary natural sciences has shifted from the description and classification of individual entities to the theoretical explanation of classes of entities forming ordered structures of events. The shift is evident both in the physical and in the life sciences. Classical physics operated on the assumption that the data to be explained are individual entities or events, located against a background of space and time, "flowing equitably through all eternity." Particular entities, such as masses moving or at rest, could be plotted against coordinates of space and time and classified on the basis of their observed similarities. Thus classical physics posited particular entities as the ultimate furnishings of this world, with classification serving the primarily heuristic purpose of building theories for calculating their behavior. Although thinkers have always recognized that no two entities can ever be exactly alike, the differences could be disregarded as long as they created no difficulties for solving the problems at hand. But contemporary physics gave up the notion of unique if classifiable masses moving against a background of eternal space and time. The new physics deals with ordered sequences of events, forming wholes, which can only arbitrarily, and usually without success in formulating exact laws, be analyzed into individual components. The general construct for these ordered wholes is *field*. Einstein summed up the shift in orientation between classical and field physics when he said, "Before Clerk Maxwell, people conceived of physical reality—insofar as it is supposed to represent events in nature—as material points, whose changes consist exclusively of motions. . . . After Maxwell they conceived physical reality as represented by continuous fields, not mechanically explicable . . ." And Einstein added, "This change in the conception of reality is the most profound and fruitful one that has come to physics since Newton."[1]

The field concept permits the solution of problems which are insoluble with the individuated particle conception of classical physics. Planck pointed out that "it is impossible to obtain an adequate version of the laws for

---

[1]Albert Einstein, *The World As I See It*, New York, 1934, p. 65.

which we are looking, unless the physical system is regarded *as a whole.* According to modern mechanics (field theory), each individual particle of the system, in a certain sense, at any one time, exists simultaneously in every part of the space occupied by the system. This simultaneous existence applies not merely to the field of force with which it is surrounded, but also to its mass and its charge."[2]

Physical phenomena are now viewed as systems, in which subsidiary events are not separate particles but subsystems: subpatterns within the overall pattern which is the object of investigation. For example, in field theory one can no longer say that an electron is "here," as assumed in classical mechanics—one can legitimately conclude only that field strength is highest in this area of the total pattern of focus called "atom," and that the probability is highest, therefore, that the electron will be found here when the instrument interacts with it.

Physics' conception of nature as forming ordered fields with complex subsidiary patterns is reflected in the new chemistry, where molecules are viewed as complex dynamic patterns formed by the sharing of the outer electrons (conceptualizable both as waves and as particles) between several nuclei. Further counterparts to these concepts are found in biology. There, organisms are no longer viewed as discrete entities made up of similarly discrete components in some mechanistic type of interaction. The new concept of organism is that of a whole which, rather than being constituted by the mechanical interaction of discrete parts (organs, cells, molecules and atoms), forms a system within which such elements are discernible as co-ordinate subsystems endowed with boundaries consisting of semi-permeable surfaces, concentration gradients and energy transfer interfaces. Moreover, even the full organism is no longer thought of as a strictly delimited, rigorously bounded system but is believed to merge with its environment over a series of further boundaries, none of which are absolute (i.e., adiabatic) walls. The organic boundaries themselves are interfaces, permeable to some—though not to all—informations, energies and substances. As a consequence we also get ordered wholes in the sciences dealing with multi-organic units: *supra*organic systems, which can be analyzed to individual organisms as component subsystems. And even the supraorganic wholes form a continuum, broken only by relative boundaries, with still larger systems, as evidenced by such systems concepts as continental ecologies and planetary economic and social systems. The remarkable fact is that contemporary science has effectively, though largely tacitly, abandoned the notion of isolated particular entities as its units of investigation. Its concern is the discernment of ordered totalities constituting, at the basic

[2]Max Planck, *Where is Science Going?*, London, 1933, p. 24.

physical level, fields, and on higher levels, systems within the fields, and systems within systems within fields—and so on, in a complex hierarchy of organization in nature.

A true scientific breakthrough comes today, in Gerard's words, "when somebody has sufficient creative imagination—and courage to follow up, which may be even more important—to say 'Let us look at the universe in terms of some new kinds of entities, some new kinds of units, or, what really comes to the same thing, in some new way of combining units'; because combining units gives a new unit at the superordinate level."[3] And the many breakthroughs which have occurred in the recent past, and are likely to occur in the next future, come when someone looks at the universe in terms of entities which are not atomic material substances but organized functional systems, related to one another by "horizontal" interactions within their own level as well as by "vertical" interactions between different levels.

Microhierarchy and macrohierarchy

The recent breakthroughs in the empirical sciences prepare the ground for a new general theory of the empirical world. The units of the diverse special sciences are now largely reconceptualized as dynamic systems, and we find ourselves with a set of systems, unclearly interrelated, postulated by different disciplines, and situated at different levels of organization in nature. One can attempt, therefore, to map these externally related systems models into a second-order general model. Conceptualizing physical, chemical, biological, ecological, sociological, and even political phenomena as systems with invariant structures and properties no longer requires a radical reformation of existing scientific languages; the systems approach is being developed, independently, in most sciences. What we first require, then, is the coordination of these independently formulated systems models in a *general theory of systems.*

At the present stage of empirical investigation, however, it is necessary to impose some limitations upon even a highly generalized inquiry. Reality *in toto* may be too large an order for any theory claiming scientific foundation, and the problem becomes to choose the most general subsection of the cosmos for systemic theory construction. The choice may be dictated by the following considerations.

Levels of dynamic systems evolve on two different planes and by means of two different types of processes. On one level we find the entities of astronomy: *galaxy clusters, galaxies, star clusters, stars, planets* and their

[3]R. W. Gerard, "Hierarchy, Entitation, and Levels," in *Hierarchical Structures,* Whyte, Wilson and Wilson, eds., New York, 1969.

subsidiary bodies. On another level there are the entities of physics, chemistry, biology, ecology, sociology, and even international relations: *atoms, molecules, molecular compounds, crystals, cells, multicellular organisms,* and *communities of organisms.* Cosmology has jurisdiction over the former; almost all of the remaining natural and social sciences investigate the latter. Little is explicitly known of the laws of the vertical build-up of level-structures in either area, other than that gravitational forces play the principal role in the astronomical sector of evolution, and electrical and subsidiary forces are instrumental in evolution on the microphysical-to-organic level, with cognitive factors playing a role thereafter, when symbolized communications supervene over energetic interactions. Now, cosmology is in a truly rudimentary state at this time, with scientists such as Harrison admitting that, although it is "the science of the universe as an organized whole," "nobody in their right mind for a moment dreams that we are anywhere near achieving even an elementary notion of the universe as an organized whole."[4] Cosmology, he points out, has not yet succeeded in explaining in general terms why the universe is fragmented into planets, stars, and galaxies—it is known neither how galaxies evolved nor what initial conditions determined their present properties and structural differences.[5] This situation contrasts with that which prevails in the microphysics-to-sociology spectrum of the build-up of level structures. There, the problem is to find a general law of development which could coordinate the enormous storehouses of relevant data, rather than to uncover the basic data themselves. The contrast between the rudimentary state of scientific cosmology and of a scientific account of micro-evolution is comprehensible: cosmology must be based on a reconstruction of past events on the basis of their redistribution in present processes, whereas all (or almost all) phases of micro-evolution are still available to investigation in one form or another. But this difficulty in cosmology precludes the inclusion of astronomic developmental processes in a general systems theory: the latter cannot undertake to advance scientific laws and theories, only to map those already available into a second-order model. And the existing theories and laws of cosmology are too uncertain, as yet, to allow the construction of a general theory of systems carrying any degree of empirical meaning and accuracy. Hence the range of that theory must be limited to the phenomena of organized complexity which are initiated with the formation of atoms, and which progress thereafter in hierarchical fashion, through molecular, cellular, organic and multiorganic forms, to increasingly large social and ecological suprasystems.

---

[4]E. R. Harrison, "The Mystery of Structure in the Universe," in *Hierarchical Structures, op. cit.*, p. 88.

[5]*Ibid.*

Now, it may be held at this point that general systems theory is hereby restricted to typically *terrestrial* phenomena, and that it says little if anything about nature or the universe as such. While the above restriction is a definite and radical one, it nevertheless does not constrain general systems theory merely to terrestrial phenomena. First of all, the complex hierarchy that evolves in the earthly biosphere may be unique in form but is most probably not unique in type. Proceeding on the assumption of the uniformity of nature, i.e., holding that like conditions bring forth like phenomena in all sectors of the cosmos, even sceptical calculations would allow that there are likely to be a considerable number of hierarchies in the cosmos which resemble that of our biosphere, i.e., which evolve life and socio- and ecosystems in some form.[6] We may be unique in regard to particular form, but not with respect to being complex, hierarchically organized natural systems. While we do not at this time possess definite evidence of extra-terrestrial life, it is clear that this does not imply that there is no such phenomenon; for even if there were, our present recording instruments could still not show any trace of it. However, with the chemical analysis of stray meteorites, and eventually through the extension of interplanetary space travel and communication techniques, ways and means become available for at least a limited testing of the uniformity of nature hypothesis. At this time it suffices to note, however, that there are no grounds on which we could dismiss that hypothesis: physics and astronomy successfully predict phenomena on its basis, and no a priori objections could be raised against extending its validity to the domains of the other natural sciences.

Secondly, general systems theory is not specifically limited to terrestrial (and hence arbitrarily chosen) phenomena in virtue of the common denominator of the two types of hierarchy. A given level of systemic complexity in the *cosmic* hierarchy serves as the basic level of the *terrestrial*-type hierarchy, when conditions permit further complexification of the modules in question. The common modules are *atoms*. They are by no means the basic modules of the cosmic hierarchy, since they are composed of electrons and nucleons, each of which may well be composed of some still simpler unit, such as quarks. These in turn may be conceived as primitive knots, condensations, or critical tensions in an underlying space-time manifold. Hence atoms are relatively high-level modules in nature's cosmic hierarchy. Moreover they are (or are believed to be) ubiquitous in the universe, occurring both in relatively loose interstellar clouds and in stellar bodies of all varieties. But atoms also appear in

---

[6]Numerous, though often divergent estimates are produced on the now rapidly growing fields of "exobiology" and "interplanetary communication."

another capacity: as the basic building blocks of a terrestrial-type hierarchy. (While it may be argued that particles, quarks, or the space-time manifold serve as such basic building blocks, to this we can always reply that these latter are readily conceptualized as subsystems, or "primitive" components, of atoms.) Given specificable conditions (defined in terms of temperature, chemical environment, density, gravitational and electric fields, etc.), some species of atoms undergo joint structural reorganization, and result in *molecules* and *crystals.* These in turn, under a still more rigorously delimited set of conditions, coalesce into higher-level structures which we identify as *polymers,* and various *giant molecules,* including *proteins.* The latter form (under further restricted conditions) *cells,* these in turn form multicellular *organisms* and ultimately *sociological* and *ecological systems.* While the force or compulsion to form such superordinate structures cannot be explained by a study of the morphology of the component substructures, the fact remains that there is structural integration on one level, followed by the superimposition of structures of the next higher level. Thus the evolutionary record takes us from atoms, to multiatomic molecules and crystals, and in progressive steps to some planet-wide dynamic self-maintaining system. And the pivotal point, i.e., the point of the intersection of the cosmic astronomical hierarchy with the terrestrial-type hierarchy, is the level of atomic nature. Although it is undoubtedly possible to advance a large number of different ways in which hierarchical organization in nature can be conceptualized, this point of intersection is likely to be preserved in most. One such conceptualization is offered, as illustration of this argument rather than as a finished theoretical product, in Figure 2.

In order to reduce the complexity of the terminology of this exposition, I shall call the terrestrial-type atoms-to-ecologies hierarchy the *microhierarchy,* and the astronomical hierarchy extending from the space-time field to the metagalaxy the *macrohierarchy.* Using this terminology, we can say that general systems theory is concerned with the microhierarchy, and intercepts (in its theoretical considerations) the macrohierarchy at the level of the atom. It focusses on those processes of organized complexity which supervene upon the atomic level under suitable planetary conditions, and which thus intervene upon the macrohierarchy's atom-to-stellar-object part-whole relation. In other words, about cosmic regions where the energetic conditions are unsuitable to multiatomic structurations of the molecular-multimolecular kind, this theory has little to say (other than noting that populations of atoms constitute stellar macroobjects). But where conditions permit multiatomic structurations to take place, there interposes between the atom-star part-whole relation an entire series of remarkably complex

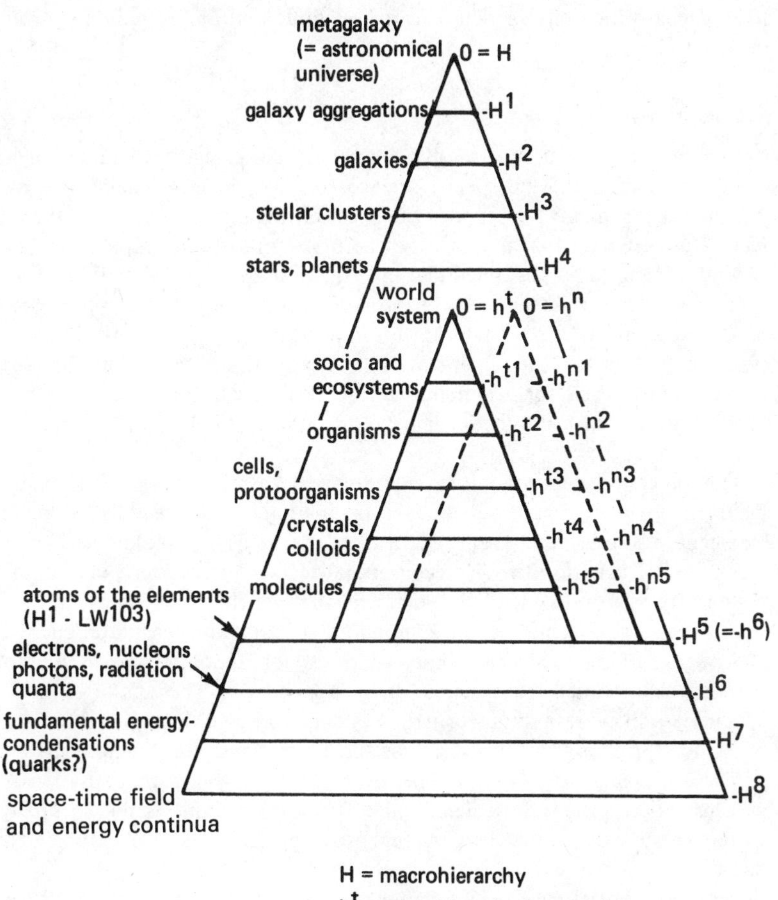

Figure 2

A hypothetical identification of the principal levels and interrelation of the micro- and macrohierarchies. Note that the emergence of each higher level out of systemic structurations of units of the lower levels is contingent upon local conditions and results in the uneven build-up and distribution of modules of intermediate levels (-H7— -H1, including -h6 – h) within the space-time continua.

---

[1]Based on the uniformity of nature hypothesis: whether the upper levels have been evolved, and what form they take, is not known.

phenomena, which offer a rich field for scientific and philosophic investigation.

Natural systems

The task, as I see it, is to evolve the conceptual categories within which the many findings of the special sciences could gain new significance by yielding the general laws and principles of organization *in the microhierarchy*. I suggest that this task can be most fruitfully attempted by conceiving of the units of scientific investigation as systems with certain definable invariant properties. The properties serve to define the behavior of the systems in their environments, including their relations to other systems. In proposing a *general theory of systems* (to be explored subsequently as the basis for a *systems philosophy*) I shall postulate properties of this kind and measure them against data gathered by the pertinent special sciences.

The suggested concept *natural system* replaces the many misleadingly disjunctive names of natural entities in the microhierarchy such as "atoms," "molecules," "crystals," "cells," "viruses," "organisms," "ecologies" and "societies." Each of these is reconceptualized as one variety of natural system. By "natural system" I understand that which is sometimes referred to as a *concrete* system, i.e., a "nonrandom accumulation of matter-energy, in a region of physical space-time, which is nonrandomly organized into coacting interrelated subsystems or components."[7] The units of such systems are likewise natural (concrete) systems, with the possible exception of some "primary" or "least hierarchical" systems, the components of which are elementary blobs of energy ("electromagnetic condensations," in Einsteinian terms). Such definition of "natural system" is not so broad to be meaningless yet general enough to replace the above enumerated names of natural entities. The definition excludes all "aggregates" or "heaps" of natural events, i.e., matter-energy agglomerations which do not interact or coact, although sharing a defined space-time region. (It is always possible that a "heap" should transform into a "natural system" by virtue of the components entering into systematic interactive or coactive relationships with one another. The gradual colonization of *heaps* as emerging natural *systems* is one form of evolution, in the macro- as well as the micro-universe.) Natural systems are characterized by the measurable nonrandom regularity of the coactions of their components. These regularities add up to conceptually discovered invariances, i.e., differential equations stating functional relationships between variables. The particular

---

[7]James G. Miller, "Living Systems: Basic Concepts," *op. cit.*

transformations of this invariance are determined by the hierarchical level of the natural system, i.e., by the number and kind of systems which it incorporates (*intrasystemic hierarchy*) and by the number and kind of systems with which it coacts in its environment (*intersystemic hierarchy*).

The here to be presented hypothesis suggests invariances extending to the full range of the atoms-to-ecologies microhierarchy. Each system within that hierarchy is capable of conceptualization as a system manifesting the postulated invariances in the coaction of its parts as well as in its total-system coaction with systems in its surroundings. The hypothesis does not suggest that the known physical laws are invalid in application to the behavior of these systems. Rather, it proposes that they are inadequate as presently formulated to grasp the invariances which underlie the dynamic-functional behavior of systems in the full range of the microhierarchy. Our hypothesized invariances describe the constraints upon the behavior of co-acting parts and systems; constraints which do not violate the laws of physics, but which are imposed upon the deterministic laws of classical mechanics, exploiting the degrees of freedom permitted by the latter. The universe here conceptualized is an emergent-holistic universe, with systemic processes supervening over the mechanical processes and resulting in the local reversal of the generally entropic trends in the cosmos.

Mapping invariances by means of creatively postulated general systems constructs overcomes the difficulties entailed by the usual approaches to the investigation of natural systems. One such approach is to ask, in regard to a system, "how does it work?" That is, how the parts are coordinated so that the laws governing their motion yield the patterns actually observed of the system as a whole. The task is one of analysis—a taking apart of the whole to find the detailed mechanisms which "make it tick." The type of effort is reductionist in intent, if not necessarily in result. The latter depends on whether or not the mechanistic laws of the primitive components can account without lacunae for the behavior of all parts, so that their combination yields the system under consideration. Such endeavors have often been frustrated; hence the prevalent belief in the fallacy of reducing biological phenomena to physico-chemical machines, and of reducing societal phenomena to interactions between biological individuals (to cite merely two outstanding instances).

Another type of problem concerning the explanation of organized systems arises when we examine them from the viewpoint of the primitive parts. Then, the deterministic laws of physics are assumed to apply to their behavior, and the problem is "how do the constraints constituting the interrelations of the parts arise?" The difficulties of attempting to derive the laws governing even a system of medium complexity from the laws governing the motion of the parts are enormous, as anyone knows who is familiar with the type of calculations required to give the laws of even simple chemical molecules with the equations of quantum mechanics. Consequently the laws applying to wholes are always simplifications (such as statistical averages), and it is often not clear how the higher-level laws, applying to sets of phenomena, are related to the deterministic laws of the parts. It is often held (e.g. by Polanyi[8]) that the laws defining the constraints by the whole are *not* derivable from the laws of the parts. This position is one of non-reductionism and can lead to dualism or pluralism, depending on how many areas of irreducibility are recognized.

In the face of such difficulties in interrelating the constraints arising in organized wholes with the degrees of freedom permitted by deterministic laws for the parts, the present approach seeks invariant systems properties as a set of constant and universal constraints. Rather than denying that such constraints exist in addition to the known laws of physics, or looking on them as irreducible emergent laws of organized entities at their own level of organization, we assume that they are formative constants of nature: properties of all "natural systems." These (the Pauli exclusion principle is a good example) are not observable below a given level of organization, i.e. they are "system properties" or "laws of organization." But just as the exclusion principle does not come into being by some external *fiat* when two or more electrons find themselves in the same orbit around a nucleus but is more cogently viewed as a constant law of nature, so the laws defining the constraints at higher levels of organization do not simply emerge when that level has been reached but should be assumed to be present all along, though not in an observable or measurable capacity.

The invariant properties of natural systems are universal constraints in this view, and they become effectively observable at particular (hierarchical) levels of organization. Thereafter they are manifest in diverse transformations corresponding to the various levels of the microhierarchy. If this is a meaningful working hypothesis, our task is a large but not an impossible one: it is to postulate the commonality underlying the manifest behavior of organized entities as a set of invariant properties of natural systems. These creatively postulated invariances are not laws of physics as these laws have been hitherto understood, nor are they laws of biology or of any other special inquiry. They are general laws of natural organization cutting across disciplinary boundaries and applying to organized entities in the microhierarchy at each of its many levels.

---

[8]M. Polanyi, "Life's Irreducible Structure," *Science* **160**, (1968), 1308–1312.

Part One

OUTLINE OF A GENERAL THEORY OF SYSTEMS

Chapter 4

THEORY OF NATURAL SYSTEMS

> The need to create sound syntheses and systematizations of knowledge . . . will call out a kind of . . . genius which hitherto has existed only as an aberration; the genius for integration. Of necessity this means specialization, as all creative effort inevitably does; but this time the man will be specializing in the construction of a whole. (Ortega y Gasset, *Mission of the University*, Princeton, 1944, p. 91)

The purpose of the here presented theory is to state those invariances which apply to phenomena of organized complexity throughout the microhierarchy. Although some relatively detailed and quantitative formulations of theories of natural systems are now emerging from the workshops of physicists, biologists, sociologists, economists and ecologists, most of these represent theories proposed in view of specific foci of attention; hence they are not necessarily reliable as a foundation for a *general systems "philosophy."* Rather than surveying and summarizing the existing theories, I shall propose one of my own, mapping explicitly but a handful of systems properties, but ascertaining that each one of these is applicable throughout the range of phenomena of organized complexity that constitutes the terrestrial microhierarchy. Whereas individually these properties may appear trivial in their utmost generality, in their combination, as properties of natural systems, they produce a significant theory of what systems in nature are like. Since man himself is one species of such system, and his environment and experience is made up of others, we get anchor points for a considered general systems philosophy, rooted in an unbiased general theory of systems.

Our theory of natural systems will first be stated, then explored both axiomatically, and in regard to some of its many empirical applications.

THEORY: $R = f(\alpha, \beta, \gamma, \delta)$, where $\alpha, \beta, \gamma, \delta$ are independent variables having the joint function R ("natural system").

THE INDEPENDENT VARIABLES[1]

$R = f(\alpha)$ (SYSTEMIC STATE PROPERTY)

Natural systems: wholeness and order

An *ordered whole* is a non-summative system in which a number of constant constraints are imposed by fixed forces, yielding a structure with mathematically calculable parameters.

The concept *wholeness* defines the character of the system as such, in contrast to the character of its parts in isolation. A whole possesses characteristics which are not possessed by its parts singly. Insofar as this is the case, therefore, the whole is other than the simple sum of its parts. (For example, an atom is other than the sum of the component particles taken individually and added together; a nation is other than the sum of individual beings composing it, etc.) However, no mysticism is implied or involved in this assertion. Traditionally, wholes were often considered to be qualitative and intrinsically unmeasurable entities because they were seen as "more than the sum of their parts." This conception is spurious. Wholes can be mathematically shown to be other than the simple sum of the properties and the functions of their parts. Consider merely the following basic ideas. Complexes of parts can be calculated in three distinct ways: (i) by counting the *number* of parts, (ii) by taking into account the *species* to which the parts belong, and (iii) by considering the *relations* between the parts.[2] In cases (i) and (ii) the complex may be understood as the sum of the parts considered in isolation. In these cases the

---

[1]The present, mainly axiomatic discussion of the four independent variables, will be followed by descriptions of particular subclasses of natural systems (physical, biological, social: Ch. 5), as well as of cognitive systems and its known subclass (the mind: Ch. 6 and 7).

[2]Cf. v. Bertalanffy, *General System Theory*, *op. cit.*, Chapter 3.

complex has *cumulative* characteristics: it is sufficient to sum the properties of the parts to obtain the properties of the whole. Such wholes are better known as "heaps" or "aggregates," since the fact that the parts are joined in them makes no difference to their functions—i.e. the interrelations of the parts do not qualify their joint behavior. A heap of bricks is an example of this. But consider anything from an atom to an organism or a society: the particular relations of the parts bring forth properties which are not present (or are meaningless in reference to) the parts. Examples of this range from the Pauli exclusion principle (which does not say anything about individual electrons), through homeostatic self-regulation (which is meaningless in reference to individual cells or organs) all the way to distributive justice (likewise meaningless in regard to individual members of a society). There are many natural complexes which are not heaps, but constitutive complexes: wholes which are *other* than the sum of their parts. Mathematical examples for both summative and constitutive complexes can be readily found.[3]

The noteworthy point for our consideration is that the mathematics of non-summative complexes apply to physical, biological, psychological,

---

[3]I shall not burden the reader with the mathematics of wholes, but for those interested give merely an equation for each type of complex (following v. Bertalanffy).

(a) non-summative complex of interdependent elements:

$$\left.\begin{aligned} \frac{dQ_1}{dt} &= f_1(Q_1, Q_2, \ldots Q_n) \\ \frac{dQ_2}{dt} &= f_2(Q_1, Q_2, \ldots Q_n) \\ &\ldots\ldots\ldots\ldots \\ \frac{dQ_n}{dt} &= f_n(Q_1, Q_2, \ldots Q_n) \end{aligned}\right\} \quad (1)$$

In this system of simultaneous differential equations, change of any measure Q_i is a function of all Q's and conversely. Thus an equation governing a change in any one part is different in form from the equation governing change in the whole.

as well as sociological systems. These systems form ordered wholes, i.e., constitutive complexes in which the law-bound regularities exhibited by interdependent elements determine the functional behavior of the totality. The demonstration that this, in fact, is the case, will be attempted subsequently. I shall merely emphasize here that the construct "ordered wholeness" is not identical with the mystical interpretation of the principle "a whole more than the sum of its parts," but is an acceptable and indeed an often used construct in natural, anthropological and social scientific literature.

$R = f(\beta)$ (SYSTEM-CYBERNETICS I)

Definition of "system-cybernetics"

In any description which interrelates input-output functional analysis with internal state description, the control processes whereby the input is channeled through the system and results in the modification of some (or all) of its existing parameters—and eventually issues in an output—become the foci of attention. Now "cybernetics" is the term coined by Wiener to denote "steersmanship" or the science of control. Although current engineering usage restricts it to the study of flows in closed systems, it can be taken in a wider context, as the study of processes interrelating systems with inputs and outputs, and their structural-dynamic structure. It is in this wider sense that "cybernetics" will be used here, *to wit,* as *system-cybernetics,* understanding by "system" an ordered whole in relation to its relevant environment (hence one actually or potentially open).

Two kinds of system-cybernetics will be distinguished. The *first* is that which is most directly associated with the name of Wiener: it is the study of self-stabilizing controls operating by error-reducing negative feedback. The *second* has received attention primarily since Maruyama called attention to the importance of error (or "deviation") *amplifying*

(b) summative complex:
Assume that equation (1) can be developed into Taylor series: (2)

$$\frac{dQ_1}{dt} = a_{11}Q_1 + a_{12}Q_2 + \ldots a_{1n}Q_n + a_{111}Q^2{}_1 + \ldots$$

Let the coefficients of the variables Q_j ($j \neq {}_i$) become zero; the complex becomes summative (change of each element being unaffected by its relation to others): (3)

$$\frac{dQi}{dt} = a_{i1}Q_i + a_{i11}Q_i^2 + a_{i111}Q_i^3 + \ldots$$

control processes, which function by means of positive feedback (the "second cybernetics"). These two forms of system-cybernetics give us a qualitatively neutral conceptual tool for assessing processes which result in the pattern-maintenance, as well as the pattern-evolution, of ordered holistic systems through periodic, or constant, energy or information exchanges with their environment. Instances of "system-cybernetics I" and "II" can be found in the realm of atomic processes (*microphysical cybernetics*), organismic processes (*bio-cybernetics*), societal processes (*socio-cybernetics*), and even in the area of cognitive processes (*psycho-cybernetics*[4]). These terms refer to the most general characteristics of the controlled flows which result either in the maintenance of a typical structure over a period of time in a dynamic environment (stationary or steady states), or in the progressive modification of that structure in response to inputs from the environment (evolution). The use of the general cybernetic framework in studying qualitatively divergent self-stabilizing and self-organizing processes permits the discovery of invariances, appearing under qualitatively divergent transformations. Such invariances would be masked over by the use of special languages with limited connotations. (Hence the justification of such neologisms as may be involved in the above mentioned hyphenated identifications of particular cybernetic processes.)

Natural systems: adaptive self-stabilization

If a system is entirely governed by its fixed forces, the constant constraints these impose bring about an unchanging steady state. If, however, together with the fixed forces some unrestrained forces are present in the system, the system is capable of modification through the interplay of the forces present *in*, and acting *on*, it. The presence of some fixed forces is sufficient to bring about a state which persists in time ("steady state") when all the flows in the system which correspond to the unrestrained forces vanish. And any fluctuation in the system gives rise to forces that tend to bring it back to its stable configuration due to the fact that the flow caused by the perturbation must have the same sign as the perturbation itself.[5] Hence the flow will tend to reduce the perturbation and the system will establish, and return to, its stationary (or steady) state. If the fixed forces are conserved, the stationary state is characterized by their parameters (since the unrestrained forces vanish). If both the fixed and the un-

---

[4]There is no connection in this use of the term with the meaning attributed to it in popular psychological literature (e.g., in Maltz's book of that title).

[5]Cf. A. Katchalsky and P. F. Curran, *Nonequilibrium Thermodynamics in Biophysics*, Cambridge, Mass., 1965, Chapter 16.

restrained forces are removed, the system is reduced to a state of thermodynamical equilibrium.[6] "Ordered wholes" are always characterized by the presence of fixed forces, excluding the randomness prevailing at the state of thermodynamical equilibrium. Thus ordered wholes, by virtue of their characteristics, are self-stabilizing in, or around, steady states. Given unrestrained forces introducing perturbations which do not exceed their threshold of self-stabilization, they will tend to return to the enduring states prescribed by their constant constraints.

The above is a contemporary statement of Le Chatelier's principle, advanced in 1888 in reference to closed systems in chemical equilibrium. The principle states, "Every system in chemical equilibrium undergoes, upon variation of one of the factors of the equilibrium, a transformation in such a direction that, if it had produced itself, would have led to a variation of opposite sign to the factor under consideration."[7] Le Chatelier's principle was adapted by Prigogine to open systems and elaborated by Onsager, Katchalsky, and others in nonequilibrium thermodynamics. In various forms, it reappears in biology (Cannon, v. Bertalanffy, Weiss, *et al.*), in sociology (Pareto, Parsons, *et al.*), in economics (Ricardo, Schumpeter, ...), political theory (Taylor, Kaplan...), and has gained such importance that Miller was led to state that no concrete system could exist without such self-regulation.[8]

"Cybernetic stability" in systems, i.e., the capacity of self-regulation by compensating for changing conditions in the environment through coordinate changes in the system's internal variables, is stated here as a form of adaptation by concrete systems which are ordered wholes. It can be restated by a simple formula advanced by Weiss to symbolize the characteristics of systems in general:[9]

> Let us focus on any particular fractional part (*A*) of a complex suspected of having systemic properties, and measure all possible excursions and other fluctuations about the mean in the physical and chemical parameters of that fraction over a given period of time. Let us designate the cumulative record of those deviations as the variance (v_a) of part *A*. Let us furthermore carry out the same procedure for as many parts of the system as we can identify, and establish their variances v_b, v_c, v_d, ... v_n. Let us similarly measure

[6] *Ibid.*

[7] H. Le Chatelier, "Recherches experimentales et théoriques sur les equilibres chimiques," *Annales des Mines,* 8eme Série, Paris, 1888.

[8] James G. Miller, "Living Systems: Basic Concepts," *General Systems Theory and Psychiatry, op. cit.*

[9] Paul Weiss, "The Living System: Determinism Stratified," *Beyond Reductionism,* A. Koestler and J. R. Smythies, eds., *op. cit.*

as many features of the total complex (S) as we can identify, and determine their variance (V_s). Then the complex is a system if the variance of the features of the whole collective is significantly less than the sum of variances of its constituents; or written in a formula:

$$V_s \ll \Sigma(v_a + v_b + v_c + \ldots v_n).$$

Weiss regards as the essential character of a system its invariance as a whole, compared with the more variant fluctuation of its constituents. Such systemic invariance is a general property of ordered wholes of interdependent parts, governed by fixed internal constraints and exposed to externally introduced perturbations. Ordered wholes of this kind constitute *systems*: within a given range of perturbations they return to the steady states characterized by the parameters of their constant internal constraints.

Inasmuch as systems of this kind reorganize their flows to buffer out or eliminate the perturbations (and thereby conserve a relative invariance of their total complex compared with the greater range of variation in their coacting components), they are *adaptive* entities. This is one form of adaptation in systems; not the only one. A more striking form of it, involving the reorganization of the fixed forces themselves, is discussed next.

$R = f(\gamma)$ (SYSTEM CYBERNETICS II)

Natural Systems: Adaptive Self-Organization

We have shown that ordered wholes, i.e., systems with calculable fixed forces, tend to return to stationary states following perturbations introduced in their surroundings. It is likewise possible to show that such systems *reorganize* their fixed forces and acquire new parameters in their stationary states when subjected to the action of a physical constant in their environment.

This conclusion follows if we consider Ashby's "principle of self-organization" with some modifications. The latter concern the substitution of "ordered steady state" for Ashby's "equilibrium state" in reference to *natural* systems. Undertaking the pertinent substitutions, Ashby's principle of self-organization reads as follows.

We start with the fact that natural systems in general go to ordered steady states. Now most of a natural system's states are relatively unstable. So in going from any state to one of the steady ones, the system is going from a larger number of states to a smaller. In this way it is performing a selection, in the purely objective sense that it rejects some states, by leaving them, and retains some other state by sticking to it. Thus, as every determinate natural system goes to its steady-state, so does it select.[10] The selection described by Ashby involves not merely the reestablishment of the parameters defining a previous steady state of the system after perturbation, but the progressive development of new steady states which are *more resistant* to the perturbation than the former ones.

Ashby suggests the following example. Suppose the stores of a computer are filled with the digits 0-9. Suppose its dynamic law is that the digits are continuously being multiplied in pairs and the right-hand digit of the product is going to replace the first digit taken. Since even *x* even gives even, odd *x* odd gives odd, and even *x* odd gives even, the system will "selectively evolve" toward the evens. But since among the evens the zeros are uniquely resistant to change, the system will approach an all-zero state as a function of the number of operations performed.

Ashby concludes that this is an example of self-organization of the utmost generality. There is a well-defined operator (the multiplication and replacement law) which drives the system toward a specific state (Ashby's "equilibrium state"). It selectively evolves the system to maximum resistance to change. Consequently all that is necessary for producing self-organization is that the "machine with input" (the computer+dynamic-law system) should be isolated. Adaptive self-organization inevitably leads toward the known biological and psychological systems. *"In any isolated system, life and intelligence inevitably develop."*[11] Or, to quote his more general conclusion, *"every isolated determinate system obeying unchanging laws will develop 'organisms' that are adapted to their 'environments.'"*[12]

The above argument applies to the present thesis with the suggested two modifications: (1) it is restricted to natural (as opposed to artificial) systems; and (2) the operator drives not toward a state of equilibrium in the system, but toward stationary or quasi-stationary *non-equilibrium* states. The reasons are potent for discarding the concept of the equilibrium state in favor of that of a non-equilibrium steady state in *natural* systems: (i) equilibrium states do not have available usable energy whereas natural systems of the widest variety do; (ii) equilibrium states are "memoryless," whereas natural systems behave in large part in function of their past histories. In short, an equilibrium system is a dead system—more "dead" even than atoms and molecules. Thus, although a machine may go to

[10] W. Ross Ashby, "Principles of the Self-Organizing System," *op. cit.*
[11] Ashby, *op. cit.*, p. 272. (Italics in original.)
[12] Ashby, *op. cit.*, p. 270. (Italics in original.)

equilibrium as its preferred state, natural systems go to increasingly organized *non*-equilibrium states.[13]

The modified Ashby principle shows that in any sufficiently isolated system-environment context, the system organizes itself as a function of maximal resistance to the forces which act on it in its environment. A number of other investigators have shown that, starting with an arbitrary form and degree of complexity (i.e., anything above the thermodynamically most probable equilibrium distribution), systems will complexify in response to inputs from the environment.[14] Such "natural selection" leads from populations of relatively simple systems to hierarchically organized sets and clusters, determined by adaptive traits and common origins. The evolution of any arbitrary complex system is always in the direction of merging some characteristics, differentiating others, and developing partially autonomous subsystems in a hierarchical sequence. Levins concludes that it is not necessary to postulate some unique rule for the synthesis of complex systems: the starting point can be "arbitrary complexity." "The dynamics of the system itself and the action of evolutionary forces on populations of such systems produces structure, merges some subsystems, subdivides others, reduces total interaction on some parts, gives spontaneous activity, organizes hierarchy, and transforms discrete systems and fields into each other."[15]

The principles of self-organization can be stated in the formula,

$$\frac{\textit{external}}{\textit{forcings}} \rightarrow \frac{\textit{internal}}{\textit{constraints}} = \frac{\textit{adaptive}}{\textit{self-organization.}}$$

However, an adaptively reorganized system is not necessarily a more *stable* system: adaptation is not synonymous with structural stability. An adapted system is optimally resistant to the kind of forcings which elicited the process of self-organization; it is not thereby more resistant to *all* factors in its general environment. In fact, normally just the opposite is the case: to the extent that adaptive self-organization occurs by means of a complexification of structure, the system becomes thermodynamically more "improbable" and hence structurally unstable and prone to physical

[13]Ashby himself seems to be aware of the difficulty in regard to living systems. He suggests that intelligence, as well as such phenomena as conditioning, association, learning, etc., may be the result of a process where the input comes to a system with *many* equilibria. But such a system would be cogently defined, it would seem, not as an equilibrium system, but as one with a number of fixed constraints.

[14]E.g., Sewall Wright, S. A. Kaufmann, R. Levins, *et al.*

[15]Richard Levins, "The Limits of Complexity," *Hierarchy Theory,* H. H. Pattee, ed., New York, 1973 (International Library of Systems Theory and Philosophy).

disorganization. Its increased adaptive potential derives from its higher functional capacity, afforded by the greater degree of freedom of the higher organizational structure. Hence systems evolve toward increasingly adapted, yet structurally unstable states, balancing their intrinsically unstable complex structure by a wider range of self-stabilizatory functions (Figure 2′).

Increased complexity of organization in a self-organizing system is measurable both as negative entropy, and as the number of "bits" necessary to build the system from its components. Every system produces entropy relative to time. Disorder in systems grows at a rate $\frac{ds}{dt}$. This is the dissipation function, Ψ.[16] Ψ may be positive, negative, or zero. If Ψ is zero, the system is in a stationary state. If Ψ is positive ($\Psi > 0$), the system is in a state of progressive disorganization. But if Ψ is negative ($\Psi < 0$), the system is in a state of progressive *organization*, that is, it actually decreases its entropy or, what is the same thing, gathers information ($\Psi < 0 = \frac{d \text{ info}}{dt} > 0$). The positive, negative, or zero entropy change (dS) is governed by the relative values of the terms in the Prigogine equation

$$dS = dS_e + dS_i,$$

where dS_i denotes entropy change through the input and dS_e entropy change due to irreversible processes within the system. Whereas dS_e is always positive, dS_i may be positive as well as negative. If the latter, the system "imports negentropy" (Schrödinger) and can not only offset disorganization by work performed within its boundaries, but can actually use the excess free energy to organize itself. Thus there is nothing supernatural about the process of self-organization to states of higher negative entropy: it is a general property of systems, regardless of their materials and origin. It does not violate the Second Law of thermodynamics since the decrease in entropy within an open system is always offset by the increase of entropy in its surroundings. The net change of entropy in every open-system—environment complex is positive

$$\frac{dS}{dt} = \frac{dS_e}{dt} + \frac{dS_i}{dt} > 0$$

since within the complex as a whole,

$$\frac{dS_e}{dt} > \frac{dS_i}{dt}.$$

Self-organization conduces systems toward more negentropic states; self-stabilization maintains them in their pre-existing state of organization. In

[16] Cf. Katchalsky and Curran, *op. cit.*

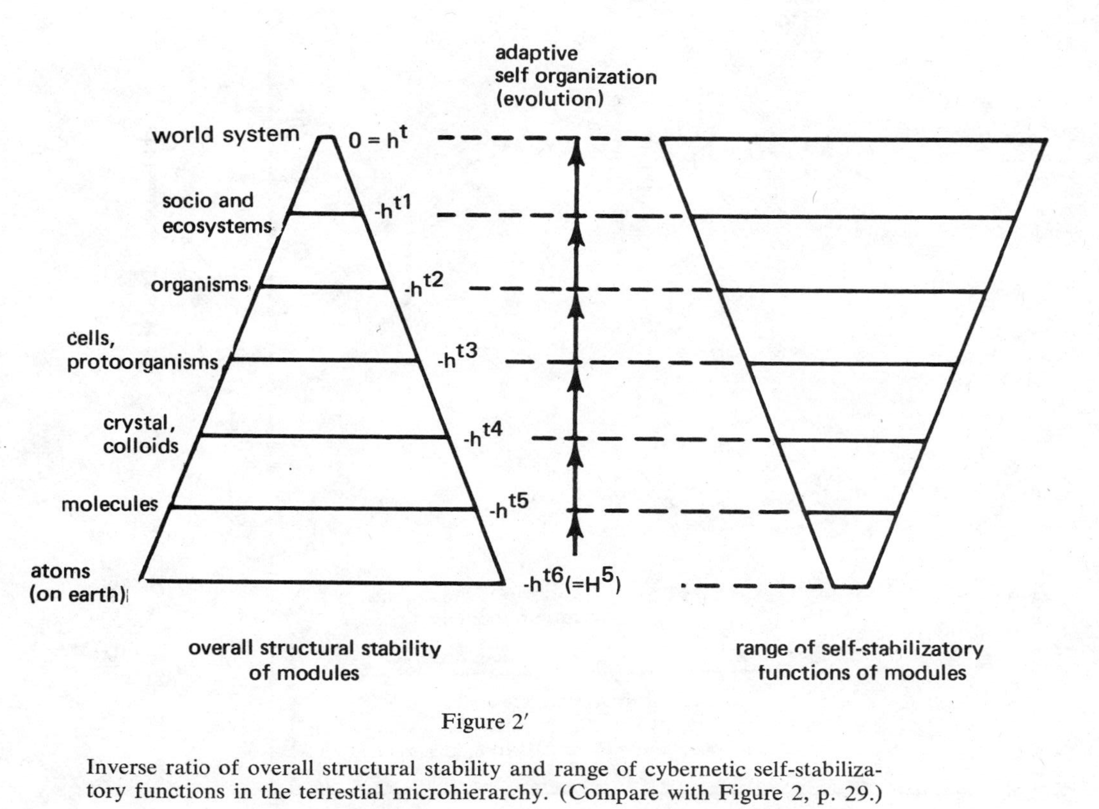

Figure 2′

Inverse ratio of overall structural stability and range of cybernetic self-stabilizatory functions in the terrestial microhierarchy. (Compare with Figure 2, p. 29.)

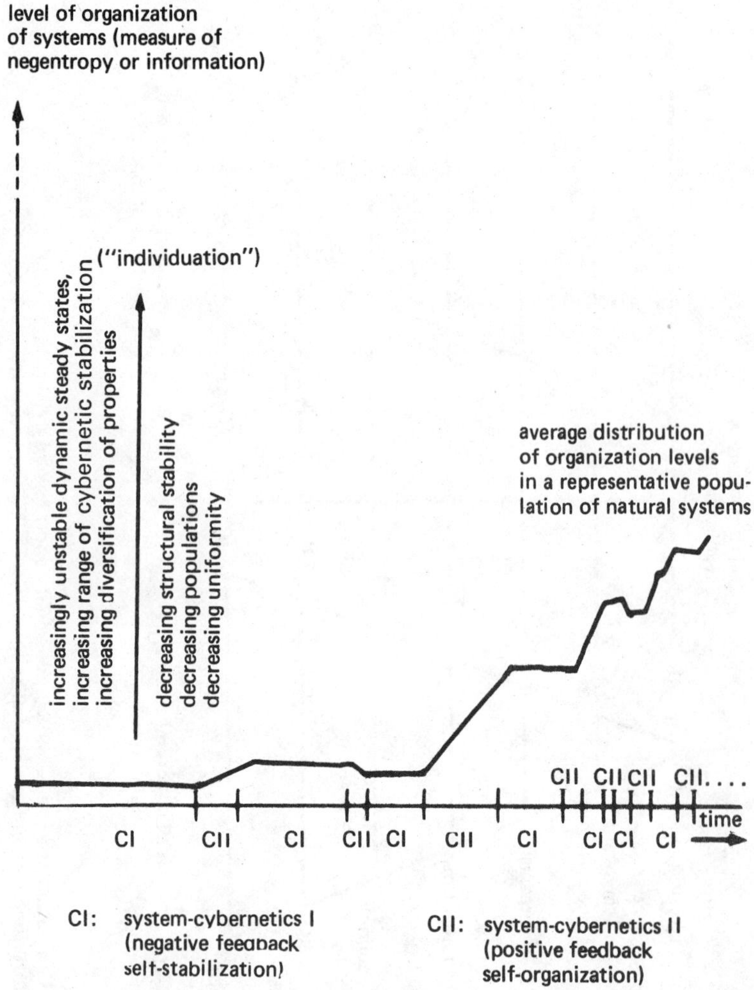

level of organization
of systems (measure of
negentropy or information)

("individuation")

increasingly unstable dynamic steady states,
increasing range of cybernetic stabilization
increasing diversification of properties

decreasing structural stability
decreasing populations
decreasing uniformity

average distribution
of organization levels
in a representative popu-
lation of natural systems

CII CII CII CII. . . .

CI CII CI CII CI CII CI CI CI CI

time

CI: system-cybernetics I
(negative feedback
self-stabilization)

CII: system-cybernetics II
(positive feedback
self-organization)

Figure 3

Typical evolutionary patterns of natural systems. Cycles of self-stabilization and self-organization conduce populations of systems to higher levels of organization as a function of time, and trade uniformity in large populations of structurally stable systems for diversity and differentiation in small populations of structurally unstable but cybernetically self-stabilizing systems. Duration of cycles diminishes with the increased sensitivity of the more organized systems.

an environment in which forces are constantly operative, and the perturbations they occasion are within the range of correction by self-stabilization, systems not only survive, but evolve. The development of systems in such environments can be conceptualized as a sequence of parallel, or irregularly alternating, *stabilization* around the parameters of existing fixed forces, and *reorganization* of the fixed forces as a function of increasing resistance to perturbing forces in the environment. (Figure 3)

Self-organization radically modifies the existing structure of a system and puts into question its continuing self-identity. A new system appears to emerge, synthesized from one or more older systems (as one helium atom is synthesized from four hydrogen atoms minus the radiated excess energy), or the new system seems to start a new line of reproducing entities (e.g., a new species brought about by mutation and natural selection). Also, the detailed mechanism whereby a change is induced in a system leading to its reorganization is often difficult to understand when that system is considered in isolation. It becomes necessary to choose a more suitable strategic level in the examination of the processes and mechanisms of self-organizing systems, and this level is that of the next higher suprasystem, i.e., the system formed by the given species of systems in a shared environment. Such a system can be seen to evolve notwithstanding the change of identity of all its constituent subsystems, and it can exhibit causal factors in the system-environmental relations—leading to reorganization as a response to environmental constants—which are obscured in the individual system view. Thus whereas the processes of self-stabilization can, in general, be clearly apprehended in reference to an isolated system viewed in relation to its environment, the processes of self-organization require that the strategic level of the next higher suprasystem be chosen for clear conceptual grasp. This is not to deny that self-organization takes place in a given system in relation to its environment, by means of the above mentioned processes of selective reorganization through the use of imported negative entropy; it is only to suggest that self-organization is better amenable to conceptualization from the viewpoint of a population of systems than it is from that of the self-organizing single system itself. (Details of the advantage of this strategy will come to light in the ensuing discussion of inorganic and organic systems.)

$R = f(\delta)$ (HOLON-PROPERTY)

Natural systems : intra- and inter-systemic hierarchies

In systems which constitute ordered wholes, adaptively stabilizing themselves in their environment around existing steady-states as well as evolving themselves to more adapted, and normally more negentropic (or informed) states, development will be in the direction of increasing hierarchical structuration.

The above conclusion follows directly from Simon's hypothesis, according to which complex systems evolve from simple systems much more rapidly if there are stable intermediate forms than if there are not.[17] The resulting complex systems are, as he points out, hierarchic. The reason for the decreased time-span required for a hierarchical form of complexification is that any failure in the organization will not destroy the system as a whole, but only decompose it to the next stable subsystem assembly. Then, instead of starting all over again, the process of complexification starts from the stable hierarchy-level and reconstitutes the loss in a relatively short time. Hence the observed predominance of hierarchies among natural systems are explained, on Simon's hypothesis, by the fact that hierarchical systems are the ones that have the time to evolve.[18] Given the self-organizingly adaptive property of natural systems, their mutual adaptations are more likely to result within any given time-span in a multi-level hierarchy than in a non-hierarchical structure. Phrased qualitatively, it is "simpler" for systems to cooperatively constitute higher systems than to do the job of complexification alone. And here simplicity is equal to efficiency and is measured by the time required for the process.

If, in favorable environments of enduring order, natural systems not only maintain their already attained level of organization, but intermittently or continuously evolve toward more highly organized states, *then*, given the relative rapidity of hierarchical over non-hierarachical organization, systems sharing a common environment will tend to form hierarchical suprasystems, rather than set forth their self-organizing processes on their individual level. Consequently an environment composed of such systems itself becomes a system, with the subsystems introducing the fixed forces which define its parameters. The medium of dense populations of natural systems, taking on the characteristics of a system on the next level of the hierarchy, *stabilizes* itself in the face of varying perturbations and *organizes* itself when confronted with constant forces in its environment. Thus such systems adapt themselves to their own (more inclusive) milieu. Provided that milieu is likewise populated with other systems of the corresponding level of organization, the suprasystems adapt to one another and jointly constitute systems of the 2nd level order. And so on, in a sequence of adaptive organizational interaction limited only by the number, extension, and density of the systems available for inclusion.

This concept of a natural hierarchy is an entailment of the concept of

[17]Herbert A. Simon, "The Architecture of Complexity," *Proceedings of The American Philosophical Society*, **106** (1962).

[18]*Ibid.*

self-stabilizing and self-organizing ordered wholes in common environments. It suggests a vertical law of organization by deduction from horizontal laws of interaction. It has a wide range of potential applicability to natural phenomena. Before exploring these it should be remarked that the concept of a hierarchy composed of systems defined by invariant general properties does not connote reductionism and is well capable of accounting for the manifest diversity of functions and properties in nature. Fresh "qualities" can emerge in the hierarchy in the form of new transformations of invariant properties. Such *nova* are accounted for by the consideration that systems at each level contain systems at all lower levels plus their combination within the whole formed at that level. Hence the possibilities for diversity of structure and function increase with the levels, and one need not reduce the typical characteristics of higher-level entities to those of lower levels but can apply criteria appropriate to their particular hierarchical position. The higher we raise our sights on the hierarchy, the more diversity of functions and properties we are likely to find, manifested by a smaller number of actualized systems. (Cf. Figure 2″) Thus atoms exist in greater numbers than molecules but have fewer properties and variations of structure; organisms exist in smaller numbers than molecules but have an enormously wide repertory of functions and properties and are capable of existing in untold variety of structural forms (the roughly ten million species of plants and animals are but a fraction of all the possible species which could be capable of existence), and the number of ecologies and societies is smaller than that of organisms, but already within their small population manifest greater diversity and flexibility than biological phenomena. It is evident that both the numerical and the functional differences are due to the hierarchical relations of the systems on the various levels: many systems on one level constitute one system on a higher level, consequently higher level systems are less abundant and have a wider repertory of functional properties than systems on lower levels. Thus to claim that all systems exhibit invariant properties and types of relationships does not entail reductionism: the invariances express themselves in specific non-reducible transformations corresponding to the degrees of freedom proper to each level of the hierarchy.

Now, the concept of "hierarchy," while much used in the contemporary natural scientific and philosophic literature, is seldom defined rigorously, and when it is so defined, it is often inapplicable to the phenomena for which it is most often used.[19] A rigorous definition implies a governing-

[19]Cf. Bunge's rigorous definition in terms of a set equipped with a relation of domination or subordination which, in his own admission, is inapplicable to physical and biological phenomena. Mario Bunge, "The Metaphysics, Epistemology and Methodology of Levels," in *Hierarchical Structures, op. cit.*

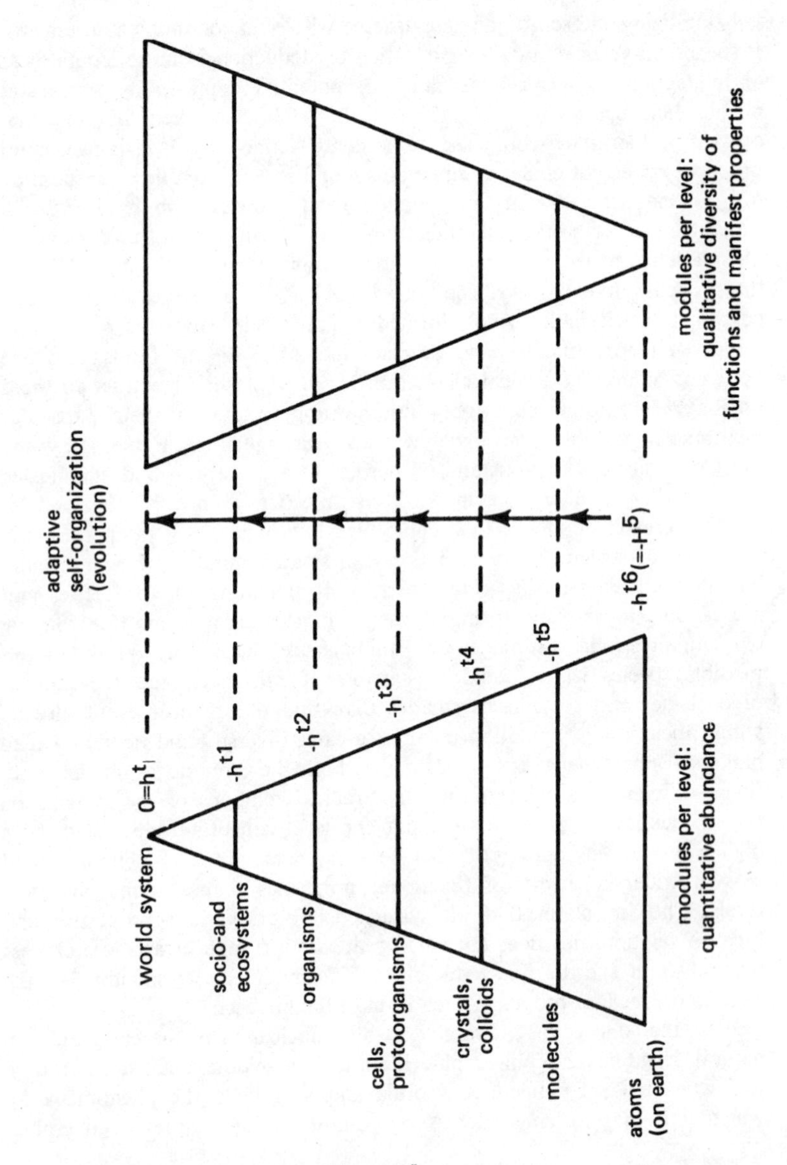

Figure 2″

Inverse ratio of quantitative abundance and qualitative diversity in the terrestrial microhierarchy. (Compare with Figure 2′.)

governed or "bossing" relation between levels, so that a diagram of a hierarchy becomes a finite tree branching out of a single point, without loops. Such hierarchies apply at best to military or quasi-military organizations with established non-reciprocal chains of command. But hierarchies have found their most fruitful application in nature, where rigorously unidirectional action is hardly ever the case. Hence in the present use the concept of "hierarchy" will not be given its rigorous meaning but will denote a "level-structure" or a "set of superimposed modules," so constituted that the component modules of one level are modules belonging to some lower level. Using the term 'system' for 'module,' we can speak of a hierarchy as a "multi-holon" type structure in which the systems functioning as wholes on one level function as parts on the higher levels, and where the parts of a system on any level (with the exception of the lowest or "basic" level) are themselves wholes on lower levels.

Systems belonging to a level below that of any chosen level are called "subsystems" in relation to the system of the chosen level, and the system belonging to the next higher level is a "suprasystem" in relation to it. The relativity of these terms is evident: a given system a may be a subsystem in relation to b and a suprasystem in relation to c. Merely that $(c \subset a) \subset b$ is required (where $\subset$ is a symbol of relative inclusion). Then b is a suprasystem in relation to a, and c a sub-subsystem in relation to b.

A given system a has component subsystems $c_1, c_2, c_3 \ldots c_n$ in determinate association, the sum of which is expressed as R. System a is part of a hierarchy when its components are likewise systems and when it is itself a component in a more encompassing system:

$$[a = (c_1, c_2, c_3 \ldots c_n)^R] \subset b$$

where a is a subsystem (component) of b, and all c's are comparable systems in relation to which a is a suprasystem.[20]

The resulting hierarchy is of the "Chinese box" variety, and is theoretically infinite. However, infinite hierarchies lack empirical interpretation, and our task is therefore to propose a finite-level hierarchy and identify each of its levels with one class of observable (or constructable) natural systems.

Attempts of this kind have often been made and, until relatively recently, came under the heading of an ontological category scheme (one of the latest major systems of this kind being that of the German philosopher Nikolai Hartmann). More recently, this type of endeavor has been taken over by general systems theorists, who wish to establish similarities as well as dif-

---

[20] Based on Clifford Grobstein, "Hierarchical Order and Neogenesis," *Hierarchy Theory*, H. H. Pattee, ed., *op. cit.*

ferences between systems encountered in various empirical domains. Thus Boulding supplied key notions of a "hierarchy of systems" which von Bertalanffy tabulated as a table of system levels, theories and models, and empirical descriptions.[21] It includes both natural and artificial systems (e.g., both atoms, molecules and organisms; and clockworks, control mechanisms and symbolic systems). The hierarchy we are concerned with is less inclusive than this, dealing only with *natural* systems, and more rigorous in one basic regard: its levels follow the hierarchical scheme of relative inclusion without gaps or redundancies. Thus we seek to order natural phenomena into a "vertical" order wherein any given system, with the exception of those on the lowest or *basic* level, and that (one) on the highest or *ultimate* level, is both a supra-system in regard to its components, and a subsystem in regard to the totality which it forms together with other systems in its environment. Hence, from the viewpoint of a system of level n, there is an *intrasystemic hierarchy* of its structure-functional constitution, made up of the hierarchically ordered series . . . $[(a \subset b) \subset c] \subset n$, as well as an *intersystemic hierarchy* consisting of the structure-functional wholes constituted by its environmental co-ordinations with other systems, $[(n \subset x) \subset y] \subset z$ . . . Since n is situated at the intersection of the intra- and inter-systemic hierarchies, the number of levels in each defines n's specific position within the objective level structure of a given microhierarchy.

This hierarchy of rigorously inclusive (but not necessarily rigorously governed) natural systems is akin to Gerard's hierarchy of "orgs." For Gerard, "org" denotes "those material systems or entities which are individuals at a given level but are composed of subordinate units, lower level orgs."[22] If a hierarchy of this kind could be empirically confirmed, a basic ideal of science would be realized: the many entities investigated by the diverse empirical sciences would be plotted on a map of hierarchical organization and the theories applicable to them could thereby be interrelated. But such confirmation encounters serious difficulties: the identification of different levels with empirically known entities is often

[21] In *General System Theory*, pp. 28–29.

[22] R. W. Gerard, "Units and Concepts of Biology," *Science*, **125** (1957). Gerard subsequently broke down "org" hierarchies into two classes: those consisting of non-living *orgs* (from atoms, molecules, crystals and larger units, to geographical, geological, astronomical units); and those consisting of living *animorgs* (unicellular, multicellular organisms and epiorganisms, i.e., multi-individual living systems such as anthills, villages and the like). Cf. "Hierarchy, Entitation, and Levels," *op. cit.* His classification contrasts with the present, which distinguishes two types of hierarchies, the astronomical "fields-to-galaxies" *macrohierarchy*, and the physico-bio-sociological "atoms-to-ecologies" *microhierarchy*. The former consists entirely of non-living "orgs," whereas the latter bridges the gap between "orgs" and "animorgs."

problematic or unclear. These difficulties will engage our attention when we confront the present hypothesis with empirical evidence. (Chapter 5) But empirical difficulties of identification do not adduce evidence for the inapplicability or falsity of the concept of natural hierarchy, only for the methodological and observational problems of identifying its rungs with classes of empirical entities. The hierarchy concept conserves its appeal on the basis of its supreme potential as an ordering principle, relating, through invariant hierarchical relationships, empirically dissimilar phenomena.

Chapter 5

EMPIRICAL INTERPRETATIONS

The principal propositions of the hypothesized theory of natural systems have been formulated; all further theorems should be derivable from these through the rules of formal logic. However, such rigor remains an ideal for the present; we must content ourselves with showing that cogent interpretations can be derived from the above propositions for the principal varieties of empirical phenomena. The interpretations do not involve modifications of existing theories, merely their translation into the synoptic framework offered by our theory. There can be no question of adjudicating among scientific theories or of revising them; the task is to map them into a common, internally consistent framework wherein their particular propositions become mutually reinforcing as descriptions and explanations of one reality with a rationally knowable, overarching species of order. Such order is an assumption concerning nature and is not evident when the theories of the special sciences are examined one by one. But if these theories, which in themselves exhibit but limited types of order in nature, should prove to be capable of integration with a second-order theory—a model of the empirical-theoretical models—the assumption of general order (or orderability) in nature will have been vindicated: orders are not "givens" of immediate experience but hypotheses of the rational mind, and if experience is orderable, there are no valid reasons for including special orders but excluding a general one.[1]

Nature offers a vast field of events and processes for observation and experimentation. It is not possible in a work smaller than an encyclopedia to survey all phenomena of nature, or even a majority of them. What we can do instead, given the limited space of a single book and the modest knowledge of one investigator (who is particularly conscious of glimpsing but a small slice of the spectrum of contemporary knowledge), is to select scientific theories in key areas of investigation and consider whether they can be integrated within the aforementioned second-order framework. If areas which are ordinarily viewed as being very different in themselves can be encompassed within a general theory of systems, we will have

[1]Cf. the "Introduction," above.

evidence for the cogency of arguing for general order in nature, demonstrated by means of a second-order "meta-model."

The vast panorama of organized complexity offered in nature can be categorized in numberless ways. However, when general categories are sought, the concepts of *physical, biological* and *social* phenomena acquire paradigmatic colors. Within each of these broad categories many different entities are subsumed; yet here, too, some gain paradigmatic value over others. *Atoms* within the range of physical systems, *multicellular organisms* within that of biological entities, and *human societies* within the scope of social systems stand out as primary foci of investigation. Hence, without underestimating the importance or denying the reality of a host of other phenomena of organized complexity, and of intervening categories and hierarchical levels, we shall undertake an empirical interpretation of the foregoing general theory of natural systems in regard to *atoms, organisms,* and human *social systems*. The relevant scientific theories will be those of atomic physics, biology (and theory of evolution), and sociology. Our task is to explore, whether or not these key entities on the physical, biological and social levels can be reconceptualized as subclasses of natural systems, characterized by the same independent variables, and hence sharing invariant fundamental properties.*

---

*Readers who are more interested in the philosophical implications of the theory than in its empirical foundations in contemporary science, may prefer to skip the present chapter and continue on p. 119.

(i) PHYSICAL SYSTEMS (Atoms)

THEORY: $R_i = f(\alpha, \beta, \gamma, \delta)$, where $\alpha, \beta, \gamma, \delta$, are independent variables having the joint function R_i ("atomic natural system").

$R_i = f(\alpha)$ (SYSTEMIC STATE PROPERTY)

Atoms: wholeness

Theories of the atom have undergone rapid modification in this century, and have come up with state descriptions which exhibit atomic structure more and more as an instance of what Warren Weaver said was the subject matter of *biology,* namely "*organized complexity.*" Gone is the mechanical "solar system" model of the atom with indivisible entities exercising well-defined deterministic effects on one another. In its place we have the conception of a system of fields within which are embedded particle-like energy and spin concentrations. The character of the entire structure is not the simple sum of the character of what appears within it as elementary particles, or even of the forces between them. Rather, the character of the whole equals the sum of the relations of the components in the exact ordering in which they are found. Mathematically, the non-summativity of atoms is expressed by the change in form of the equation describing changes in the whole and change in any given atomic component. In terms of thermodynamics, there is a difference between the sum of the entropy of the particles in equilibrium and the entropy of the atomic structure as a whole.

The difference appears likewise in information theory, where it is measured in "bits." The entropy of the whole atom is less, and its information content more, than that of the sum of its parts at equilibrium (electrons + nucleons). This is further emphasized in the modern conception of the atom as constituted by functionally interacting nuclear and electronic fields, rather than by mechanically interacting parts. In fact, the mechanical (summative) model of the atom has been definitely discarded in quantum theory.

Atoms: order

Despite the well-known uncertainties associated with the state of the electron, the notion of mathematically calculable order within the atomic structure has not been surrendered. Merely stochastic probability laws describing statistical regularities have come to replace exact dynamic laws. Atomic physicists no longer think of the electron as a small sphere revolving around the nucleus, but they have not, for all that, ceased to think of the relation of the electron to the nucleus (as well as of the particles of the nucleus among themselves) as highly ordered.[2] In fact, the opposite is the case: atomic structure reveals a most remarkable variety of mathematical order. Let us consider the order embodied in the complexification of atomic structure as shown in the periodic table.

The central nucleus of the atom is surrounded by something like an "electron-fog," calculated in terms of relative probabilities as to the location of the electron. The fog is "densest" in the region of the highest probability and thins out toward the edges, making up in its full range the permissible energy band of the given electron. The simplest atom, hydrogen, has a single electron-fog associated with its nucleus. Helium has two positive charges (or protons) in its nucleus and a corresponding two negative charges (i.e., electrons) in the orbital "fogs." "Paired" electrons, such as those of helium, can be distinguished in reference to the concept of *spin*, which allows a distinction to be made between particles having negative and those possessing positive spin. Any one spin is readily complemented by its counterpart, and both spins together complete the electron fog in a manner which, as a rule, gives rise to short-range forces with other orbitals and produces very little chemical activity. Consequently, surrounding the helium nucleus we have an electron fog composed of a matched set of electrons of low combinativity; this explains why helium is ordinarily found as a gas.

The three-electron atom introduces a new structural pattern into the picture. This atom is lithium, and it is made up of the paired electrons of helium plus an extra electron. The additional electron is excluded from the lowest shell (or orbital) around the nucleus and must occupy a position further removed from it.

When the nucleus of an atom contains four positive charges, it can accept four negatively charged electrons. Then, the third electron of lithium is paired with a fourth, occupying the same orbit but having opposite spin. The atom we thus get is beryllium. Adding still another electron, we encounter boron. Here a new structural principle appears. The possibilities of negative and positive spin on the first and second shells have already been used up by the four electrons of beryllium. The extra electron of boron is forced into an unsymmetrical pattern. The spatial orientation of the fifth electron will be perpendicular to the plane of vibration of the other electrons. The dynamic form of this electron is the

[2]An example of the pervasiveness of the contemporary belief in "elegant" mathematical order is the discovery of mesons. Rather than agree to the presence of entirely different orders and laws within the nucleus, Yukawa postulated the *meson*, which permitted the application of consistent equations to electronic and nuclear fields. (Mesons were subsequently confirmed in cosmic ray researches.)

p orbital. It is a two-loop "vibration pattern" which occupies one of the three independent orientations permitted by space: horizontal, vertical, or transverse.

Without going into details of chemical combinativity, we can simply note the ordered structural patterns we get as we add electrons to existing structures. One more electron, corresponding to an extra proton in the nucleus, gives us carbon. Another electron, corresponding to an additional proton in the nucleus, gives us nitrogen. Oxygen contains eight electrons, with the new electron pairing with one of the three occupying the outermost shell. The ninth electron of fluorine pairs with the second such electron and the tenth electron of neon with the third. Consequently neon has a closed pattern and a very small potential for combining with other atoms.

As we build up the various kinds of atomic structures, the three-loop pattern allows us eight chemical elements, and each additional shell, with its correspondingly wider range of possibilities of spin and orientation, successively produces the structures of the heavier elements. The complexification of structure can be determined directly from the equations of wave mechanics. We get the first family of 2 elements (hydrogen and helium), followed by the next where we have $2+2\times3$, and then $2+2\times3+2\times5$, and finally $2+2\times3+2\times5+2\times7$, and so on, leading to the heaviest elements with the number of electrons approaching the one hundred mark. Here we have a clear and beautiful example of the way in which the building up of atoms mirrors the structure of mathematics. The complexification of increasingly heavy atomic species follows the laws of wave-mechanics, abundantly testifying to the applicability of concepts of order to the realm of atomic actuality. (Figure 4)

There are further remarkable exemplifications of mathematical order in atoms. There is, for example, a clear-cut relationship between the abundance of a species and the structure of its nuclei: nature appears to favor even numbers both as regards the number of electrons and protons (the atomic number) and the number of neutrons in the nucleus. In the earth, elements of even atomic numbers are estimated to be ten times those of odd numbers, while in meteorites the ratio is about seventy to one. Even more striking is the fact that 99 per cent of all atoms in the earth possess an even number of neutrons in their nuclei. Moreover nuclei of even mass numbers are more abundant than those with odd numbers. Of the 272 known stable nuclei, 162 have even numbers of protons and neutrons, 55 even numbers of protons and odd numbers of neutrons, 49 odd numbers of protons and even numbers of neutrons, and just 6 have odd numbers of both protons and neutrons. The complexities of atomic structure resolve themselves into relatively simple mathematical relations of supreme balance

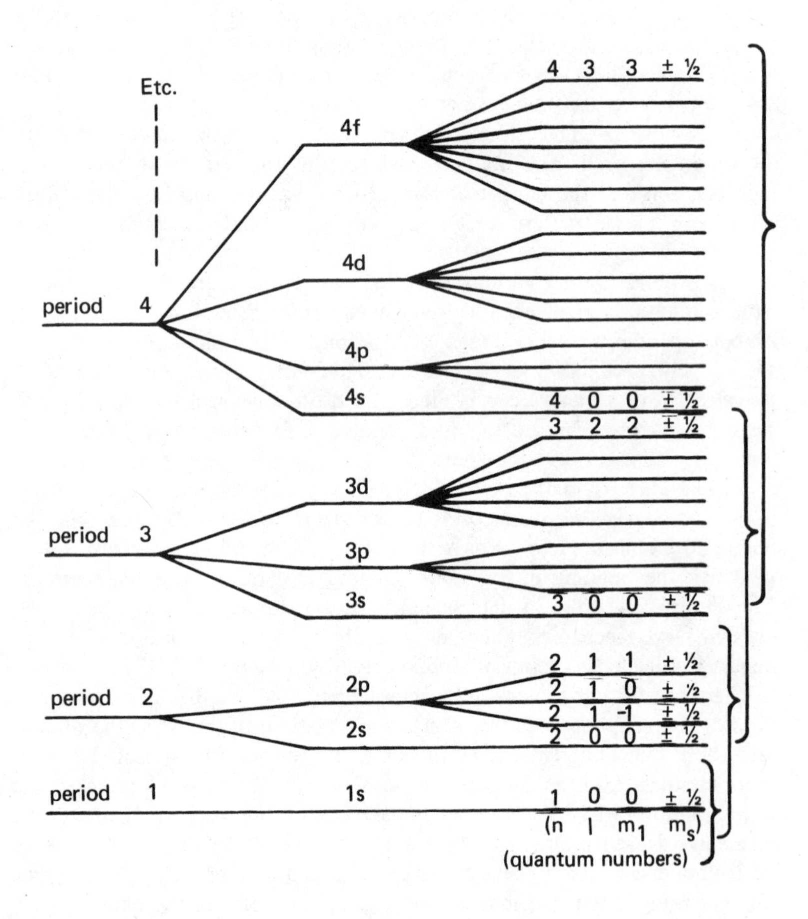

Figure 4

The complexification of the elements, based on the Mendeleyev periodic table. Given in Harold G. Cassidy, "Meaning of the Generalization of Mendeleyev's Table. Summary of Theoretical Issues," in E. Haskell, ed., *Extension of the Structure of Mendeleyev's Periodic Table to Physical, Biological, and Social Sciences,* Proceedings of the Twenty-first Anniversary Symposium of the Council for Unified Research and Education, AAAS, Boston, December, 1969.

and symmetry. If some mathematical physicists feel moved to exclaim that if there was a Creator, he must have been a great mathematician (essentially the position Plato took in the *Timaeus*), it is because there are a number of fixed forces introducing constant constraints within the atom, and these forces (electronic attraction and repulsion, exclusion, nuclear exchange forces) introduce conditions with highly ordered parameters. As a result atoms are not random heaps of unordered parts but ordered wholes, fully satisfying our theory's relevant postulate.

$R_i = f(\beta)$ (MICROPHYSICAL CYBERNETICS I)

Definition of "microphysical cybernetics"

The conclusion that atoms form ordered totalities is relatively clear-cut and un ontroversial. By contrast the next two postulates, those namely of cybernetic processes amounting to sequences of *self-stabilization* and *self-organization*, appear highly controversial in regard to the atom. In fact, at first sight they may appear to be contradicted by empirical evidence. It is my belief that this assessment is spurious and is only due to a misconception of the meaning attributed here to the concepts "self-stabilization" and "self-organization." Self-stabilization means that the ordered whole in question maintains its parameters when disturbed, provided the perturbations are within the limits of its capacity for correcting for them. Self-organization, then, stands for the process whereby the typical parameters of the system undergo modification due to the process of constant perturbations acting upon the structure in its surroundings. It is clear that both self-stabilization and self-organization presuppose interactions with the environment and can, therefore, be only performed by *open* systems.

Theoretically, systems can be fully closed, i.e., they can be equipped with adiabatic boundaries which prevent the exchange of both matter and thermal energy between the system and its environment. Such systems cannot interact with their surroundings otherwise than through mechanical work not accompanied by an exchange of heat or matter. Systems can also be less rigorously closed: they may have diathermal boundaries which allow the exchange of thermal energy but prevent that of other forms of matter and energy (other than mechanical work). Finally, systems may be "open," i.e., have boundaries which are permeable to certain forms of matter and energy, although not to all, and hence act as sieves rather than as walls. It is evident that if self-stabilization and self-organization presuppose interactions between system and environment, closed systems are excluded from this phenomenon. In general, the greater the range of effective interaction between system and environment, the greater the effect of perturbations introduced by external vectors in the system. Consequently the greater is the requirement on enduring systems for correcting for these perturbations, i.e., for adapting to conditions in the environment through cybernetic processes. Physics shows that atoms, as enduring systems in dynamic environments, do indeed interact with the surrounding medium from time to time, and that these interactions manifest functional properties which signify the here discussed cybernetic processes on the level of microphysics ("microphysical cybernetics").

Atoms : adaptive self-stabilization

Stable atoms in ground energy states are closed systems with diathermal boundaries. Perturbations reaching them from the environment must be in the form of thermal or kinetic energies. In the absence of such vectors, stable atoms maintain themselves in their medium without actual interaction. They dispose over a structural-dynamic balance of energies in virtue of the integration of the particles within the atomic fields. For example, from the viewpoint of an electron, the atom behaves as the center of an attractive force which exists at all points but whose magnitude decreases proportionately to the distance from the nucleus. By means of this negative potential energy, electrons are bound as "standing waves" to the nucleus.

It is when the normally closed dynamic pattern of the atom is momentarily opened that we could encounter the phenomenon of "adaptation through self-stabilization." A temporary opening of the structure does occur under electron bombardment. When an atom is bombarded with electrons of well-defined velocity, or with electromagnetic radiation of a defined wave length, it is found that unless a given minimum amount of energy is supplied by the bombarding agent an electron cannot be dislodged from the atom. The interaction between the electron in the atom, and the one shot into it, is an elastic one, since the latter bounces off without loss of energy. However, when the threshold of energy is surpassed, effective interaction can occur : the atom absorbs the entire energy of the bombarding agent and ejects one of its own electrons. The ejected electron possesses the absorbed surplus energy in the form of kinetic energy.

The threshold of elasticity is usually exceedingly sharp. When it is transcended and an electron is completely knocked out of the atom, the atom can capture another electron and readjust itself by one, or by a series of, quantum jumps. The atom thus returns from an excited state, occasioned by the removal of one of its outer electrons (whereby it becomes a positive ion) to its normal (neutral) state. This is the stable state of lowest energy. The readjustment, by means of radiating off the surplus energy (excitation potential), may be effected extremely quickly, with some excited states lasting no longer than 10^{-8} seconds.

The above amounts to a process of effective self-stabilization in the atom, with an output (radiation) balancing an input (bombarding agent) as a function of re-establishing the atom's appropriate steady (ground energy) state. The self-stabilization is momentary rather than continuous, as in typically *open* systems. However, the phenomenon does occur and evidences that atoms satisfy the postulate of adaptation to their surroundings through self-stabilization whenever perturbations are effectively introduced from the medium. The range of perturbations which can thus be corrected for encompasses at its lower limit the atom's threshold of elasticity, and at the upper boundary the temperatures under which nuclear fission occurs. Any input of energy below the threshold of elasticity bounces off without loss of energy by the atom and produces no structural changes in it; and every input over the "nuclear cooking boundary" disrupts the existing atomic structure. But between these limiting values disturbances from the environment are normally compensated for through proton capture and the radiation of the surplus energies. Since the atom maintains itself in its typical steady (ground energy) state by means of these processes, the latter can be meaningfully considered adaptations to the temporary disturbances from the surrounding medium. In this sense, atoms *de facto* adapt to their environment by means of a pattern-maintaining cybernetic process.

$R_i = f(\gamma)$ (MICROPHYSICAL CYBERNETICS II)

Atoms: adaptive self-organization

Self-organization, similarly to self-stabilization, presupposes an interaction between system and environment. In the case of the atom, effective interaction can occur above the threshold of elasticity. Above this level and below that where nuclear fission takes place, atoms react by the above outlined processes. These amount to self-stabilization; not, however, to self-*organization*. The latter could occur, therefore, only at temperatures *above* the threshold of nuclear fission. It is into these processes that we shall inquire.

With the exception of artificially created conditions in reactors and nuclear fission devices, the temperatures required for nuclear chemistry obtain only in stars. Recent astrophysical theories suggest that the heavy elements are progressively built up from the lighter ones in stellar interiors, in sequences of nuclear transmutation processes. The theories tend to disregard cosmogonical questions concerning the origin of matter in the universe and assume that the lightest and simplest elements, hydrogen or hydrogen nuclei, constitute the raw material of all further astrophysical processes.

Perhaps the first state of the universe which can be inferred with any amount of certainty is that wherein some 10^{69} hydrogen nuclei swirled loosely bound in interstellar gases. These primordial hydrogen clouds, carried about by vast cosmic currents, formed vortexes and condensed into protostars under gravitational pressure. Thus the older stars of Population II were born. Stars of this population began to undergo recognizable life-cycles, including formation, evolution and eventual explosion (novation), and the reconstitution of younger Population I stars from the scattered matter of the older stars.

The immediately pertinent and most remarkable fact concerning these processes of stellar evolution is the essentially one-way synthesis of progressively heavier atomic structures. The evolution of stars is cyclic; their life histories resemble the histories of biological *individuals*: they include birth, development and death, with new stars forming out of the remains of the old ones. But the evolution of atoms is linear, analogously to the evolution of biological *species*. It progresses from the simple to the complex, from the light to the heavy elements, notwithstanding occasional local reversals.[3]

Relevant work, especially by Fowler and M. and G. Burbidge,[4] suggests that as many as eight nuclear transmutation processes are involved in the full-scale building up of all known naturally obtaining elements in stars. The first and most sustained of these is the conversion of hydrogen into helium, either through the carbon-nitrogen cycle or through the proton-proton chain. After most of the star's hydrogen has been converted into helium, rapid structural changes bring about the so-called helium-burning process, whereby various intermediate weight elements are synthesized. The presence of nuclei produced by these processes permits the synthesis of the majority of the less abundant isotopes of carbon, oxygen, neon and nitrogen by means of the ordinary hydrogen-helium transmutation reactions.

Thus hydrogen and helium conversion processes account for a considerable number of the elements known to exist in stars. But the synthesis of *all* such elements requires the postulation of no less than six additional processes of atomic transmutation. These include the α-process, the *e*-process, the slow and rapid neutron-capture processes, the *p*-process (a modification of the process of proton-capture), and a process accounting for the synthesis of some light elements, such as deuterium, lithium, and boron, in nuclear reactions occurring in the atmosphere of stars and in the shells of

---

[3]If the analogy to biological evolution is significant, the reader may wish to conclude that the development of matter in the cosmos is more fruitfully studied by taking atoms, rather than stars, as basic units of investigation.

[4]Cf. Margaret and Geoffrey Burbidge, "Formation of Elements in the Stars," *Science*, Vol. 128, No. 3321 (1958), also Fowler *et al.* in *Astrophys. J.* 1955 & seq.

supernovae. These eight types of nuclear reactions present a cogent theoretical framework for the quantitative analysis of the observed cosmic abundance of the various species of atoms. While a full account is dependent upon a number of further factors (including the rate of star formation and the manner in which synthesized matter returns to interstellar space), much of the evidence points to the fact that all the known naturally occurring species of atoms have been built in stars in successive stages from hydrogen. Since some of the species are produced in stellar interiors and others are synthesized during the extremely high temperatures reached in a supernova explosion, it is clear that, whereas the life-cycle of stars includes the phases of initial condensation, structural transformations and partial or total eruption, the evolution of atoms is a continuous and on the whole irreversible process. The very event which destroys a star, namely a supernova explosion, actually represents a further step in the evolution of atoms: it provides the temperatures needed for the synthesis of some of the heavier elements.

The evolution of progressively heavier atoms in astrophysical processes exemplifies an ongoing process of structural development in the realm of atomic nature. Atoms are enduring structures open to perturbation by thermal and kinetic energies in their environment. Stable atoms exist in ground energy states and can maintain themselves in such states if thermal and kinetic forces in the medium are below the nuclear fission values. If energies in the environment exceed these thresholds, the closed pattern of the atomic fields is disrupted (e.g., under extremely high temperatures). Then nuclear fission occurs, conducing, under certain conditions of density and temperature, to the fusion of hitherto separate nuclei into one nucleus of greater structural complexity.

Nuclear transmutation processes may be seen as reorganizations of existing atomic structures toward higher negentropy states when not the individual atomic structure, but a *population* of atoms in a certain region in a stellar body, is considered. Such populations transform simpler structures into more complex ones and radiate off the energy differential. Thus the negentropic trend of atomic structures does not contradict the Second Law : the level of organization of the stellar substance as a whole decreases, notwithstanding the building up of the atoms of the elements in synthetic processes.

Atomic evolution is at best a paradox when viewed on the level of the individual atom : the evolving structure transforms into another kind of structure through a disruption of its existing forces. Thus the individual atom disappears in the process, to reappear within another structure in which other, formerly individual, atoms also participate (e.g., a helium atom composed of four formerly distinct hydrogen atoms). But viewed on the level of an atomic population making up a layer of a stellar substance, atoms appear to evolve: the transformation of the hydrogen content of the central layers of the sun, for example, into helium, represents the self-organization of the population of solar atomic systems toward a more organized state, notwithstanding the higher entropy of the sun as a whole. Energy is channelled into systems that can maintain themselves under conditions where less organized systems would disintegrate. However, the increased adaptive capacity of the resulting systems does not connote a higher level of structural stability. Just the opposite is the case: the more organized heavy atoms are the structurally less stable ones. Nevertheless, they are capable of self-maintenance under stellar conditions which result in fission for the less organized lighter elements. Adaptive organization is synonymous with compatibility with actual environmental forces, not with structural stability. Thus atomic evolution drives toward functionally more fit, if structurally less stable, forms. In so doing it parallels evolution in other sectors of systemic organization (cf. Sections (ii) and (iii), below).

Atomic *populations,* rather than individuals, evolve functionally adapted (even if inherently unstable) structures in response to conditions in their particular environment. This hypothesis does not contradict the current theories of nuclear and astrophysics, but translates their propositions into the language and the conceptual framework of a general theory of natural systems.

$R_i = f(\delta)$ (HOLON-PROPERTY)

Atoms : intrasystemic hierarchy

The property of constituting Janus-faced "holons" (wholes in regard to the components and parts in relation to certain sets of nexūs in the environment) is not an independent property of natural systems, but a resultant of the sum of the above discussed properties. As illustrated in Figure 3, natural systems, when subjected to varying perturbations in their environment, not only maintain their existing level of organization in an invariant steady state, but tend to organize toward new steady states, which are usually more organized than the previous ones. In stellar environments of changing temperatures and densities, atomic populations tend to evolve toward more complex forms through nuclear fusion (Figure 5). Each type of atomic structure represents a hierarchically organized whole of its nucleonic and electronic subsystems. The shells are built up according to the periodicities mapped in the Mendeleyev table of the elements. Each new shell is superimposed on the filled shells of the simpler atoms, paralleling the addition of neutrons in the nucleus. The addition of a new shell modifies the previous shells through the increased charge in the nucleus,

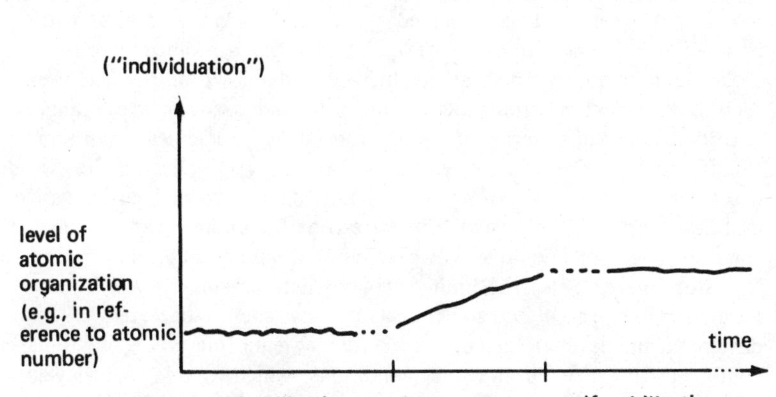

CI	CII	CI
(readjustments to ground-energy states [in individual atoms])	(nuclear fusion processes [in populations of atoms])	(readjustments to ground-energy states [in individual atoms])
(e.g., in hyrdogen)	(e.g., hydrogen → helium)	(e.g., in helium)

Figure 5: patterns of development in physical systems.

as well as through the presence of new electrons. Thus each higher period comprises all the structural characteristics of the lower periods plus new characteristics, which modify the rest. As Cassidy points out, such an arrangement clearly exemplifies a systems hierarchy.[5]

Atoms: intersystemic hierarchy

Although an *intra*systemic hierarchy is definitely exemplified in all atoms, *inter*systemic hierarchies are contingent in their regard. In environments where energetic conditions are relatively constant and permit the formation of higher-order structures, molecular bonding obtains, whereby atoms share unpaired outer electrons in a common structure. The jointly formed molecular system belongs to a higher level than any of the component atomic subsystems (since it includes each subsystem plus the relations between them), and, from the viewpoint of a given atom, the rest of the molecule can be regarded as its hierarchical "environment." However, such hierarchical environments occur only under a restricted range of conditions, defined in terms of temperature, density, chemical environment, etc. If the conditions obtain, a microhierarchy is initiated and can lead, if further requirements on favorable conditions are satisfied, to an evolved multilevel structure, of which the terrestrial hierarchy is an instance. If conditions are *not* favorable, however, multiatomic structuration can come to a stop, or regress, or be limited to weak interactions between relatively isolated atoms (e.g., in interstellar space). Hence the presence, as well as the developmental level (if given), of an *inter*systemic hierarchy of atoms is contingent upon environmental conditions which decide the development of a microhierarchy, or its total or partial absence.[6] But if a microhierarchy is even primitively evolved, some atoms in its region will be components in larger suprasystems. Hence such atoms exhibit the here discussed holon-property: they appear as wholes in regard to their subatomic parts, and parts in regard to their supra-atomic (molecular, crystalline, etc.) environment.

(ii) BIOLOGICAL SYSTEMS (Organisms)

Physical nature was long considered radically different from the organic world. Until as recently as the 1930s, it was common practice for both physicists and biologists to make casual reference to each other's domains

[5]Cassidy, *op. cit.*

[6]Cf. Figure 2, where the presence of a microhierarchy is mapped as supervening upon the part-whole relation of atoms and stellar bodies (level −H5 of the macrohierarchy).

in terms of areas lying entirely outside their fields of interest and having little if any connection to them. The emergence of biophysics and biochemistry brought the two fields closer together, but did this at the expense of raising the unity-duality problem within biology itself. There it crystallized as the controversy over mechanism and teleology. Those who approached the categorical divide in the hope of bridging it by extending the laws and concepts of physics into the realm of the living organism were accused of "physicalist reductionism" by their opponents, who championed either the emergence of radical novelty in nature in bringing about biological organization, or wished to see biological concepts extended into the realm of physical nature. Thus on the one hand biological organisms were viewed as specially complex physical and chemical systems, and on the other, physical entities were attributed entelechy and, at least the seeds of, "spirit" or "soul."

The radical distinction between physical and biological nature can be overcome without entailing the labyrinths of the mechanism-vitalism controversy by the simple expedient of analyzing physical as well as biological phenomena in terms of a general theory of natural systems. Specific differences are not overlooked by such systems-analysis, but are accounted for in terms of structural *isomorphisms* (rather than identities). At the same time these isomorphisms guarantee the unity of the established phenomena in the significant organizational respects. This new and immensely high-potential approach is the one we shall explore here by subjecting the extensive class of biological organisms to the same kind of treatment applied to atoms. That is, we shall inquire whether current biological theory permits an affirmative answer to the question of the applicability of the general theory of natural systems to "living organisms."

THEORY: $R_{ii} = f(\alpha, \beta, \gamma, \delta)$, where $\alpha, \beta, \gamma, \delta$ are independent variables having the joint function R_{ii} ("organic natural system").

$R_{ii} = f(\alpha)$ (SYSTEMIC STATE PROPERTY)

The first query concerns the validity of the concepts "*whole*" and "*order*" in relation to the living organism. The answer is so evident that a few remarks will suffice to establish it.

Organisms: wholeness

The non-summative nature of the organism is demonstrated in the difference between the entropy of its components at thermodynamic equilibrium and the entropy of the whole organism. The entropy difference can also be expressed in terms of "bits" of information (Dancoff and Quastler), i.e., as *information-entropy*. In fact, the very essence of biological organization lies in the mutually constitutive interrelation of the organic components —the mere summation of their characteristics does not give a biological organism, much less an individual. Relationships are so important that the most minute change in the interrelation of an identical number of components of an identical set or species may produce entirely different results. This is evident already on the biological level where any change in the order of the amino-acid strands in a DNA helix can completely change the characteristics of the whole. Moreover, even complex structures are integrated to a high degree, with many reactions involving the entire organism (e.g., the stress-reactions of the General Adaption Syndrome). I do not wish to enter here into the controversy of a metaphysically absolutized "mechanism" and "holism," and shall not argue that some organic reactions are "meaningless" apart from consideration of the totality.[7] In the present perspective it suffices to establish that the organism is a "constitutive" (non-summative) totality of interdependent components. And contemporary biology fully affirms this proposition.

Organisms : order

What about, then, the *ordered* nature of the organism? Order refers to both spatial and temporal configuration and sequence of events. Spatial order is grasped as the "structure" or "morphology" of the organism; temporal order as its "function" or "physiology." There is no antithesis between these concepts. As von Bertalanffy points out, "The antithesis between *structure* and *function, morphology* and *physiology* is based upon a static conception of the organism. In a machine there is a fixed arrangement that can be set in motion but can also be at rest. In a similar way the pre-established structure of, say, the heart is distinguished from its function, namely rhythmical contraction. Actually this separation between a pre-established structure and processes occurring in the structure does not apply to the living organism. For the organism is the expression of an everlasting orderly process, though, on the other hand, this process is sustained by underlying structures and organized forms. What is described

---

[7]As, for example, fear, which is sometimes said to be a meaningful organic reaction only if attributed to the total organism, and meaningless on the endrocrine level, where the fear reaction actually takes place.

in morphology as organic forms and structures, is in reality a momentary cross-section through a spatio-temporal pattern."[8]

When we consider the morphology of living organisms, then, we consider a momentary cross-section (a "slow-running wave") upon which are superimposed dynamic functions or patterns (fast process-waves). Order in structure and order in process are not different species of orders but different manifestations of the basic orderliness of the spatio-temporal pattern of the organism. Let us consider, first, the kind of orders characterizing the spatially exhibited "slow-wave" patterns, i.e., the *morphological* order of living organisms.

The morphology of living organisms exhibits remarkable forms of symmetry. The earliest Protozoa show *spherical* symmetries: their bodies have the shape of a sphere with the parts radiating from, or being arranged around, the center. *Radial* symmetry is found in all species adapted to an attached existence, including most plants. Here the body has the general form of a cylinder or bowl with a central axis from which radiate, or around which are arranged, the parts. The axis is heteropolar: it is unlike at the two ends. In animals, one end is known as the oral or anterior end and it bears the mouth; the other is the aboral or posterior end and it bears the rear organs (e.g., the anus). In plants, one end may bear the roots, the other the flowers. *Biradial* symmetry occurs in certain phyla such as Ctenophora and differs from radial symmetry in that, in addition to the anteroposterior axis, it also has two other symmetry axes at right angles to the main axis and to each other. (These are known as the sagittal and the transverse axes.) Finally, in *bilateral* symmetry the same three symmetry axes occur as in biradial forms, but the organism has only one pair of symmetrical sides (the lateral sides) since the dorsal and ventral sides are differentiated. Such symmetry is characteristic of the great majority of animals. Bilateral symmetry signifies the differentiation of the anterior end as a potential head (in a process of "cephalization"), bearing eyes and other sensory organs and a concentration of nervous tissue. Cephalization becomes progressively more pronounced as one ascends the evolutionary scale. Although these symmetry forms are established for most species, variations may be possible through relative growth, e.g., the same symmetry element may be expressed in long, short, flattened or sharpened forms.

If, in the living world, structure expresses process, symmetries in the

[8]Ludwig von Bertalanffy, *Problems of Life*, New York, 1952. In a similar vein, Weiss holds that "the features of order, manifested in the particular form of a structure and the regular array and distribution of its substructures, is no more than the visible index of regularities of the underlying dynamics operating in its domain." Paul Weiss, "The Living System: Determinism Stratified," *op. cit.*

morphology of organisms must have functional significance. Weyl suggests that if the "over-all symmetry inherent in the laws of nature" is not limited for a given body by anything other than the accident of its position P, then it will assume the form of a sphere around the center of P.[9] Hence Protozoa tend to be spherical if suspended in water. Weyl accounts for radial symmetry by the consideration that, for bodies attached to the bottom of the ocean, the direction of gravity enters as determining factor and narrows the set of symmetry operations, from all rotations around the center P, to rotations about an axis. Biradial symmetry is the logical form for organisms capable of self-motion in a medium since, for them, both the anteroposterior direction and the direction of gravity are of decisive influence. Bilateral symmetry appears to be the most complex symmetry form which has functional value for organisms, since factors in evolution which tend to introduce inheritable differences between left and right (the only sides between which the distinction is arbitrary in bilateral forms) tend to be held in check by the advantage offered by symmetry in the organs of motion: an asymmetrical development would result in a screw-wise motion in the medium. At the same time, bilateral symmetry offers the advantage of cephalization, so that it may be expected that it represents the most accomplished morphological design for the living world. Viewing the processes of evolution in terms of the symmetry of morphological form, it appears that there is a clear-cut correlation between degree of organization and complexity of symmetry form. The correlation is explained in Weyl's argument by the fact that to maintain complex organization requires motility, and motility in turn requires complex symmetries for efficiency and control.

If order of the exacting variety of symmetry is present in the morphology of the organism, order of equally remarkable varieties must characterize its *functions*. (These concepts, I repeat, "morphology" [structure] and "function" [process] are expressions of spatio-temporal patterns of relatively slow, respectively rapid, kind.) In fact, the very processes of growth and development exemplify varieties of symmetry. Examples of progressive growth incorporating symmetry elements are the spiral arrangement of leaves, the concentric or conical accretion of growth rings of trees, the spiral forms of snail shells and the sunflower spiral, and the spiral climbing action of the morning glory. These relatively "slow-wave" patterns are expressions of more rapid patternings in the manifest functioning of the organism. Examples of the latter fill the textbooks of biology and physiology. Rather than attempting to review them, let us merely note, with Sinnott, that all living organisms, whether plant or animal, appear to be

---

[9]Hermann Weyl, *Symmetry*, Princeton, N.J., 1952, pp. 27–28.

exactly programmed in their manifold actual and emerging functions to reach the form typical of their species. The final form is encoded in each cell or seed; these are able to control the many cell divisions whereby the developing organism grows to the form typical of its species with great precision. The manifold functions in self-maintenance and development constitute, in Sinnott's words, "an organized and self-regulating system, well named an 'organism.'" [10]

$R_{ii} = f(\beta)$ (BIO-CYBERNETICS I)

Definition of "bio-cybernetics"

We have now raised the question of *self-regulation* in organisms. The presently relevant proposition of our theory is that natural systems (hence also organisms) adapt to their environments through sequences of self-stabilization around steady states. The phenomenon of organic self-regulation is thus interpreted as an instance of adaptation to the environment through the maintenance of the steady states of the organism.

I shall argue that viewing organisms as open cybernetic systems stabilizing themselves in their environments provides a consistent viewpoint from which to view organic as well as inorganic natural organizations. It is worthy of note that, contrary to this premiss, the observation of self-regulation in organisms suggested a fundamental duality in nature to Sinnott as recently as in 1955. He assumed that the fact that living things are organized self-regulating systems is the "basis for regarding the life sciences as *distinct* from the physical ones" (my italics).[11] The distinction reappears even in the argument of von Bertalanffy (who would certainly not wish to be committed to a dualist view of nature) when he holds that it is the *openness* of organisms that is their fundamental characteristic.[12] The theory espoused here, however, removes this dualism, leaving only specific differences among basically isomorphic systems. The differences arise due to the high level of organization on the upper levels of the hierarchy of nature: the systems become thermodynamically increasingly improbable, and rely to an increasing extent on precisely controlled environmental relations. The self-maintenance of complex systems in highly unstable steady states requires a constant openness of the systems' boundaries, for

---

[10]Edmund W. Sinnott, *The Biology of the Spirit,* New York, 1955,. p. 16

[11]*Ibid.*

[12]Take, for example, the following passage: ". . . the fundamental characteristics of life, metabolism, growth, development, self-regulation, response to stimuli, spontaneous activity, etc., ultimately may be considered as consequences of the fact that the organism is an open system." *General System Theory, op. cit.* p. 149.

purposes of the importation of free energies from the environment. This presupposes the general mechanism of a metabolism—the breaking down and assimilation of the negentropic input, and the output of the entropic waste products. Thus we view organisms as continuously open systems because of their need for cybernetic stabilization, and not as highly organized because of their continual openness.

The degree and constancy of the openness of natural systems is a consequence of their steady-state maintaining self-regulation in a state of thermodynamical disequilibrium. Since cybernetic self-regulation is a universal system property, this conceptual frame exhibits the unity of nature through the isomorphy of bio-cybernetics with cybernetic processes in other (physical, social) systems.

Now, the steady states of most organisms are not entirely stationary: their parameters are not fully time-independent. In function of time, the organic system adopts new steady states. The time-bound changes are those which differentiate the normalcy states of the embryo, the young, the mature, and the aging organism. At each period of its life, the organism maintains the appropriate steady states. Since these states are themselves changing in time, they are but *quasi-stationary*. They are maintained by means of a wide variety of processes, some of which fall under the general heading of "homeostasis," others under that of "learning." The principal distinction between them is that whereas homeostasis maintains the existing organic structure through a genetic programming of behavior, learning maintains that structure by evolving behavior patterns based on individual experience. Hence both varieties of processes are classified here under the general heading of "bio-cybernetics I," and are surveyed in turn.

Organisms: adaptive self-stabilization

(I) *Varieties of homeostasis*

The organism lives in a potentially hostile environment, and it must guard against inputs which may inflict damage upon its delicately balanced components. Excessive quantities, even of needed energies and stimuli, can be toxic, and the prime survival task for organisms is the reduction and avoidance of deviations from their required norms of environmental interactions. Highly specialised cells and organelles may succumb even to small changes in temperature, pH, oxygen saturation and other parameters. The self-regulation of organisms, whereby they maintain their needed constancies by adaptively balancing potentially noxious environmental energies and stimuli, is denoted by the term originally proposed by Cannon, *homeostasis*.

Homeostatic self-regulation can be understood as a cybernetical process involving the organism in an ongoing transactional relationship with its environment. The parameters of the particular distribution of constraints which add up to the normalcy state of the organism (its appropriate quasi-stationary steady state) are encoded in a control-center within it. In evolved species, the function of the latter crystallizes in the autonomic nervous system. Self-regulation by the organism with the goal of maintaining itself in or around these states presupposes active behavior *vis-à-vis* the environment, and this in turn requires an autonomous power supply. The latter is furnished by the metabolism of the body, as it takes in and breaks down substances and uses the liberated energy for its own regulative purposes. Signals must reach the control-center of the organism concerning the relevant states of the environment, since it is by means of regulating itself in reference to these environmental states that the organism can attain and maintain its own quasi-stationary steady states. The informative function is fulfilled by the sense receptors. Since the organism commands a built-in energy source, it does not require of the signals conducted through the receptors to furnish the energies needed for self-regulation. An external signal of low energy level can be converted by the organism into a high-energy reaction. This represents an "informative or sense-organ coupling," realizable also in artificial servo-mechanisms.[13] Minute signals can be coupled to major responses, with the latter determined not by the energy carried by the signal, but by its correspondence to a "code" (or norm) within the system. A radar signal can be coupled to heavy-duty motors operating steering mechanisms; a low-energy stimulation of the optic nerve in the eye can be coupled to intense muscular and neural activity. In each case it is the relation of the input to the system's codes which determines the type of output produced, and the output may be amplified through incorporated power-sources to bring about corresponding changes in the input. In homeostasis, as in servomechanisms, the output is chosen so as to minimize the deviation of the input from the system's organizational requirements. These are deviation-reducing negative feedback processes, governed by codes ("programs") and having as their goal a specific match between input and organization. The match is brought about by the appropriate behavioral response of the organism to its stimulus. Feeding when hungry, drinking when thirsty, mating, sleeping, and so on, are deviation-reducing negative feedback processes, goal-directed toward the matching of the signal (input) to the normalcy codes (genetically programmed quasi-stationary steady states) of the organism.

In addition to behavioral responses acting directly on the environment,

[13]Cf. A. Rosenblueth and N. Wiener, "Purposeful and Non-Purposeful Behavior," *Philosophy of Science*, **17** (1950).

some homeostatic responses compensate for changes in environmental states through a partial and temporal reorganization of the states of the organism. These are the classic examples of homeostasis. (Figure 6) For instance, if the organism finds itself at high altitudes, its response may not be to descend to lower altitudes (which may be impracticable), but to compensate for the decrease of the oxygen content of the air in the lungs through the increase of the supply of red corpuscles through the hypertrophy of the erythron. When the normal oxygen content of the air is re-established (perhaps through a later behavioral response taking the organism back to its usual habitat), the erythrocytes decrease. Similarly with environmental conditions of drought. When the organism's water input is reduced, pituitrin is secreted which has an anti-diuretic effect and checks the output of water through the kidneys. When the water supply is replenished, pituitrin secretion is inhibited and output through the kidneys becomes accelerated. If any of several substances (such as protein, fat, calcium, etc.) are not supplied through the usual channels, stores of these substances are released. If skin-temperature drops, body-heat is generated by a higher rate of heart-beat through increased blood sugar and sympathetico-adrenal action; the same effect is also produced by automatic muscular action such as shivering. If skin-temperatures increase, sweat is produced, the evaporation of which reduces body-temperature. By means of such stimulus-response patterns essential constancies are maintained in the organism's *milieu intérieur*. For the human being these include blood pressure, the oxygen content of blood (Cannon enumerates no less than fourteen homeostatic processes directed toward this goal alone), water content (which is independent of actual water intake and output up to three days), the supply of sugar, salt, protein, fat, and so on. In homeostasis, negative feedback activates response patterns which bring about a minimum error-margin between the sensed input of the organism and the requirements of its organization.

The threshold of homeostatically deviation-reducible inputs is considerable, especially in organisms possessing an evolved nervous system. Within the threshold fall inputs which represent, for example, a sharp object penetrating some distance into some part of the organism. If the wounded part is not a vital organ (such as the heart or brain), the threshold of self-stabilization is not overstepped. In that event, homeostasis involves the processes of healing and perhaps regeneration. Complex responses are coupled to the input, including, for example, the instantaneous snatching of one's hand from a hot surface. Further signals activate cells in the area bordering the wound. Through division, a sufficient number of cells is produced and, by means of still imperfectly understood processes, channelled along to the wounded area, to provide a covering layer which is

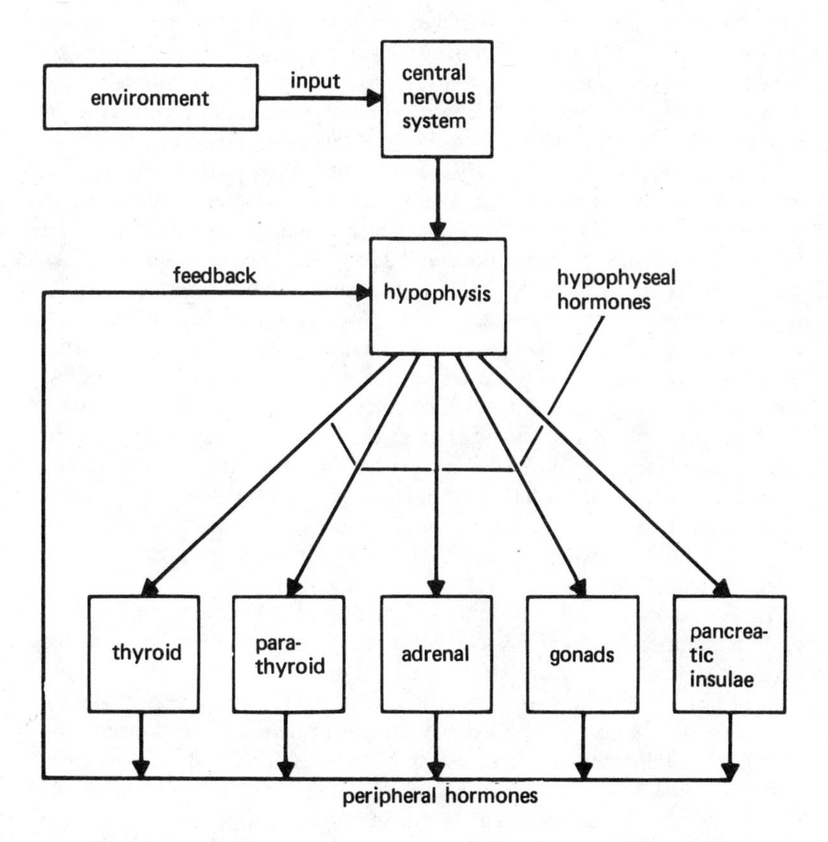

Figure 6

The endocrine feedback: an example of biological homeostasis (as given by A. Sollberger, *Biological Rhythm Research,* Amsterdam-New York, 1965). The organism adaptively maintains its essential constants through feedback-stabilizing itself in reference to its environmental input, transmitted through the central nervous system. Examples of biological feedback are found at all levels of organization: in the body chemistry, the endocrine system and the nervous system; different varieties of it are present in plants as well as animals.

thicker than will be eventually needed. When the cuticle has formed, this layer thins out by cell degeneration to the correct density. The healing cycle is then completed.

Strains produced by the environment may also be reduced to correspondences with the requirements of the organism through a non-localized *general* adaptation process. The disturbing input in this case is defined as the "stressor" and the homeostatic response to it is "the state manifested by a specific syndrome which consists of all the nonspecifically induced changes within a biologic system."[14] Two types of responses are distinguished in the "General Adaptation Syndrome": the primary change, nonspecific both in form and in causation, and a secondary change, which manifests the specific pattern of the general adaptation syndrome. The first type of change triggers the second. The latter consists of three stages: the "alarm reaction," the "stage of resistance," and the "stage of exhaustion." Virtually every organ and chemical constituent of the organism participates in this tri-phasic stress reaction. The process occurs in the presence of a "stressor," which represents a disturbing variety of input; the reaction is a tri-phasic feedback process adapting the organism to it, through sensing and responding. For example, if the body is exposed to cold (a stressor), the adrenal cortex first discharges the available fat-granules which contain the cortical hormones (alarm reaction); then it becomes laden with an unusually large number of fat-droplets (stage of resistance). But, if exposure continues, the fat-droplets are eventually lost (stage of exhaustion).

The stress-syndrome represents a time-limited homeostatic process. It reduces the margin of error between a specific input (the stressor) and the appropriate quasi-stationary steady states of the organism. It accomplishes a match in virtue of a partial reorganization of organic processes, tenable for a limited time. Thereafter the stressor proves to be beyond the threshold of self-stabilizing homeostasis of the organism.

Perhaps the most remarkable form of homeostasis is that where the entire organism is reconstituted from some of its specialized cells and the remainder of the parent organism eventually decays ("dies"). Individual mortality is the price complex species pay for their adaptation. Reproduction, however, is but a special manifestation of homeostasis. The existing organization of a species is maintained over a time-span beyond the reach of its individual members by an intricate programming of processes which involves the cyclic degeneration and regeneration of individuals. Aging and death are balanced by birth and growth, and the species maintains the organization characterizing its genotype. The control over these

[14]Hans Selye, *The Stress of Life*, Toronto, 1956, p. 54.

processes is exercised by information carried within the genes, and only its actualization is conditioned by the environment. The genes carry the full set of "instructions" for rebuilding the mature organism from the specialized reproductive cells; hence the growth-process, while providing a vehicle for phylogenetic reorganization, is basically homeostatic in character. In order that a new structure may emerge by the reconstitution of the old one, information must be generated and then applied. If the old structure contains the necessary information, the new structure represents the carrying out of the program built into the old one, and we would be mistaken in thinking of the emerging structure as signifying a *reorganization.* The structure which comes about may be new from the standpoint of the particular system which thus comes into being; but the type of organization which is generated on the basis of a set of instructions encoded in a parent system is not new. Hence, although growth serves as a mechanism of evolutionary adaptation (since the carrying out of the genetic instructions is conditioned by the nature of the environment), by itself growth is a conservative (stabilizing) rather than an innovative (reorganizing) process.

Using Waddington's concepts, we may categorize growth and development as *homeorhesis* (from the Greek *Rheo*, to flow). In a homeorhetic process, what is constant is not a stationary state but a flow-process. For example, if disturbances cause a homeorhetic process to deviate from its normal course, its negative feedback control mechanisms bring it back not to where it was when disturbed, but to where it would have progressed if left undisturbed. Flow-processes of this kind follow relatively fixed trajectories, which Waddington calls *chreods* (*Chre,* fated or necessary, and *Hodos,* path). The many intricate processes of growth, from fertilization and embryogenesis, through birth and maturation, represent homeorhetic processes following more or less fixed chreods, programmed into the organic system in virtue of its genetic structure, and conditioned in its process of actualization by the environment it finds. Thus the phenotype is built up from the genotype through the interaction of instructions carried in the latter and the relevant conditions encountered by the former. Natural selection, for example, acts on the phenotype, but what it selects is the genotype (since it is that which is handed down from generation to generation).

When we place homeorhetic developmental processes within the context of the existential functions of an organism, they appear as special cases of the latter's homeostasis. If self-maintenance of a complex and inherently time-limited system involves alternating cycles of degeneration and regeneration, then the constructive regenerative processes must be under the system's control, and these controls must be homeorhetic in character, i.e., they must buffer out disturbances by bringing the processes back to their normal developmental pattern (rather than to the values which characterized them when disturbed). A re-establishment of the previous steady state would arrest regeneration and disrupt the overall homeostasis of the reproductive cycle. Thus the intricate processes of dynamic multivariable regulation within a developmental system must be seen in the functional perspective as homeostatic systems-phenomena. If not the detailed mechanisms, then at least their function, can be clearly understood in that context.

(II) *Varieties of learning*

We distinguish (i) genetically programmed, and (ii) empirically acquired processes of self-stabilizing *bio-cybernetics I.* The latter can be subsumed under the general concept of *learning.*

Traditionally, under "learning," one understood the acquisition and recall of "facts," attitudes and beliefs, together with the correlated ways of responding. Today, the meaning of "learning" has been extended and placed in a holistic-organismic context. For example, Clausse in a recent paper defines learning as a very general process which affects all aspects of the personality: knowledge, capacities, interests, attitudes, affective responses, social and ideational adjustments, techniques of thought and action. The common element in all varieties of learning is that learning is obtained by a certain form of activity in which the organism pursues a conscious or unconscious goal and attempts to escape a situation or a problem. By that token the situation of the organism is modified. Hence learning is "a modification of behavior attained by the 'solution' of a problem posed for the individual by his environmental relations." [15]

The process of learning is readily studied through artificial models which, though on a scale of lesser complexity, exhibit the types of operations characterizing even sophisticated instances of learning in animals and humans. From the cybernetical viewpoint, learning is a modification of programs (or norms) incorporated in the system by means of signals informing the latter of conditions in its environment. Such modification in "learning machines" permits the systems to adapt to these conditions by the appropriate reprogramming of their codes. Learning machines differ from ordinary non-learning machines by having what Wiener called a "flexible" (as contrasted with a "rigid") personality: they "learn from experience" rather than keep making the same mistakes all over. Here, of course, "mistake" refers to an interaction which fails to accomplish the

---

[15] A. Clausse, "Formalisme et Réalisme Pédagogiques," *Scientia*, CV (1970), pp. 80–81 (my translation).

operations programmed in the system, due to maladaptation to the actual conditions in the environment. Flexible coding (or "dynamic programming") allows the systems to reorganize their programs to fit the information fed back through their receptors. Thus learning machines can carry out the intended operations even under changed conditions in their milieu, conditions which frustrate non-learning machines.

The scale of programming from "rigidity" to "flexibility" can be illustrated by drawing on familiar examples. The following grades can be distinguished:[16]

1. Fixed program of a single value. Example: a refrigerator or air conditioner set to maintain a given temperature.

2. Fixed time-curve for a given value. Example: gradual heating or cooling systems for chemicals, tissues, etc.

3. Fixed time-curve for a series of values, in a given sequence. Example: automatic washer with a program consisting of wash, rinse, and spin cycles, in definite order.

4. Program with a temporal sequence determined by the feedback of information. Example: industrial control circuit, coordinating phases of a manufacturing process according to the success of each step in the process.

5. Program with alternate subroutines with the choice determined by the feedback of information. Example: industrial control circuit of "managerial" capacity, i.e., one capable of stopping a sequence, or switching in emergency procedures, or accelerating it, depending on the existing situation in the plant.

6. Optimizing dynamic programming with capacity to reprogram according to need signalled in information feedback. Example: checker- or chess-playing machines which revise their strategies according to their success on the playing field.

The six grades given above take us from a system with a fully rigid personality (such as a preset air conditioner) to one with a fully flexible one (e.g., a self-improving chess-playing machine). Each system carries out a set of operations, as given in its basic program, but the more flexible the system is, the more leeway it has to perform successfully under changing conditions. Programs 1 through 3 are non-feedback systems, unable to compensate for changes in their surroundings ("fields of operation"). Programs 4 and 5 are negative-feedback systems maintaining a given program

[16]Following Hans Sachsee, *Die Erkenntnis des Lebendigen*, Braunschweig, 1968, p. 55.

under changing circumstances by varying their subroutines. Program 6 couples positive feedback to this in evolving new programs for carrying out the operations beyond the limits of feasibility, or efficiency, of the original programs. Such a system embodies a high-level decider, or program-selector, which not only picks out the most fitting of the existing programs but writes out new programs, within the "degrees of freedom" of the system, to meet the challenges of the changing environment. Learning, in the form in which one encounters it in higher species of organisms, is of this variety. Man excels at it more than any other species; in fact, it may be argued that it is this capacity which differentiates man from other species and assures his control and domination over them. In the past few years artificial systems with analogous (though much simpler) capacities have been developed, and it is the possibility of evolving machines with higher learning capacities that feeds the imagination of those writers who foresee an era of machines in which robots eventually enslave their own creators.

Learning, even in the sophisticated form of writing out new programs, is not necessarily what is ordinarily termed a "conscious" process, i.e., it does not presuppose a high-grade awareness of one's own mental events. (It does presuppose *some* degree of self-awareness, on the other hand, since information must be received in the system of the success of its various subroutines before new routines can be evolved to improve them.) In some cases, learning may be a "physiological" as contrasted with a "psychological" process: the body can perform it even in the absence of a conscious awareness of what takes place. Here is the lower end of the scale of learning in organisms, and it is present—of course, conjoined with the higher varieties—also in man. The most striking example of such physiological learning is furnished by the now discovered instances of "biofeedback learning": the adaptive reprogramming of certain homeostatic norms. Heart-beat, kidney-function, blood-pressure, blood-flow, intestinal and stomach contraction, and even brain waves, may be amenable to regulation in function of an adaptive response of the organism as a whole to its environment. The experiments in this regard have been pioneered by N. Miller of Rockefeller University and use the principle of positive and negative reinforcement. Starting with rats, whose brains have been "wired for pleasure" (i.e., electrodes implanted in the nerve clusters near the cerebral "pleasure sites"), and graduating to humans who are likewise hooked into a reward-producing network (which informs them when they have reached the goal), Miller, and his colleagues at Cornell and Harvard, have shown that organisms can reorganize their genetically coded homeostatic norms as a means of optimizing the frequency of the reward. Although it is not clear by what processes, rats as well as human subjects can slow down, as well as speed up, their heart-beat rate, decrease and increase their blood-pressure and the blood-flow in the stomach walls. Visceral learning is a new and striking example of *bio-cybernetics I* in complex warm-blooded organisms. When the environment changes to reward certain types of variation in the steady-states of the organism, these states are gradually brought about : the organism thereby maximizes the rewarding (matching) input and avoids the punishing (mismatching) kind.

For man, biofeedback learning is a newly discovered nonconscious form of learning, yet analogues of it have existed in many forms among organisms of all kinds. Learning, as physiologists and biologists now contend, is as old as life itself. It is the adaptation of a self-regulating system to changes in the surrounding medium. As the number of organisms increases, as species and genera multiply, conditions of life become more varied and its risks greater. The inherited structure of organisms becomes increasingly less adequate to insure its survival. Learning is necessitated, in the form of an adaptive response effecting the reorganization of inherited behavior patterns to fit the existing environmental situations. Learning represents, in Thorpe's words, "the process of adjusting more or less fixed automatisms or patterns of behavior and more or less rigid releasing mechanisms to the changes and chances of life in the world."[17] It is based on a large preprogrammed network, in the autonomic nervous system and its sensory-motor reflex-arcs, which activates many varieties of reflex patterns. But learning signifies the supervention of empirically acquired patterns upon these genetically coded ones. The biological range of learning extends from organisms with relatively primitive nervous systems capable of no more than minor adjustments of instinctive patterns, to organisms with highly evolved nervous systems, in whom learning can take the form of rational and aesthetic insights. Thorpe classifies the varieties of learning under five headings, which, as we shall see, cover the ground from primitive organisms to man.

The first kind of learning process distinguished by Thorpe is *habituation*: a process whereby a behavioral attitude drops from a hereditary pattern under changing conditions which render it superfluous or irrelevant. For example, if repeated noises do not harm birds, they become habituated to them and no longer fly away. Behavior is kept economical and efficient through habituation; surplus patterns are thereby effectively dropped.

Secondly, there is *conditioning*: the establishment of a behavioral response to one relevant feature of the environment through reaction to another feature which repeatedly precedes it. Two types of conditioning are

[17]W. H. Thorpe, *Learning and Instinct in Animals*, London, 1963.

distinguished: one is possible subcortically and is effected by the substitution of one sign stimulus for another in activating an innate release mechanism, the other is not possible subcortically and is effected by the establishment of a particular series of acts as leading to the appropriate releaser ("instrumental conditioning").

Thirdly, there is *trial-and-error learning.* It involves the presence of a problem (typically in the appetitive phase of behavior, by varying the conditions under which the consummatory act can be performed), the setting up of "hypotheses," and the trial and error activities of confirmation. Bruner and Postman hold that all cognitive processes involve such "hypotheses" (expectancies) set up by organisms in the course of their behavioral activity. The hypotheses are not consciously entertained in lower biological species (and not always in man) but they condition the receptivity of the organism and demand further exploratory activity to confirm them. If the "input of information from the environment" matches the "hypothesis," the latter is confirmed; if it is not, the hypothesis is modified, giving rise to the "consequent hypothesis." Hypotheses may be strong or weak, according to the extent of their confirmation in the past, the extent of competitive hypotheses, the immediacy of their relevance to the satisfaction of organic needs, and, if explicit, their agreement with the hypotheses of other people. In any case, the behavior pattern resulting from confirmed hypotheses represents an adaptation of the organism to the new (or merely newly discovered) conditions of its environment.

Fourthly, there is *latent learning*: typical of trial and error learning but without direct motivation by a problem blocking the way to a consummatory act. Rather, it is an exploratory activity which does not need to be reinforced by confirmation. Many animals indulge in such exploratory activity (wasps and bees in learning the topography of their environments, rats in a maze without food in the box, monkeys without expectation of food, and finally men, without expectation of reward of any kind), and they exhibit evidence of having learned by it, for their performance in repeated instances is superior to what it was the first time. Thorpe holds this type of learning to involve "insight," which he defines as the apprehension of relations. He goes on to say that "there is implied a highly organized background of the familiar against which something new stands out requiring investigation. The investigation completed, this new object is 'built in' to the perceptual world as something which is now familiar and which can henceforth be ignored or which can be utilized later, by 'transfer of training,' in a different context."[18]

Fifthly, there is direct *"insight-learning"*: the "sudden production of a

[18] *Ibid.*

new adaptive response not arrived at in trial behavior; the solution of a problem by the sudden adaptive reorganization of experience." Examples of it range from insects to men: bees display it in direction-finding and orientation; ants by the circumvention of novel obstacles on their paths; birds reveal it in their tool-using and pre-linguistic numbering ability; rats in finding short-cuts in mazes which cut through previously blocked passages, while mammals, through the higher apes to men, provide countless examples which are familiar to everyone. In all these processes of learning there is evidence of a gain in insight; of an increasing apprehension of the relations between objects in the environment in view of their relation to the environmental needs of the organism. In Thorpe's words: "the work of recent years has, on the whole, confirmed . . . that all learning is in some degree the manifestation of a process basically identical with insight." Adaptation leads to learning, and learning to some insight into the network of relations in which the organism finds itself. Insight, if sufficiently pronounced, becomes the foundation of empirical knowledge.

The chain of perfected adaptation thus leads, through perception, innate release mechanisms, motivated behavior and learning, to *intelligence*. Piaget emphasizes that intelligence is not a separate and unique function of the human mind but is continuous with biological adaptation to, and behavior within, the environment. Intelligence, according to Piaget, is not separable from instinct, nor opposed to trial and error learning: it is "the most highly developed form of mental adaptation, that is to say, the indispensable instrument for interaction between the subject and the universe when the scope of this interaction goes beyond immediate and momentary contacts."[19]

Piaget accounts for the development of intelligence in man through the study of its unfolding in the child from basic perceptual and behavioral processes.[20] In the development of the infant, new factors are assimilated into the innate schemata of reflex action, and by such means higher-order schemata are evolved. These grow by mutual association to form comprehensive and adaptable patterns of behavior. (E.g., innate sucking mechanisms are perfected and adapted to different objects; these are correlated with bodily movements leading to manipulation and affording more satisfactory responses to varying environmental conditions.) Intelligence is displayed when the child coordinates perception with behavior and repeats bodily movements in anticipation of a desired result. Stages of the development include accommodation of the perceptive organs to follow objects, and learning to distinguish them despite motion and fore- and background patterns. Thus awareness of shape and size evolves by means

---

[19]J. Piaget, *The Psychology of Intelligence,* London, 1950.
[20]See Chapter 7, below.

of the development of perceptual constancies through incorporating "secondary schemata" in the apprehension of the relations between innate motivating factors and environmental conditions. Through trial and error further schemata are assimilated, and the stable space-time world gradually develops as an environment within which the self becomes progressively detached.

In this account of the development of intelligence from basic sensorimotor responses to perceptual stimuli, trial-and-error learning plays a major part, involving elements of problem, hypothesis and test. (Even the most elementary of learning processes, that connected with the sucking of the infant, is triggered by a problem, posed by the lacking of one element in the structure of a total situation whereby the infant obtains its nourishment.) An elementary hypothesis as to the stimuli of the consummatory act appears to be operative, and it is tested in the first movements of trial and error. According to Piaget, success in trial-and-error learning is measured by the subject's awareness of the implicative relations of the elements of the system which demands completion. Success is thus an element of insight: a grasp of the relevant factors interconnecting the organism and the environment. It is an outcome of appetitive behavior impelled by internal motivating factors and seeking satisfaction in the environment through the consummatory act.

Intelligence may be defined as the insight into, or grasp of, the relations which are relevant to the compatibility of the organism with its environment. This definition is supported by Piaget's conclusion: "In fact, every relation between a living being and its environment has this particular characteristic: the former, instead of submitting passively to the latter, modifies it by imposing on it a certain structure of its own. . . Mental assimilation is . . . the incorporation of objects into patterns of behavior, these patterns being none other than the whole gamut of actions capable of active repetition."[21] Thus, through perception and behavior, objects are incorporated into patterns of behavior, and the relations of objects (features of the environment) to the organism become explicated. As Thorpe repeatedly emphasized, insight is the apprehension of relations, and the relations here involved are those which are relevant to the compatibility of the intelligent organism with its environment.

The present approach has been shown to be capable of accounting for the grasp of mathematico-logical operations as well as for desire-motivated thought and action-processes. Bühler, Janet, Rey and Piaget have collected a rich variety of examples showing how schemata are assimilated which make possible the rounding off of perceived systems through an awareness

---

[21]Piaget, *op. cit.*

of their combinativity, reversibility, associativity, identity and tautology. In the child these "groupings" are developed from sensori-motor and perceptual processes, progressing through representation in perceptual thinking to concrete action requiring the combination of groupings in a unitary schema. As a result, developing insight imposes upon innate reaction-patterns the kinds of action indicated in the light of the particular relations binding the observed objects between themselves as well as to the subject. Insight thus subserves the final end of perfecting appetitive behavior through trial-and-error learning by means of problem, hypothesis and test, to lead to the consummatory act (or series of acts) through varying degrees of immediacy. This, of course, is more evident in behavior directly motivated by organic needs than in the often complex thought-processes leading to the conceptual grasp of the scheme of relations themselves. (In fact, disinterested thought processes can and do occur, which use for their own ends insight and reason originally evolved in function of organic self-maintenance.) Intelligence, to judge from our species, crystallizes gradually, and evolves in continuous elaboration from basic sensori-motor insight involved in manual and bodily skills, through symbolic-representative insight present in mechanical and technical reasoning, to abstract reasoning, subserving the survival needs of the organism at a considerable remove, if at all.

However varied the processes of learning may be, they lead in a continuous series of gradations, from minor shifts in instinctive behavior patterns, to examples of profound intellectual and aesthetic insights in human intelligence. It would be to commit the genetic fallacy to infer from this that intelligence is "nothing but" a highly evolved form of empirical adaptation to the environment through the reorganization of genetic behavior patterns.[22] But it would be equally fallacious not to see the evident: that all varieties of learning evolve as forms of the progressive adaptation of the organism to its environment. Here not a passive adaptation is meant, a kind of accommodation of the subject to environmental constraints. Rather, adaptation, in learning as well as in biology in general, is an active process, of creatively responding to the challenges of the environment by a suitable modification of endogenous activity patterns. It involves *both* "accommodation" and "assimilation." "Any biological adaptation implies two poles. . . . On the one hand, it is an 'accommodation' i.e. (by definition) a temporary or lasting modification of the organism's structures under the influence of external factors. But [. . .] any adaptation, even momentary, implies a complementary pole which, in very general terms, could be called the 'assimilation' pole, and which has the task of

---

[22]Cf. Part Two, Chapters 10 and 11.

integrating external factors into the organism's structures; this necessarily implies a continuity between earlier and later structures. This is why any reaction or response is the expression of its continuous structuralization due to the organism as much as it is due to pressures from the stimuli, the environment."[23] The two poles of *adaptation* and *assimilation* are present on all levels of development, organic as well as cognitive, Piaget adds, and concludes that "it is in the strictest sense of the word that *knowledge is a special case of biological adaptation.*"[24]

Learning, then, extends the range of biological adaptation into the conscious cognitive realm. Yet learning is not an evolutionary adaptation of a species, since the knowledge an organism acquires in its lifetime, however valuable it may have proven to be, cannot be passed on to the succeeding generations. Learning of the here discussed varieties effects a *temporary* reorganization of the parameters of the organism's essential functions, physiological as well as behavioral. The reorganization attained by means of learning normally lasts as long as there is positive reinforcement for the new pattern. Not being heritable, the modifications leave the genotype unaffected and, when viewed in the perspective of a species, they belong to the general domain of conservation rather than innovation.

Oparin was characterizing all homeostatic processes of *bio-cybernetics I* when he said, "not only are the many tens and hundreds of thousands of chemical reactions which occur in the protoplasm, strictly coordinated with one another in time, harmoniously composed into a single series of processes which constantly repeat themselves, but the whole series is directed toward a single goal, toward the uninterrupted self-preservation and self-reproduction of the living system as a whole in accordance with the conditions of the surrounding medium."[25] His words reaffirm the present thesis: the organism, an ordered whole, adaptively maintains itself in the states proper to its level of development, correcting for environmental perturbations both by a purposive behavioral response to the environment, and by a limited temporal reorganization of its organic and behavioral parameters to compensate for persistent disturbances.

$R_{ii} = f(\gamma)$ (BIO-CYBERNETICS II)

Organisms : adaptive self-organization

Homeostasis signifies the organism's adaptation to the environment by means of negative-feedback self-regulation, controlled by genetic codes,

---

[23]J. Piaget, "Afterthoughts," *Beyond Reductionism,* A. Koestler, ed., *op. cit.*
[24]*Ibid.*, p. 157 (my emphasis).
[25]A. I. Oparin, *The Origin of Life,* London, 1957.

handed down from generation to generation. These codes represent the norms of the organism, the parameters imposed by its fixed internal constraints. Learning can effect a temporal reorganization of the codes through new and flexible behavior patterns; but the reorganizations are not heritable. They represent new operational programs which last as long as they are successful, and function to maintain the existing organic structure in its ambient. In the machine analogy, they are instances of "optimizing dynamic programming" with the programs written and rewritten to enable the machine to function effectively. With *phylogenesis,* however, we come upon a different bio-cybernetic process. Here the reorganization, while likewise adaptive, affects not merely the program, but the very structure of the system. A fully self-organizing machine would evolve itself, and not only its programs: it would create a new, better adapted machine, by a progressive revision of its own basic organizing principles. Such machines are as yet in the realm of theory rather than in that of practice. But living systems are machines of this kind. The process whereby they reorganize their fundamental structure is called *evolution.* It consists of mutations exposed to the test of natural selection and it procures changes not merely in the behavior pattern (and "insight") of the organism, but in its *genotype.*

The processes of evolution, like those of learning, are positive feedback deviation-amplification processes activated in response to a need for adaptation to conditions in the environment. *Short*-term adaptations in organisms with evolved nervous systems yield the phenomena of *learning*; *long*-term adaptation, in the case of all species with a sufficiently rich genetic information pool, results in phylogenetic *evolution.* Maruyama lists no fewer than four different varieties of deviation-amplifying positive feedback processes occurring in phylogenetic evolution.[26] First, there are positive feedback processes between mutations and the environment. The selection of a certain type of environment, whether by accident or design, favors certain types of mutants. These in turn favor certain features of the environment, and these features determine the viability of new mutants. Thus a kind of vicious circle obtains which amplifies the original deviation. Second, there are interspecific deviation-amplifications. The increased protective ability of one species of mutants calls for increased detection ability and hunting techniques of its predators. Such ability in turn favors mutants with still greater protective abilities, and so on. The respective abilities of the species in the food chain amplify each other and increase generation after generation. Third, intra-specific selection can be

[26]M. Maruyama, "The Second Cybernetics: Deviation-Amplifying Mutual Causal Processes," *American Scientist,* **51** (1963). Note that our "system-cybernetics II" corresponds to Maruyama's "second cybernetics."

deviation-amplifying. Certain individuals may prefer stimuli of specific kinds, leading to the selection of mates and collaborators of particular kinds. By giving more responses to these stimuli, the members of a species can amplify the original deviation, found in the deviating individuals chosen for mating and cooperation, and produce more offspring of such characteristics. Such preferences may be inborn, as well as culturally conditioned (in the biological vein, Maruyama gives as an example of the latter the contemporary preference in American culture for females with abnormally long and slender legs). Finally, inbreeding, too, can be deviation-amplifying since the characteristics of the inbreeding population can be amplified in successive generations. If families would not intermarry, each family would develop into a separate species, amplifying its distinguishing characteristics. Interbreeding, on the other hand, has a stabilizing effect, in eliminating preponderant tendencies toward specific characteristics.

It is readily noted that all such positive feedback deviation-amplifying mutual causal processes refer to *groups* of individuals, and that the causal processes involve mutual deviation-amplification either between successive generations, between the individuals and the environment, or between members of different species of populations. No genetic deviation-amplification occurs by or within a given individual. Any given individual is merely one part of a mutual causal process, one component in a set of components which jointly produces the amplification. The individual as such has no control over the processes; it is the population which is evolving, although the effect of the evolution is demonstrated in each of its individual members.

The overall course of phylogenetic evolution leads to the emergence of better adapted species. In the phylogenetic perspective, organic species adapt themselves to their environments by bringing forth those types of organizations which are optimally fitted to the conditions of the surrounding medium. Phylogenetic evolution is a selective progression toward the organization capable of handling all possible types of fluctuations in the environment. It is effected through mutations (imperfect genetic replications), which are exposed to the test of survival (natural selection), with the result that the fittest—i.e., the best adapted to existing conditions—survive and propagate. Here the terms "fittest" and "natural selection" take on specific meanings. The "survival of the fittest" is not merely the "survival of the survivors," i.e., a tautology, because under "fittest" we understand an organism with a specific level of adaptive organization from which it *de facto* results that it tends to survive longer, and propagate itself more extensively, than less adapted organisms. And "natural selection" is not a brute external force weeding out less adapted organisms, as in classical Darwinism, but denotes an adaptive system-environment process.

It is comparable, in Dobzhansky's words, "not to a sieve but to a regulatory mechanism in a cybernetic system. The genetic endowment of a living species receives and accumulates information about challenges of the environment in which the species lives. The evolutionary changes are creative responses to the challenges of the environment."[27] Natural selection is the preservation of the creatively responding mutant over other species, who have not met the challenge to an equal degree. Adaptation, then, is the state of meeting the challenge of the environment: of mapping within the organisms the codes that control the conditions under which it must exist.

The adapted state does not maximize the structural equilibrium of energies within the organism. Adaptation leading toward states of increasing thermodynamical equilibrium could account neither for the emergence of organic 'ife from inorganic matter (the latter being more stable and equilibrated than the former), nor for the observed and on the whole irreversible course of organic evolution (where the simpler organisms are generally more stable than the thermodynamically fantastically improbable complex ones). Rather, in the present use, "adaptation" (and hence "fitness") means a maximization of the *cybernetic* stability of the organism: its ability to cope with the environment.

With few exceptions, adaptive evolution involves a gain in structural complexity, measured in terms of negative entropy or information content. Evolution, in most cases, is not merely a change of organic form, but a *complexification* of it. As von Bertalanffy points out, "A general progression of evolution toward higher organization . . . is a phenomenological fact, that is, a matter of the paleontological record. This transition toward higher organization is not an expression of a subjective value judgement, nor connected with vitalism, a drive toward perfection or other metaphysical supposition; it is a statement of fact describable at any length in anatomical, physiological, behavioral, psychological, etc., terms."[28]

Thus, since adaptive reorganization tends to result in the complexification of organic structures, it involves both an overall decrease in the level of structural stability of the typical evolving organism, and a greater functional capacity to cope with the changes and challenges of its environment. (This tendency toward decreasing structural and increasing cybernetic stability is illustrated in Figure 2′ as a general characteristic of the evolution of natural systems.[29]) As a result we get smaller populations of more vulnerable, but more efficient organisms, compensating for their inherent instabilities of

[27]Th. Dobzhansky, *The Biology of Ultimate Concern*, New York, 1967.
[28]L. von Bertalanffy, "Chance or Law," *Beyond Reductionism, op. cit.*, 66-67.
[29]Cf. p. 45, above.

structure through highly evolved cybernetic functions. Expressed in terms of the latter, evolution may be seen as a process of *bio-cybernetics II*, perfecting the deployment of processes of *bio-cybernetics I* in the organism's actual environmental situation. (Figure 7) Hence one can maintain that evolution involves a generally irreversible trend without exposing himself to the fallacy of teleology.

The so-called "typing error" theory of random mutations and random natural selection, giving the gist of the statistical theory of evolution, and the contrary "teleological" or "inner-directed" view of evolution, are extreme cases resulting from an undue bias for one or the other aspect of the phylogenetic adaptive process. They can be replaced by a theory based on the basic general formula,

$$\frac{\textit{external}}{\textit{forcings}} \rightarrow \frac{\textit{internal}}{\textit{constraints}} = \frac{\textit{adaptive}}{\textit{self-organization}}$$

which applies to natural systems of the widest variety. This is a holistic systems view, applicable to populations of adaptive systems in changing environments. Adaptation becomes the key function of evolution, without assuming the randomness of the process as a whole. Rather, evolution is the outcome of the interaction of populations of systems and their environments controlled by the adaptive capacities of the former. The basic concepts of random mutations and subsequent natural selection are not discarded, but integrated within the holistic context of evolving systems populations.

Indeed, there are good reasons for believing that the basic categories of random mutations and resulting natural selection are unfalsifiable.[30] These Darwinian concepts may be looked upon as the necessary components of any scientific theory of evolution. Their unfalsifiability is due to the fact that every such theory must take into account the effect of the environment on the survival of the organisms, and that the varying survival potential of different organisms is evidently due to variations in the herited genetic structure. Such variations cannot be ascribed to causal factors in the environment, since otherwise a teleological process would have to be acknowledged: the effect (mutation) would be prior to the cause (selective advantage). However, what is sometimes disregarded by Darwinists is that the randomness of mutations with respect to selective advantage is by no means an inefficient procedure for bringing about better adapted species: random mutations replenish the stores of variation contained in a population and prevent the latter from running out of usable mutants even if

[30]A. R. Manser, "The Concept of Evolution," *Philosophy*, XL (1965) 151.

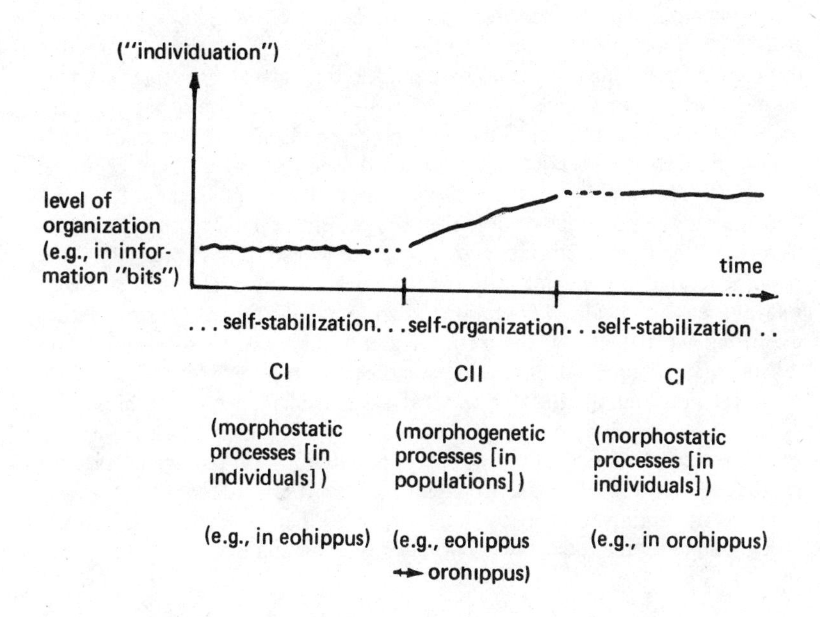

Figure 7: patterns of development in biological systems.

selection operates for a very long time. Thus random mutation is the safeguard of continued variations and, combined with natural selection, genetic recombination, and the evolution of capacities to deal with environmental problems,[31] proves to be a very efficient instrument indeed, of systems-controlled adaptive evolution.

The "science of developmental pattern" requested by Huxley (1953) appears to take shape now in the population, rather than individual oriented mathematical theories pioneered by Sewell Wright, Waddington, René Thom, and others. Here the fitness of entire populations of organisms is taken into account in terms of complex topological concepts, such as "fitness surface" or "epigenetic landscape." The interaction of the environment ("fitness space") with the phenotype, and of the latter with the developmental "epigenetic space" and through it with the "genotype space" (Waddington) provides the dynamics of a complex adaptive process whereby populations of organisms creatively respond to changing conditions in their environment, evolve internal constraints expressed in the phenotype and in behavior patterns, and become more able to cope with relevant environmental fluctuations than previous organic forms. Exact mathematical models of the process still remain to be worked out in satisfactory detail, and the understanding of the detailed mechanisms is as yet outstanding. But the general systems-phenomenon of biological evolution is accessible to clear conceptualization. It appears to be a process of adaptive self-organization whereby populations of biological systems fit themselves into their environment and fit the environment to their intrinsic constraints, and in so doing organize themselves to progressively more complex, vulnerable, but cybernetically stable states.

$R_{ii} = f(\delta)$ (HOLON PROPERTY)

Organisms : intrasystemic hierarchy

The complexity of even the simplest form of living organism is enormous. Dancoff and Quastler computed the number of bits of information in organisms by taking atoms as basic building blocks and considering how much information is needed to (i) select all the atoms which make up the organism, and (ii) specify the position of the chosen atoms within the limits

[31]Waddington could experimentally confirm "genetic assimilation" of responses to environmentally induced stress. He found that not the individually acquired characters, but the capacity to acquire such characters, is inherited. These characters, if selectively bred, become assimilated into the genotype, although they are induced by stresses produced by the environment. Cf. C. H. Waddington, "The Theory of Evolution Today," *Beyond Reductionism, op. cit.*

of precision allowed by thermal vibrations at body temperature. They found that *Escherichia coli,* a simple but self-sufficient organism, contains an estimated 2×10^{11} atoms and that the selection and specification of each atom requires 24.5 bits. The total number of bits required to build the bacterium as a whole is thus 5×10^{12}. This, however, is an upper bound since it neglects all constraints resulting from the atoms being arranged into molecules. Correcting for these constraints, the information content of *E. coli* comes to 1×10^{10} bits. Using a similar technique, the information content of the human organism can be placed at 2×10^{28} bits. As a lower bound, a self-sufficient organism without the redundancies which characterize all natural organisms can already be built on the basis of 10^3 bits of information. But even this corresponds to a sequence of 1,000 binary choices or, if the organism as a whole were to be selected from alternate possibilities of atomic arrangement, a single choice among $2^{1,000}$ possibilities. The probability of making such a choice in a random process of shuffling the components is 10^{-301}!

This raises the question, whether the time-span allowed by physical conditions on the earth's surface is sufficient to permit the evolution of complex organisms on the basis of chance associations in a primordial chemical "soup." The unqualified random shuffling hypothesis runs into difficulties at this point, since the probability of achieving systems of the order of complexity of living beings far exceeds the order of the time-span indicated by astronomical and geological evidence.[32] However, when we take systems as ordered wholes, adapting to their environment in function of compensating for changes, and realize that atoms and molecules are systems of this kind, we get a more conservative time estimate. Using Simon's hypothesis, according to which the time required for the evolution of complex from simple systems depends critically on the numbers and distribution of potential intermediate stable forms,[33] organic evolution fits into the physically available time period. While the exact time is not readily calculable due to many unknown variables in the origin and branching of forms of life, it becomes evident that nature can build complex structures relatively rapidly by using existing stable systems as building blocks, and combining them in superordinate stable associations under favorable conditions. If conditions change, these associations break down, but the disorganization comes to a stop at the level of the sub-

[32]Since, as von Bertalanffy points out, a primeval soup composed of nucleic acids, various molecules and enzymes is unlikely to form open systems which remain in a state of high thermodynamical improbability; instead, they would break down into nucleic acids again.

[33]H. A. Simon, "The Architecture of Complexity," *op. cit.*

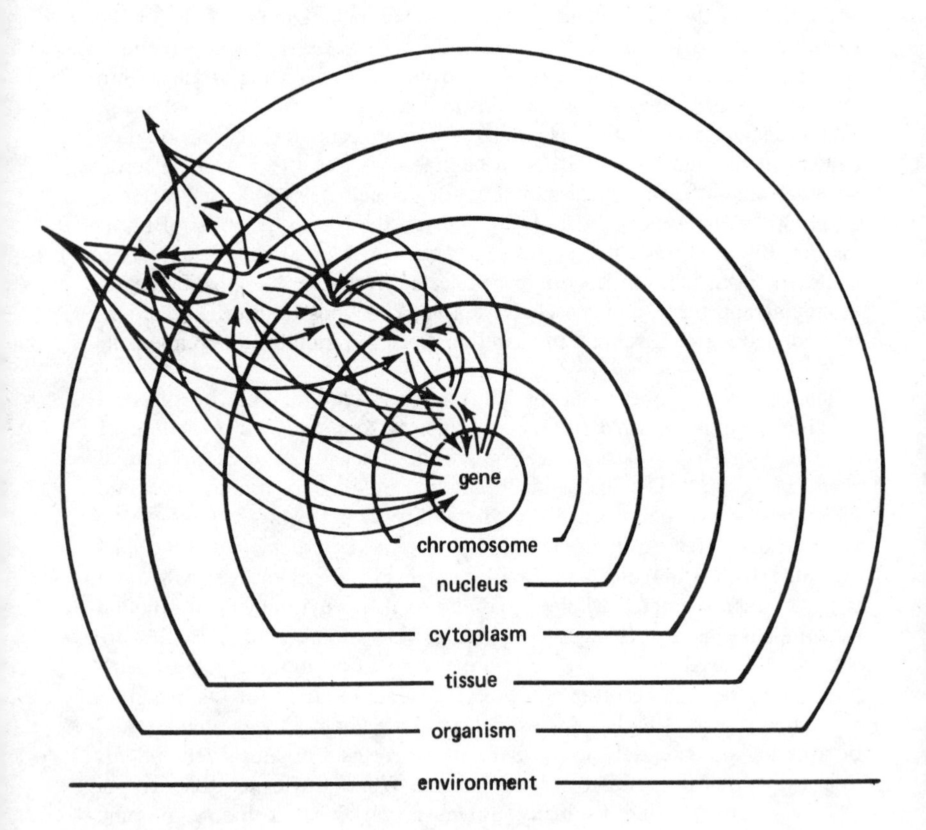

Figure 8

Mutual relationships of the hierarchically ordered subsystems of organisms. (From Paul Weiss, *Dynamics of Development: Experiments and Inferences,* New York, 1968.)

system which is resistant to the disturbance; and from there new intersystemic associations can subsequently be formed.

One would expect, based on these purely pragmatic considerations, that each organism one cares to examine will exhibit hierarchical order in its constitution. And such indeed is the case. Organisms of all kinds are built by the integration of systems into superordinate systems, and these again into still higher level systems, until we encounter the organism as a whole. Each subsystem finds constraints imposed on its behavior by the higher system, with the result that the total organism's functional behavior dominates the behavior of all its parts, through successive hierarchically organized steps.[34]

A remarkable aspect of the hierarchical organization of organisms is the emergence of new properties at each new level. The properties appear only if a given structural configuration at one level is placed within the systemic environment which defines the next: e.g., the enzyme activity of proteins is dependent not only on the amino-acid sequence, but on the establishment of a suitable context within the organism—the presence of appropriate templates. Grobstein concludes that the emergence of new properties relates to set-superset transition within the hierarchical order of organisms. In the higher sets (or superordinate level systems), the determinate association of components constitutes configurations with relationally transformed information, not resident in the lower-level sets. Thus "hierarchical organization in biological systems . . . is characterized by an exquisite array of delicately and intricately interlocked order, steadily increasing in level and complexity and thereby giving rise neogenetically to emergent properties."[35] (Figure 8)

Organisms : intersystemic hierarchy

The systemic hierarchy of nature does not stop at the boundaries of the individual organism: it continues into the environment. The latter may include interpenetrating sets of *socio-* and *ecological systems,* in which the subsystems are coacting individual organisms. All species of organisms integrate their life patterns to some extent in social and ecological systems; these range from primitive communities in desert and arctic regions to the complex socio-ecosystems wherein man plays a dominant role. These, as I shall now show, may be cogently interpreted as natural systems in their own right, manifesting wholeness, self-regulation and hierarchical relationships at their particular level of organization.

---

[34]Cf. the detailed work of the systems biologists, e.g., von Bertalanffy, Paul Weiss, C. Grobstein, R. W. Gerard, *et al.*

[35]Clifford Grobstein, "Hierarchical Order and Neogenesis," *Hierarchy Theory, op. cit.*

(iii) SOCIAL SYSTEMS (Human Societies)

The systemic aspects of the environment into which organisms are integrated have received increasing recognition in the last decades. We have witnessed the rise of ecology and ecosystem analysis, seen sociology espousing functionalism as its method and conceptual paradigm, and cultural anthropology dealing with transformations of "structures." An adequate overview of these developments is well-nigh impossible within the scope of one section of a work which, like the present one, ranges over a great variety of topics. Hence, in support of the proposed theory, I shall restrict the discussion to one area, namely the field of human *social relations*. (Evidence concerning the systemic nature of *ecosystems* has been provided in profusion in the current literature; these topics are touched upon in Chapter 14, below.)

THEORY: $R_{iii} = f(\alpha, \beta, \gamma, \delta)$, where $\alpha, \beta, \gamma, \delta$ are independent variables having the joint function R_{iii} ("social natural system").

Social philosophers have long speculated on the nature of human society. Some have thought it to be an association entered into by free agents for mutual aid and benefit (e.g., the social contract theories of Rousseau and Locke); others saw in it a voluntary surrender of individual sovereignty in exchange for safety and the satisfaction of needs (e.g., Hobbes); still others assessed it as essential relationships formed by basically social human beings (e.g. Aristotle and Marx).[36] With the rise of sociological theory in the work of Comte, society was recognized as something of an entity in itself, and various structural and equilibrium models emerged, describing its endogenous and exogenous states and variables. One's ordinary common sense baulks, however, at treating what appear to be merely associations between individual human beings as entities in themselves. Thus the first models of "social entities" were defended, understandably enough, on the basis of an analogy with the biological organism. Social Darwinists imported into the sphere of social relations concepts developed in the area of biology. Although Spencer stressed that "there exist no analogies between the body politic and a living body, save those necessitated by that mutual dependence of parts which they display in common," he persisted in talking of the "social organism" as the most convincing way in which to regard society as an individual entity. He fought against the "nominalist" position which holds that a society is but a collec-

[36]See Ervin Laszlo, *Individualism, Collectivism and Political Power*, The Hague, 1963.

tive name for a number of individuals. "Carrying the controversy between nominalism and realism into another sphere, a nominalist might affirm that just as there exist only the members of a species, while the species considered apart from them has no existence; so the units of a society alone exist, while the existence of the society is but verbal." Spencer, however, wished to "consistently regard a society as an entity, because, though formed of discrete units, a certain concreteness in the aggregate of them is implied by the general persistence of the arrangement among them throughout the area occupied."[37]

Such relatively naïve reasoning from spatial cohesion and temporal persistence to "entitivity" in the realm of social phenomena gave way to a less realistic but just as entity-oriented *methodological* approach in later sociological theory. Rice points out that "Scientific method conceptualizes the perceptual data and treats them *as if* they were real and exact entities. This methodological process, like a great deal more of scientific method, is essentially fictional. Its justification is to be found in the results to which it leads. Hence it is as valid for the scientist to speak of a social group as to speak of an ounce of ether, provided he can do something further with the idea."[38]

Contemporary social theory is still divided on the ontological issue, whether social entities are methodologically conceptualized theoretical entities or concrete, real existents "in nature." However, an increasing number of investigators now recognizes the relativity of perception, which leads to seeing only medium-sized entities as "things" and micro- and macro-entities as at best theoretical entities (Warriner, Campbell, *et al.*), and "entitivity" is seldom denied to social units by any social scientist. Benefiting from the development of a general systems theory, which views *systems as such* as the entities to be investigated, independently of their substance, origin, and identity of components, contemporary sociology gave birth to *functionalism*, and the related Continental school, *structuralism*.[39] Gouldner points out that "The intellectual fundament of functional theory in sociology is the concept of a 'system.' "[40] The ontological interpretation of this concept may still be open to debate, as shades of nominalism and

---

[37]The above quotations are from H. Spencer, *Principles of Sociology*, Part II, "The Inductions of Sociology," New York, 1897.

[38]S. A. Rice, *Quantitative Methods in Politics*, New York, 1928.

[39]The close relation of these two schools is discussed in Ervin Laszlo, "Ludwig von Bertalanffy and Claude Lévi-Strauss: Systems and Structures," *Unity Through Diversity, op. cit.*

[40]Alwin W. Gouldner, "Reciprocity and Autonomy in Functional Theory," *Symposium on Sociological Theory*, Llewelyn Gross, ed., New York, 1962.

realism survive in modern theory.[41] However, it is clear that, unless one allows that there are events and relationships in the field of empirical phenomena to which system-models are applicable, the latter conserve but a purely formal character. Thus, if sociologists claim, for example, that their model has properties of interdependence and self-maintenance, then they must mean that the network of human relations for which the model is interpreted exhibits these properties. In other words, model and phenomena must be isomorphic in the strategic respects in the light of which the model was advanced. I shall conclude, therefore, that it is meaningful to speak not only of theoretical, but also of concrete systems in sociological inquiry, although it may be, as e.g. Lévi-Strauss holds, that no direct observations can be made of the latter. The working tool remains the theoretical system, but its validity is determined by the concrete system for which it is interpreted.

$R_{iii} = f(\alpha)$ (SYSTEMIC STATE PROPERTY)

Societies: wholeness

We are asking for the internal state variables of social systems: their endogenous characteristics. We can most clearly answer it by showing what is entailed by the contrary assumption. A set of objects related merely by spatial or mechanical adjacency is a heap, or conglomeration, of externally related, mutually independent units. It is a summative complex, for it is properly described by the summation of the individual parts. Taking away one part, or adding some, makes no difference to the character of the whole. Of this character are dumps, untidy attics, pages from two unrelated books glued together, etc.[42] Each part can be understood by itself; reference to the rest is not helpful or indicated. Pure summativity excludes "systemness." As we already noted, it does not apply to any natural organization on the atomic or organismic levels. But does it apply, perchance, on the level of *society*?

A society, if it is a non-summative entity, is totally intelligible by an analysis of its parts, namely human individuals as such. No reference to group-structures, organizations, economic, legal, political and other entities and networks, should be helpful or indicated. Moreover, adding or taking away members makes no difference in the results of the analysis, regardless of the numbers involved. But the entire trend of contemporary social theory

[41]An overview of current debates in functionalist sociology is given in *System, Change and Conflict,* N. J. Demerath and R. A. Peterson, eds., New York and London, 1967.

[42]These examples are from Pitirim Sorokin, *Social and Cultural Dynamics,* New York, 1937, "Spatial or Mechanical Adjacency."

flatly contradicts all these assumptions. We can contrast them with Parsons' statement, "The most essential condition of successful dynamic analysis is continual and systematic reference of every problem to the state of the system as a whole . . ."[43] Social behavior is always interpreted in the context of a social *system*, having some systemic, "irreducible" properties of its own. The most often voiced of these are "interdependence of parts" and "equilibrium-maintenance" in their relationship. "The most general and fundamental property of a system is the interdependence of parts or variables . . . This order must have a tendency to self-maintenance, which is very generally expressed in the concept of equilibrium . . ."[44] Societal non-summativity, or *wholeness*, is adequately documented in contemporary social theory, and insistence on the usefulness of concepts such as "system," and "irreducibility of properties" is a fair index of its acceptance.

Societies : order

Sociology in all of its forms, like other sciences, grasps essentially pattern and order. The pattern may be a synchronic, "spatial" one of a set of relationships jointly enduring in time. In that case it defines the (relatively static) *morphology* of the social system under investigation. Or it may be a diachronic order of events involving successive variations reflected in the changes of state of its component subsystems; these patterns define *dynamic* order in the system's *function*. Now, if social systems were unordered, in time and in space (here *social* time and *social* space is meant), social theorists would be creating imaginary orders when discussing morphological and dynamic properties and would have to conclude that such schemes are inapplicable to actual social systems. But whereas it is readily admitted that any system model of a social entity is an idealization, it does not follow that it is inapplicable to the aspect of social reality for which it is interpreted. Its interpretation can only be valid if the postulated morphology of the (theoretical) system is significantly isomorphic with enduring patterns in the relevant set of social relations, and if the system's dynamic orders are likewise isomorphic with patterns in processes of change in human relationships. There can be no science of a phenomenon in a constant state of flux : *some* parameters must remain constant, or invariant under transformation. These constancies and invariances furnish the systemic elements in reference to which theoretical structures can be built, mapping the fluctuating phenomena under investigation. Again, since the mappings are ordered and interpreted for the phenomena, the phenomena

---

[43]Talcott Parsons, *Essays in Sociological Theory Pure and Applied,* Glencoe, Ill., 1949, p. 21.

[44]T. Parsons and E. A. Shils, eds., *Toward a General Theory of Action,* Cambridge, Mass., 1951, p. 107.

are either ordered in the relevant respects or the mapping is invalid. It is the assumption of science, including current social theory, that *some* theoretical mappings are valid. And if so, then the phenomena must manifest strategically isolable elements of order.

The specific patterns of morphology and dynamic function in a social system are the subject of inquiry in the various theories of social relations. The fact that the theories share, explicitly or implicitly, the assumption that they are dealing with a set of (relatively) *ordered* phenomena, is the point to be emphasized here. It establishes the cogency of the here discussed systemic state property in reference to social systems.

$R_{iii} = f(\beta)$ (SOCIO-CYBERNETICS I)

Definition of "socio-cybernetics"

Social systems can be analyzed from two distinct points of view. First, morphologically : here the *structure* of the system is under consideration. "Structure" "designates the features of the system which can, in certain strategic respects, be treated as constants over certain ranges of variation in the behavior of other significant elements of the theoretical problem."[45] To define the structure of the system, therefore, is to give an internal *state* description of it. The other mode of analyzing social systems centers on its *dynamic functions*. Here the internal set of constraints which give the system its specific morphological structure are related to sets of external factors, "inputs" and "outputs" (or *disturbances* and *responses*).

The primary theoretical significance of functional analysis consists in relating external to internal factors: "those imposed by the relative constancy or 'givenness' of a structure, and those imposed by the givenness of the environing situation external to the system."[46] Hence to define the dynamic function of a system is to give a description which relates a state description to an input-output function description. Such descriptions constitute the societal variant of systems-cybernetic analysis, or "*socio-cybernetics.*"

Implicit in socio-cybernetic analysis is the concept of the *openness* of the investigated system. A closed cybernetic system must be defined in terms of its intrinsic sets of constraints: all processes are controlled by these. However, an open system deals with environmental disturbances, and processes within it are determined by its interaction with the relevant factors

---

[45]T. Parsons, "A Paradigm for the Analysis of Social Systems and Change," *Theories of Society*, T. Parsons, E. A. Shils, K. D. Naegele, and T. R. Pitts, eds., New York, 1961.

[46]*Ibid.*

in its environment. Empirically, Parsons points out, social systems are conceived as open systems, engaged in complicated processes of interchange with environing systems.[47]

The question arises then, concerning the *environment* of social systems. What are those entities with which social systems interact and in regard to the effects of which they function? To cite Parsons again (who consistently furnishes the clearest systems-analysis in social theory), the environment is systemic in nature, including cultural and personality systems, the behavioral and other subsystems of the organism, and, through the organism, the physical environment.[48] The immediate boundary interfaces of social systems are cultures, personalities and organisms, not the physical environment: the latter is mediated by the biological organisms which are the basic social subsystems. Thus defined, the social environment is never constant: as cultural norms evolve, personalities and organisms replace one another in a continuous succession in which traits are imperfectly replicated. Hence the social system finds itself in a sea of change, and is faced with the alternatives of either maintaining itself, or suffering possibly irreversible alterations. But control-mechanisms are obviously present in social systems, since they maintain themselves in their dynamic environment. These mechanisms have well-defined capacities for dealing with change: up to a given point they can maintain the parameters of the existing structures, and beyond that the system is launched on a cumulative process of change, taking it progressively farther away from its institutionalized patterns.[49] We are dealing with socio-cybernetic processes of *negative* and *positive* feedback: within the threshold of the system's control-resources negative feedback mechanisms maintain its existing pattern, beyond it, positive feedback takes over and evolves the system in new directions.

The above are the two key socio-cybernetic functions. Parsons defines them as processes of *equilibration* (or pattern-maintenance), and processes of *structural change*. The former corresponds to our general systems variable (β): adaptability by means of self-stabilization around dynamic steady-states. The latter is relevant to variable (γ) and will be discussed subsequently.

Societies: adaptive self-stabilization

The concept *equilibrium* is used by sociologists (e.g., Pareto, Parsons, Henderson) in the sense of a homeostatic steady-state and not usually in that of a thermodynamical equilibrium state. Thus a high level of

---

[47] *Ibid.*
[48] *Ibid.*
[49] *Ibid.*

organization, which is extremely improbable thermodynamically, would still be identified by such theorists as an equilibrium state, provided certain sets of relationships remain "in equilibrium," i.e., are stable. Reciprocal need-satisfaction between sets of individuals or subsystems provides an example of equilibrium. In fact, the sole normative sense of the concept refers to the continuously maintained institutionalized patterns of a relatively stable social system, whatever these patterns may be. The equilibrium state is thus the steady-state, which is maintained within a given range of disturbances.[50]

Inasmuch as it is possible to speak of social systems maintaining the *status quo*, in however limited a sense and in whatever time period (provided the latter is greater than the frequency of disturbances to which it is exposed), it is possible to speak of pattern-maintaining "self-equilibrating" socio-cybernetic processes in the systems. Since the persistence of social systems in the above sense is widely accepted, one of the two basic tasks of social theory is to explain the functional mechanisms whereby systems correct for factors of change. (The second task being to explain how systems *change* under the impact of disturbances.) It is not surprising, in consequence, that dynamic analysis should concern itself with the interlinked processes of pattern-maintenance and structural change, defining the forces of change, their impact, the factors according to which impact varies in the structure, and the resulting responses in and by the system.

Negative feedback pattern-maintenance can be traced to a wide variety of controls within the system, including economic factors (cf. Ricardo's "natural price" as a norm of economic fluctuation) and legal, political and moral constraints. These add up to a system of sanctions which tend to channel social behavior into pre-existing channels, reducing deviations by corrective measures. The control elements must be viewed as part of the system, lest we enter upon the infinite regress of asking what controls the controls, and so on. Thus social systems are conceived as *self*-regulating, pattern-maintaining cybernetic open systems. They may be analyzed from the viewpoint of the system as a whole, and in that event one speaks of information-flow channels, feedback, memory components, communication of parts or subsystems, receptors, effectors, and the like. Or, one can analyze control as a set of constraints on the parts, in which event one speaks of the values and goals of individuals, routines and factors of

[50]Buckley objects to the concept of equilibrium for it fails to signify that social systems are *not* closed, entropic systems which typically lose structure in going to equilibrium and, in that state, have a minimum of free energy ("Society As a Complex Adaptive System," in Buckley, *op. cit.*). However, given the above suggested neutral formulation, the equilibrium concept is harmless, if somewhat misleading in its implications.

acceptance, conformity and safety in social life. The difference between the two modes of analysis is a difference in the level at which the analysis is carried out. The system-level analysis takes society itself as, e.g., "a complex adaptive system" and individuals as the relevant parts or components,[51] whereas the human-level analysis takes individuals as socially interacting open systems and society as their relevant environment.[52] Although the labels attached to the controls by the investigators change with the level of their analysis, both sets of labels denote socio-cybernetic deviation-reducing feedback mechanisms for the maintenance of the existing values and institutions of the society. In the absence of such controls, societies would be propelled upon pathways of uncontrolled change by every variation in their environment.

Although structural stability in societies is a key phenomenon, capacities for change are likewise present. Whether a society changes, or maintains the *status quo,* depends on whether its control resources are capable of dealing with changes in the environment by buffering them out through internal accommodation, or whether dealing with disturbances entails a fundamental reorganization of the institutional and value structure. The switch from self-stabilizing *negative* feedback to self-organizing *positive* feedback parallels control processes in all levels of natural systems. For example, by substituting the generic term "established" in place of the socio-specific "institutionalized," Parsons' statement could apply equally to social, biological, and atomic systems: "The control resources of the system are adequate for its maintenance up to a well-defined set of points in one direction: beyond that set of points, there is a tendency for a *cumulative* process of change to begin, producing states progressively farther from the institutionalized [→ 'established'] patterns."[53]

---

[51]Such analysis is typical not only of some Western functionalists and systems' theorists (e.g., Cadwallader, Buckley, Deutsch) but also of some *Soviet* theorists: it accords with their Hegelian-Marxist collectivist viewpoint. Take, for example the following statement by E. Kol'man:

> "The goal of our development, Communist society, is a complex system, and from the point of view of cybernetics, is an open, dynamic one with ideal self-regulation. Since as a whole it has at its disposal the feedback of its numerous subsystems, it compensates automatically for the deviations from the state of dynamic equilibrium. For that it is necessary that the system and all its subsystems function according to an optimal program."
>
> *Voprosy filosofii* (Moscow), 1965, 10.

[52]An approach typical of social psychologists, biologists, anthropologists and some functionalist sociologists (e.g., Merton, Nadel).

[53]Parsons, "A Paradigm for the Analysis of Social Systems and Change," *op. cit.*

$R_{iii} = f(\gamma)$ (SOCIO-CYBERNETICS II)

Societies : adaptive self-organization

Buckley makes clear that the adequate model of contemporary society is an open system capable of changing its own structure as a condition of its viability.[54] Such "complex adaptive systems" are not only externally, but also "internally" open in that they dispose over regulating mechanisms which can give them self-direction, and not only self-regulation. In other words, Buckley argues for a social system model capable of both self-stabilization and self-*organization*: "the paradigm underlying the evolution of more and more complex adaptive systems begins with the fact of a potentially changing environment characterized by variety with constraints, and an existing adaptive system or organization whose persistence and elaboration to higher levels depends upon a successful mapping of some of the environmental variety and constraints into its own organization on at least a semi-permanent basis."[55] Such complex adaptive systems dispose over the necessary mechanisms for mapping into their own structure the type of conditions relevant to their existence in the environment. A well-adapted system is selectively matched to its environment; in the process it evolves and becomes increasingly complex in structure and selective in its operations.

The model of "complex adaptive system" applies (as Buckley himself points out) to systems on diverse levels of nature's hierarchy. The social system is a particular variant of it, manifesting characteristics and flows typical of its level, i.e., corresponding to the nature of its internal structure (subsystems) and its external environment (suprasystem). Sociological inquiry centers on these systems and their characteristics, and inquires in particular into the sources of the impetus originating change, the impact of these disturbances on the institutionalized structure, and on the generalized effect of adaptation on the social structure.

Parsons lists both exogenous and endogenous sources of structural change. Exogenous sources include disturbances introduced into the system from the personalities, organisms and culture systems of its members (these, as noted, he interprets as composing the system's *environment*), as well as changes originating in other social systems. For a politically organized society, the most important of the latter are *other* politically organized societies. Hence both national and international pressures count as exogenous impetuses for change, provided the national pressures come from the individuals and the culture-system, and not from within the social system

[54]Buckley, "Society As a Complex Adaptive System," *op. cit.*
[55]*Ibid.*, p. 491.

and its institutions. The latter are endogenous sources of change, i.e., *strains.*[56]

The accommodation of strains by structural reorganization is but a form of accommodation of existing systemic components to gain maximum stability ("equilibrium"). Response to a changing environment, however, constitutes the real test of the viability of a society, since the responding system does not encompass the sources of change, nor does it have control over their development. Hence the survival of a social system in a changeable environment depends first and foremost on its ability to cope with change in that environment. And coping with change means either offsetting the disturbances thereby introduced into the structure by regulative controls, or a reorganization of the basic parameters of the structure. The latter kind of process is said to be an intrinsic feature of modern societies viewed as adaptive systems. The cybernetical term for such systems is *ultrastability.*[57] Modern social systems prove to be ultrastable systems, inasmuch as they are capable of compensating for changes in the environment by changing their structure and behavior (*socio-cybernetics II*).

The parallel between ultrastable social systems and other ultrastable systems is most striking when the social system is compared with the likewise ultrastable *learning* system. A learning system—whether biological or artificial in origin—is capable of reorganizing its cognitive structure on the basis of its input ("experience"). It responds to changes in the environment not only by assimilating the disturbances to its existing structure, but by evolving itself to cope with the new set of conditions. Cadwallader suggests that many forms of social organizations exhibit ultrastability through learning and innovation. Their components can be analyzed in cybernetical terms, as feedback mechanisms, input receptors, information-flow channels, memory components, decision-making facilities, and effectors.[58] Such analysis exhibits the isomorphy of the strategically isolated features of social systems with similar features of other instances of natural systems (and even artificial ones), since ultrastability—the property of adaptation to a changing environment through self-organization—is shared by systems of a wide variety.

Structural change may be conceived as a change in the social system's normative culture (Parsons), which here plays the role of basic fixed forces, or internal constraints, specifying the system's typical parameters. A change in these forces—i.e., in the cultural and institutional norms of the system—means a change in the social structure itself. At the top controlling level

---

[56]Parsons, "A Paradigm . . ." *op. cit.*

[57]Cf. Mervyn L. Cadwallader, "The Cybernetic Analysis of Change in Complex Social Organizations," *American Journal of Sociology*, **65** (1959).

[58]Cf. Cadwallader, *op. cit.*, and Deutsch, Boulding, Rapoport, Buckley, *et al.*

of the hierarchy of a social system, change affects the paramount value system, and from here it radiates to progressively lower levels, until it reaches the role-level of the individual, affecting his behavior and motivation. Leaving aside now the intricate mechanisms which favor or impede change, and the relation to these of various types of disturbances, we pass to the decisive question concerning the general *effects* of change. Are there important generalizations to be made with respect to the *evolution* of ultrastable adaptive social systems?

Despite the manifest weaknesses of classical evolutionary theory, many sociologists believe that such generalizations can be made. The concepts "specialization" and "complexity" are now replaced by the concept "differentiation."[59] The content remains significant, however, for grasping evolutionary self-organizing trends and processes in social systems.

"The process of functional differentiation is one of the fundamental types of social change, and has evolutionary aspects and implications. In its bearing on the type of system, it involves more than increasing complexity—e.g., the fact that flexible disposability of resources depends on such differentiation. This dependence requires higher-order mechanisms of integration, substituting the more specialized processes of control associated with markets, power systems, etc., for control through embeddedness in diffuse structures."[60] Through progressive differentiation, social systems evolve from relatively primitive kinship societies, in which the major roles are allocated on an ascriptive basis and the division of labor is based primarily on family and kinship units, to complex modern societies, in which the main roles are institutionalized and organized in specific symbolic and organizational frameworks which become increasingly interdependent and symbiotic in their functioning within the integrated social system.

Differentiation is the general trend in the development of social systems. This is not to say, however, that different societies parallel the main stages of this development, or that they produce analogous institutional structures at corresponding stages. Great divergences exist between the particular structure of societies in the course of their progressive differentiation. Eisenstadt calls attention to cases of partial de-differentiation or "regression," as well as of partial or uneven differentiation.[61] However, as he points out, the variety of integrative criteria and institutional patterns at any level of differentiation is not limitless. The level of differentiation in any one sphere provides the preconditions for effective institutionalization

[59]Cf. S. N. Eisenstadt, "Social Change, Differentiation and Evolution," *American Sociological Review*, **29** (1964).

[60]Parsons, "A Paradigm . . ." *op. cit.*

[61]Eisenstadt, *op. cit.*

of certain levels of differentiation in other spheres of the social system.[62] *Within* these limiting conditions, variety is due to several reasons. First, different societies arrive at the same level of differentiation through different historical paths and through diverse structural forms. Secondly, a process such as modernization may begin in tribal groups, in caste societies, in peasant or in urban societies. Each of these differs greatly in resources and abilities for creating and implementing differentiated goals and for regulating the increasingly complex relations within the developing social system. Thirdly, different structural patterns on a given level of differentiation may be due to differences among the predominant elites. Elites may develop in different institutional spheres or in the same sphere but with different ideologies and action orientations.[63]

Nevertheless, if not the precise *form* of the institutional structure of a given social system is taken as normative of its developmental pattern but the fact of its *progressive differentiation,* then across the impressive range of structural variety we glimpse relative constancy: the general trend is *toward* progressive differentiation, notwithstanding partial and temporary reversals, uneven development, and differences in structural expression. The case is analogous to evolution in other sectors of nature: the overall trend refers to levels of organization of differentiated and coacting components, and not to the manner in which these are organized within the whole. Just as physical evolution "tries" various pathways in the building up of the elements and the constitution of multi-atomic chemical systems, and as biological evolution "explores" myriad possibilities for producing differentiated complex organisms, so evolution in the social sector ranges over a wide variety of structural forms in bringing about societies of highly differentiated but coacting subsystems. Such societies are highly adaptive—they are able to cope with disturbances which could prove lethal to the adaptively unresponding, less differentiated society. But, as in all sectors of organization, higher functional capacity, afforded by the more differentiated structure, is paid for in the currency of overall stability: the modern technological and bureaucratic society is considerably less stable than the relatively primitive tribal society. In the realm of society, much as in that of biological and physical nature, adaptation is synonymous not with structural but with *cybernetic* stability: functional efficiency in coping with actual environmental disturbances.

The highly complex, cybernetically ultrastable contemporary social system is the product of an evolutionary process; Boulding likens it to the development of a chick from the egg. "The 'egg' is the relatively undifferen-

---

[62]Cf. Eisenstadt, *op. cit.*
[63]*Ibid.*

tiated, unorganized subsistence economy of small farmers and craftsmen, without large organizations, without much in the way of complex equipment or formal education. The 'chicken' is the developed society, with large and complex organizations, complex accumulations of capital in the form of material, skill, and educated and informed intelligence, and an extensive division of labor and differentiation of function."[64] The difference between the two societies is one of degree of organization: the diffusely organized "egg" society transforms into the differentiated and integrated "chick" society, as the members of the former get jobs in larger organizations, acquire education and skill, and end up in highly differentiated roles.

Although Boulding does not pretend to have solved the problem of measuring levels of societal organization, he does suggest that the gain in the transformation from a lower to a higher level of organization can be interpreted in terms of the redistribution of entropy. Basing his model primarily on economy, he argues that consumption means reducing order to disorder (e.g., food to waste, new products to the garbage heap): it is a typically *entropic* process. By contrast production is *anti-entropic*: it imposes a greater degree of order on raw materials of a low level of organization. By virtue of its higher level of production, an evolved economy is more negentropic than an undeveloped one. However, since society reverses the cycle in the complementary process of *consumption*, one might have to seek a measure of organization (and therefore, of social evolution), not in the accumulated *stock* of a society, but in its production-consumption *flow*.

Now, even if the entropy-redistribution concept is problematic in the realm of social phenomena, that of *information content* is readily applicable. It should be possible to calculate how many bits of information it takes to constitute a given social system from its components, if not in practice then in principle and to some statistical approximation. And it should be clear that a highly differentiated society, with a pronounced hierarchical or bureaucratic structure, would yield a much higher measure for information content than a diffusely organized system. Thus it can be argued that social evolution means increasing differentiation (regardless of the structural-institutional form it takes), and that increased differentiation means a higher level of organization in some objective form of measurement. But this need not signify the operation of some teleological final cause, a common goal striven for by all societies. The more modest and testable assumption of *adaptation-capacity in social systems embedded in changing environments* can account for progressive differentiation and

[64]Kenneth E. Boulding, "Some Questions of the Measurement and Evaluation of Organization," *Ethics and Business*, H. Cleveland and H. D. Laswell, eds., New York, 1962.

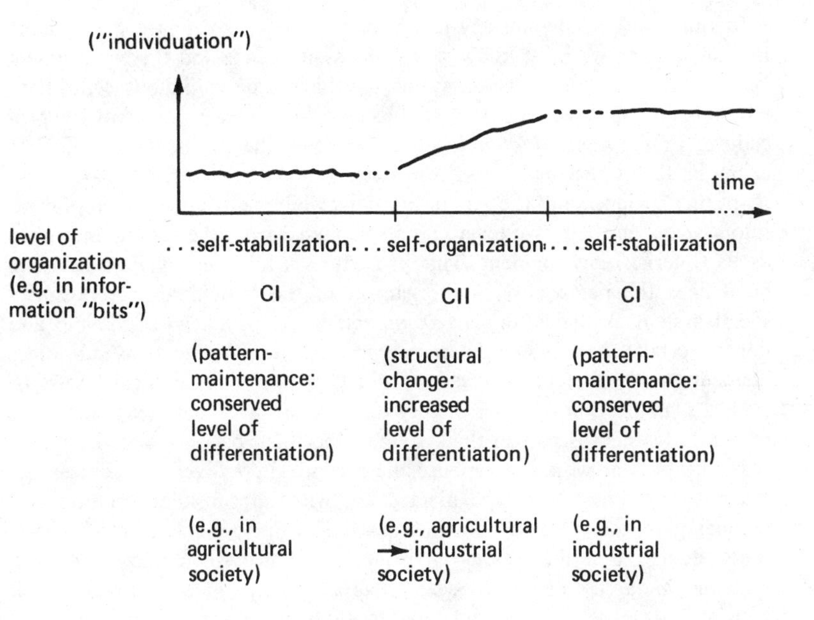

Figure 9: patterns of development in social systems.

increasing levels of organization, by showing that the structural changes which conduce to these states are necessitated as a means of solving the problems and meeting the challenges posed by changes in the environment.

$R_{iii} = f(\delta)$ (HOLON-PROPERTY)

Societies : intrasystemic hierarchy

In order to consider intrasystemic relations in a system we must define its boundaries: we must know where the system ends and the environment begins. In the case of open systems—which are continuous with their environment—this is always a problematic and somewhat arbitrary procedure. In the case of *complex* open systems the problems accrue. For example, the system may be said to extend only to the sets of relations on its particular level, and the sets of relations which constitute its components into systems on their own level may be excluded from the system and taken as its (internal) environment. This procedure is followed by Parsons when he defines the personality and organism of the individuals composing a social system as the latter's *environment*. Yet it is likewise possible and cogent to take the sets of relations composing the subsystems as subsidiary elements of the system in question, albeit on lower hierarchical levels. In such a classification, the personality and organism of individual members of a social system is part of the system, rather than part of its environment.

In the present work we espouse the second alternative: we conceive of systems as Chinese-box type hierarchies, with all subsidiary components belonging to the internal constitution—the "intrasystemic hierarchy"—of the system, rather than to its environment. Hence we identify the intrasystemic hierarchy of social systems as consisting, in successively lower hierarchical levels, of the personalities and organisms of its individual (human) members, the organs, tissues and cellular systems of these, and the constitutive molecular and atomic systems. Thus atoms, and even subatomic particles, become constituents of social systems, although it remains just as incorrect to say that social systems are systems of atoms as it is to say that biological systems are that. Nevertheless it is correct to say, in the light of this classification, that social systems are composed of atoms, which in turn compose molecules, etc., up to and including the institutional subsystems which compose the social system proper. In this classification we exhibit the *intra*systemic hierarchy of social systems, extending below the level of the "social" through the "biological" to the "physical."

The basic subsystems of the social system proper are its various institutions, constituting agencies of control of the behavior of individual members. Max Weber conceived of the distribution of institutional constraints in modern societies as the "bureaucratic machine" and emphasized its relative independence from the intentions and ideas of any group of individuals. The person engaged in the machine—the "bureaucrat"—is only a single cog in a functional mechanism which assigns to him a fixed set of routines. The official is entrusted with a set of tasks and cannot arrest and change the mechanism at will. Such directives or instructions can be issued only at the top, evincing the hierarchic nature of the mechanism. Moreover, such a mechanism cannot be destroyed by a group of persons opposed to it, such as anarchists, for it does not reside in public documents but is rooted in the orientation of man for placing himself, and obeying, a hierarchical set of constraints. Both the governing and the governed are thus conditioned. Thus, even when the formal structure of a bureaucratic mechanism is destroyed, it is reconstituted again due to the conditioning of most persons in the society.[65]

Bureaucratic mechanisms typify not only state administrations, but all social institutions where concerted action is systematically carried out. The centralization of control functions, and the chain of command from the center (the decision-making group) to the periphery (groups implementing the decisions), is a general property of social institution, whether we consider the legal, political, economic, or even the intellectual and cultural sphere. As a result the situation in which any given individual acts is composed of other individuals in ordered sets of relationship to him.[66] Moreover the individual's motivations for action are likewise socially conditioned, because his own personality structure is shaped through the internationalization of social objects and patterns of institutionalized culture.[67] Real possibilities for action, as well as motivated expectations and goals, jointly constitute the set of constraints whereby individuals fill subsystemic roles in regard to the social system in which they live. In a viable society, the constraints assure that the role-structures do not freeze into static bureaucratic machinery but remain capable of mapping the relevant factors of the changing societal environment into an evolving, dynamic social organization. The more complex and flexible the constituent parts of the society are, Deutsch points out, the greater the growth-potential of the society and the coherence and freedom of its subassemblies.[68] Essentially, growth involves the differentiation of the hierarchical institutional structure, and the functional integration of the differentiated units within a more highly articulated and flexible social whole.

---

[65]*Max Weber: Essays in Sociology,* H. H. Gerth and C. W. Wills, eds., Oxford, 1946.

[66]Parsons, "A Paradigm . . ." *op. cit.*

[67]*Ibid.*

[68]Karl W. Deutsch, "Toward a Cybernetic Model of Man and Society," in Buckley, ed., *Modern Systems Research . . ., op. cit.*

Societies : intersystemic hierarchy

It is clear that social systems exhibit a variously evolved, and in our day usually highly structured, hierarchical organization, or *intrasystemic hierarchy*. What about, then, *inter*systemic hierarchies in regard to social systems?

Parsons sometimes appears to suggest that the external environment of a given social system is the culture system which assigns the normative values of its institutions. Following this interpretation, the superordinate social system (or social metasystem) would be the culture system; it is in reference to this that we would have to define a social system's *inter*systemic hierarchy. However, it is not clear whether a culture system is a concrete system in its own right, or merely an abstract theoretical system—a gestalt in the conceptualizations of the investigator. Spengler adduced evidence for interpreting culture systems as concrete systems, with organic cycles of development and decay,[69] and Kroeber's more recent work reinforces such interpretation.[70] But Sorokin already stresses as its defining characteristic the "logico-meaningful integration" of a culture system which, unlike "causal or functional integration," makes it but a conceptual gestalt.[71]

However, evidence for the intersystemic hierarchical relations of given social systems does not hinge uniquely on the concreteness of culture systems. The relevant external environment of every social module includes, in addition to the nebulous culture system, more concrete superordinate social metasystems of various kinds. Those emerging today include federations and associations of the most diverse kind, ranging from formally constituted multi-national states, such as the USSR, through economic blocs, such as the Common Market, to more loosely federated inter-governmental organizations (IGOs) such as the OAS, its counterparts in Europe and Africa, and, at the most encompassing level, the struggling world-federation essayed in the United Nations. These supra-social (yet in themselves still *social*) systems carry forward the Chinese-box hierarchy of particular social systems in the intersystemic direction, generally losing articulation proportionately to their hierarchical level. Here we are dealing with relatively new *international* systems, superimposed on the established national and international modules and still lacking the degrees of differentiation and integration of the latter.

Indeed, surveying the theoretical literature in contemporary political science and international relations, we find increasing concern with the

---

[69]Oswald Spengler, *The Decline of the West*, London, 1926–28.
[70]A. Kroeber, *Configurations of Culture Growth*, Berkeley and Los Angeles, 1944.
[71]Sorokin, *op. cit.*

recognition and description of international systems *qua* systems. Prompted by writers such as Rosenau[72] and Modelski,[73] who note that study in these fields has not kept pace with empirical developments and who ask therefore for a "linkage" model, or for the "study of international systems as comprehensive, persistent, interdependent and self-maintaining entities," a number of writers now concentrate on outlining properties of supra-national ("world" or "global") systems.[74] One of the most noteworthy attempts of this kind is Singer's "The Global System and its Sub-Systems: a Developmental View."[75] Having noted in a previous paper the merits and demerits of a systems level of analysis (which produces a more comprehensive and total picture of international relations than does the national sub-systemic level but lacks the depth, greater detail and more intensive portrayal of the latter),[76] Singer goes on to outline the sub-systems; system and sub-system properties; inter-entity relationships and interactions; and interactions and changes in the system attributes, of *global systems,* defined as consisting of national, subnational, and extranational social entities, most of which are to some extent interdependent and all of which operate within the common larger environment. This analysis is paralleled by others, for example, by Modelski, who maintains that: "(1) the proper object of the study of international relations is the universe of international systems, past, present, future, and hypothetical. [. . .] (2) international systems are social systems; (3) international systems have structures; (4) the same functional requirements are satisfied in all international systems; and (5) concrete international systems are mixed systems."[77] While Modelski speaks in terms of a plurality of international systems, he points out that, since 1961, the Western international system is assumed to contain the entire world population. Hence there is but one global system at present, which is describable on the strategically chosen systemic level *qua* system. Concerned scientists and humanists see the emergence of a concrete, self-maintaining world system as a precondition of human survival. They tend to look upon such a system as existing in fact, and not only in the conceptualizations of the investigator. For example, Glenn T. Seaborg, Chairman of the U.S. Atomic Energy Commission, tells us that "The next thirty-odd years will see us engaged in

---

[72]"Introduction" to *Linkage Politics,* James N. Rosenau, ed., New York, 1969.

[73]George Modelski, "Agraria and Industria: Two Models of the International System," *The International System,* Klaus Knorr and Sidney Verba, eds., Princeton, 1961.

[74]E.g., C. F. Alger, M. Kaplan, R. D. Masters, J. N. Rosenau, J. W. Forrester, etc.

[75]J. D. Singer, in Rosenau, *Linkage Politics, op. cit.*

[76]J. D. Singer, "The Level of Analysis Problem in International Relations," Knorr and Verba, *The International System, op. cit.*

[77]Modelski, *op. cit.*

what may be the most crucial struggle of mankind's existence—the struggle to prove that one 'mankind,' as a physical entity and not just as a glorious idea, can be created and survive. In this sense I am talking about mankind as a global civilization—men and nations not only coexisting with each other and nature but essentially living and acting as an organic whole."[78] Research on the "world system" is on the agenda of such organizations as The World Institute, the Society for General Systems Research, the Club of Rome, and the World Future Society. Jay W. Forrester, Marshall McLuhan, Buckminster Fuller, John McHale, Oliver Reiser and Kenneth Boulding have called attention to this total system, conceived now as a global social system, now as a global ecological, and even psychological one.

Whatever the conceptual dispositions of the investigator, it is evident to all but the most prejudiced that supra-national systems are emergent in contemporary social interactions and that these form systemic environments within which smaller national as well as sub-national systems must find a measure of stability and coordination or perish. The larger systems evolve out of the smaller ones by integration within encompassing systemic structures. Here as in other sectors of the evolution of natural systems, Simon's thesis finds ample confirmation. Let us recall his conclusion: ". . . hierarchies will evolve much more rapidly from elementary constituents than will non-hierarchic systems containing the same number of elements. Hence, almost all of the very large systems will have hierarchic organization."[79]

The empirical identification of the levels of a theoretical hierarchy of social systems remains problematic. We know of the hierarchy constituted by such social systems as families, clans, villages, towns, cities, states and nations. But many intervening systems wind themselves into this pattern—economic, legal, political and cultural organizations criss-cross one another and form hierarchies in their own right. Nevertheless, some general invariances can be glimpsed across all this diversity. We see, for example, that hierarchical order is constantly enhanced in the usual process of differentiation and integration. Systems differentiate and integrate with other systems in larger functional wholes. Qualitatively new properties emerge at each level, due to the information introduced into the systems by

---

[78]Glenn T. Seaborg, United Church of Christ *Journal,* May, 1968. Two years later, in a speech before the World Academy of Arts and Sciences in New York, Seaborg again stated his belief that we are now entering a "transitional period from tribal man to truly organic mankind," and suggested that the latter may be "the acme of physical evolution on this planet." (Reported in *The Christian Science Monitor,* May 9, 1970.)

[79]The Organization of Complex Systems," *Hierarchy Theory, op. cit.*

the configuration of its particular subsystems. The emergence of the world system may likewise bring with it phenomenologically unique properties, many of which have been described in some detail by writers such as Teilhard, Fuller, Taylor, Reiser, Boulding, and others. Yet, if a general theory of systems is interpretable for empirically known social systems, the emergent qualities of the world system, as well as of the many subsidiary social systems, will prove to be analyzable as specific transformations of the invariant properties of all self-maintaining and self-evolving systems in nature. That the world system is one such system, and is endowed with the self-stabilizing and self-organizing capacities which render it persistently compatible with human life, is the hope of the future and a valid ground for rational belief in survival, not withstanding the shortsightedness and folly of many individual human beings.[80]

[80]See *The World System: Models, Norms, Applications,* Ervin Laszlo, ed., New York, 1973 (International Library of Systems Theory and Philosophy).

Chapter 6

THEORY OF COGNITIVE SYSTEMS

> It is the beauty of systems theory that it is psychophysically neutral, that is, its concepts and models can be applied to both material and nonmaterial phenomena. (L. v. Bertalanffy, *Robots, Men and Minds, op. cit.*, p. 100.)

A general theory of natural systems has been proposed and empirically interpreted. The phenomena which served to exemplify natural systems included physical, biological and social systems, as samples of a wider range of systems in nature. Such systems can also be described as systems of "physical events," where the term "physical" is not restricted to events described solely by the science of physics, but includes all events occurring in what is often called "the physical universe." Events of this kind by no means exhaust the range of empirical events, however. In addition to scientifically observable objective events, there are also events which can only be observed by introspection and which make up the immediate, felt experience of each of us. These sets of events may be denoted, in contradistinction to "physical events," "*mental events*" (or "*mind-events*"). To collapse the distinction between the two species of events is naïve and leads either to physicalist reductionism or some form of spiritualism or idealism. Yet to make a categorical separation between them has the further unpleasant consequence of dualism. Is there, then, another solution?

I believe that there is; it is the one I call "biperspectivism."[1] It requires that the realm of physical events, as well as that of mental events, be explored for systemic properties, and that these properties be comparable as invariant independent variables in a general theory of systems. Systems of physical events (" natural systems") have been explored and interpreted above; we shall now proceed to do the same for systems of mind events ("cognitive systems").

---

[1]See Chapter 8, below.

THEORY: $Q = f(\alpha, \beta, \gamma, \delta)$, where $\alpha, \beta, \gamma, \delta$ are independent variables having the joint function Q ("cognitive system").

α : coactive relation of parts resulting in *ordered wholeness* in the state of the system ("systemic state property");

β : function of adaptation to environmental disturbances resulting in the *re-establishment* of a previous steady state in the system ("system-cybernetics I");

γ : function of adaptation to environmental disturbances resulting in the *reorganization* of the system's state, involving, with a high degree of probability, an overall *gain* in the system's *negentropy* and *information content* ("system-cybernetics II");

δ : *dual* functional-structural *adaptation*: with respect to subsystems (adaptation as systemic *whole*) and suprasystems (adaptation as coacting *part*) ("holon property").[2]

SPECIAL DEFINITIONS

"Cognitive System": a system constituted by mind-events, including perceptions, sensations, feelings, volitions, dispositions, thoughts, memories and imagination—i.e., anything "present in the mind."

"Environment": realm of physical-events, signalled by perceptions and acted upon through volitions. Perceptions and volitions (the latter henceforth called "conations") constitute the cognitive system's means of communication with the environment (input and output).

"Flow of experience between cognitive systems and environment": Perceptions → any or all other elements of cognitive system → conations → environment → perceptions → . . . , in an ongoing multicyclic flow in which all elements may be simultaneously present. "Other elements of cognitive system" include cognitive items of all kinds, real and imaginary, factual and emotive. These are the "forms" which perceptions fill with "content"; or the "meanings" for which perceptions provide the "reference." They are denoted by the umbrella term "construct." Jointly, all constructs, and sets of constructs, constitute the system's "cognitive organization." Thus the flow can be restated as *perceptions* → *cognitive organization* (*construct-sets*) → *conations* → *environment* → *perceptions* → (etc.).

[2]A detailed description of these variables is given on pp. 36-53.

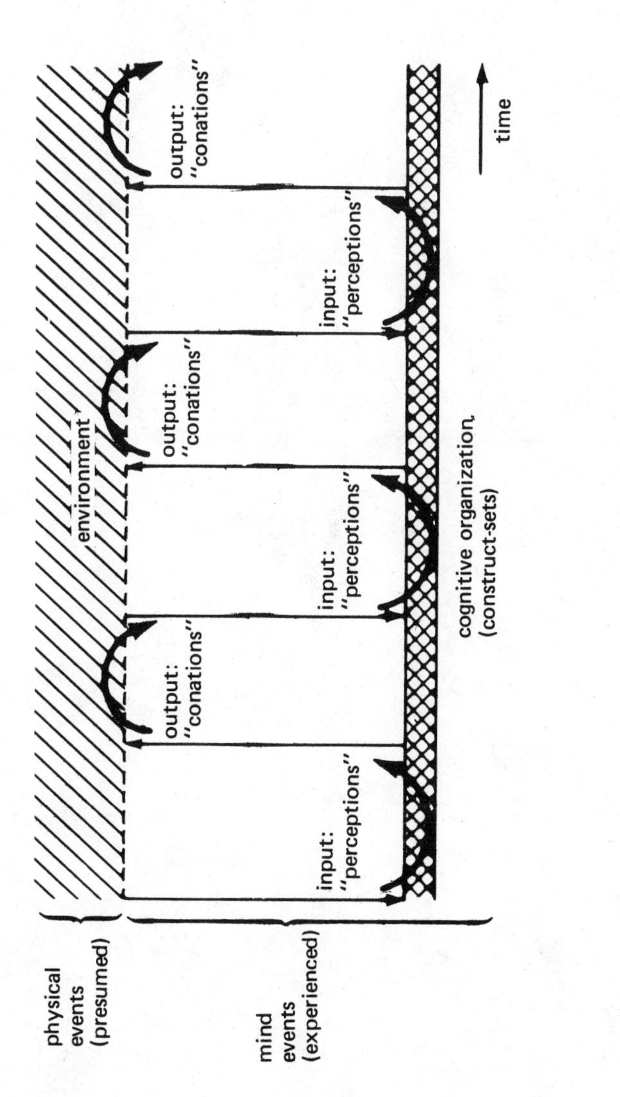

Figure 10: general information-flow in a cognitive system.

Chapter 7

EMPIRICAL INTERPRETATIONS

Cognitive systems may be as widespread in nature as natural systems. However, their identification proceeds by analogy from observed behavior, and such analogies fade in a continuous gradation as we explore species of systems progressively more unlike the human. It is both meaningful and useful to assert mind-events in regard to other human beings, since it helps us to understand them as beings similar to ourselves; and it is still meaningful and useful to do the same for some species of primates and other mammals. However, the cogency and usefulness of the concept of mind wanes with respect to systems more removed from us than these. Hence it is usually found objectionable to conceive of mind-events in regard to primitive organisms, such as slime mold, algae and plants; and with respect to groups of organisms, such as ecosystems or societies. And it appears as rampant spiritualism to speak of mind-events when it comes to cells, subcellular, molecular and atomic systems. Yet it is clear that relatively primitive systems of mind-events would be difficult to infer from observed behavior. Empirical evidence is poor, therefore, in regard both to affirming and denying the existence of mind-events for systems radically simpler than our own species. The primary grounds for decision remain theoretical, but no less important for all that. After all, theoretical principles, like the postulate of the uniformity of nature, the constancy of physical laws, and so on, are widely accepted and even human lives are staked on their truth (as, for example, in space exploration). Uniformity of nature in regard to mind-events is no less cogent a principle. In the absence of reasonable cutoff points, where we could draw a line and say, "above this level there are mind-events, and below this there are none," we must reckon with the possibility, explored by many great thinkers in the history of ideas, that mentality is a correlate of *all* physical existence.

The detailed arguments in this regard will be presented in Part Two (Chapter 8). They are announced here in order to make clear that the restriction of the empirical interpretation of cognitive systems to the human mind is not a categorical, but purely a pragmatic limitation. It is only in regard to the human mind that we have information adequate for an exploration of the empirical applicability of our theory of cognitive systems. This, however, does not entail that we deny that other interpretations of this theory are possible. The cognitive systems exemplified in cells, molecules, and plants—to cite but a few commonsensically objectionable examples—may be much simpler, but just as "real" as the cognitive system exemplified in the human mind; however, the latter remains the only such system which is accessible to empirical investigation. It thus constitutes the only instance for which the theory of cognitive systems can find empirical interpretation.

(i) THE MIND

THEORY: $Q_i = f(\alpha, \beta, \gamma, \delta)$, where $\alpha, \beta, \gamma, \delta$ are independent variables having the joint function Q_i ("human mind").

Phenomena of mind encompass an amazingly rich and varied terrain, from purely cognitive "facts" to ephemeral "moods." To attempt anything resembling a full interpretation of mind, the space of an encyclopedia is required. Our task in this section is a very much more modest one. It is to outline the general pattern of mental events in reference to the system postulates given above. More detailed explorations will be offered throughout Part Two (especially in Chapters 8, 9, and 10); the reader is also referred to this writer's *System, Structure and Experience: Toward a Scientific Theory of Mind*,[1] where the systemic information-flows of the human mind are more exhaustively diagrammed.

The purpose of this Chapter is to consider whether the set of events which constitutes the human mind (sensations, feelings, volitions, thoughts, memories, imaginations, etc.) exemplifies fundamental patterned relationships which render it isomorphic with the general information and energy-flows of natural systems. To this end we shall follow our previous procedure, and marshal evidence relevant to each general system variable in turn.

$Q_i = f(\alpha)$ (SYSTEMIC STATE PROPERTY)

Mind : wholeness and order

A few general considerations will establish the applicability of our first postulate to the human mind. It is sufficient to note that the elements of human experience constitute a non-summative system, manifesting proper-

[1]New York, 1969.

ties which are irreducible to the sum of the properties of the components. Human experience is not merely a collection of diverse sense-impressions, thoughts, fantasies, memories, etc., in an externally related heap or aggregate. Rather, it constitutes a holistic system in which these elements cohere as interdependent, mutually constitutive parts, giving rise to qualities of sensation in their mutual relationships. Of this nature are most of our experiences. Even our immediate sense-impressions tend to be complex, a fact which is adequately demonstrated in experimental Gestalt Psychology. Moreover most thought-processes operate by specific irreducible combinations of items either drawn from perception and memory, or produced by an act of abstract idealization. Neither the perception of our familiar gestalts, nor the constructs of empirical science and axiomatic mathematics, are attainable by the pure summation of the respective components.

The Humean model of the mind as a "bundle of perceptions" is definitely superseded today. Whatever the psyche may be, it is not an additive aggregate of individual elements but a system composed of mutually dependent parts. A perceptual whole or gestalt is meaningful when a demonstrable mutual dependency exists among its parts (Wertheimer); moreover the whole is very different from the sum of its parts. (In fact, once the many components of a perceptual pattern cohere in one's experience into a whole, the emergent quality—e.g., the face of a woman or a bear climbing a tree—persists and cannot be made to disappear.)

The holistic nature of the mind is not restricted to perceptual experience; it encompasses the entire range of psychic functioning. Maslow advances the concept of "personality syndrome" defined as "a structured, organized complex of apparently diverse specificities (behaviors, thoughts, impulses to action, perceptions, etc.) which, however, when studied carefully and validly are found to have a common unity . . ." He makes the point explicitly: "The whole is something else than the additive sum of the parts. So also the syndrome is something else than the additive sum of its isolated reduced parts."[2] Likewise Allport states, "Whatever else personality may be, it has the properties of a system."[3]

It would be to set up strawmen to argue against the atomistic additive conception of mind. Although cognitive psychology is not agreed on any single model or line of research today, it is agreed on rejecting the atomic "bundle of perceptions" concept and affirming something like a systems conception. It is likewise unnecessary to argue for the ordered nature of the cognitive system; it is sufficient to recall that psychology speaks of the

---

[2]A. H. Maslow, ed., *Motivation and Personality,* New York, 1954, (new edition: 1970).

[3]G. Allport, *Pattern of Growth in Personality,* New York, 1961.

organization of mind and personality, and that psychiatry combats syndromes which it terms *disorganizations*. Evidently, the cognitive mind is not a random array of elements but a structured whole. To note this suffices to establish the applicability of the "systemic state" postulate of our theory. Thus we can turn to the more complex and problematic issues raised by the other postulates.

$Q_i = f(\beta)$ (PSYCHO-CYBERNETICS I)

Mind : adaptive self-stabilization

The typical external description takes the system as a black-box phenomenon and seeks to correlate patterns of input with patterns of output. Functions of the system can thus be clarified by proposing formal descriptions of the mechanisms required to bring about the observed input-output correlations. In the case of a cognitive system, input-output correlations are given by the behavioristic S-R scheme. Although enormously useful in uncovering relatively direct, reflex-type stimulus-response connections, it has proven to be inadequate as an experimental tool for gaining a better understanding of the complex functions of the human mind, operating as the higher nervous system with its astronomical numbers of cells and looped connections. Weiss points out that "The structure of the input does not produce the structure of the output, but merely modifies intrinsic nervous activities, which have a structural organization of their own."[4] In consequence, the function-oriented external description must be intimately related to some understanding—hypothetical or even formal—of the internal state of the cognitive system which, we come to realize, determines its behavior as much as or more than its externally induced stimulation.

In the theory of *cognitive* systems we conceive of the mind as constituted of mental events, rather than of neurophysiological structures and processes. (The correlation of "mind-events" and "brain-events" will be attempted in Part Two, below.) Hence the empirical interpretation of variable β in regard to the human mind might give us the following hypothesis. "The human mind, a cognitive system constitued by the heterogeneous mind-events of introspectively disclosed human experience, tends to re-establish an internal organization made up of constructs of specific kinds in specific relationships, when exposed to new patterns of perceptual experience." This is the self-stabilizing property of mind, which expresses itself in many forms and in diverse areas of experience. Maslow defined it in its most general terms in saying that "from the point of view of functional psychology the unified organism is seen as always facing problems of cer-

[4]Paul Weiss, in *Hixon Symposium*, ed. Jeffriess, New York, 1951.

tain kinds and attempting to solve them in various ways permitted by the nature of the organism, the culture, and the external reality."[5] The cognitive system, conceived as an open system "thrown into" an external environment and constantly interacting with it, forms an organized, relatively determinate whole, which deals with external change through a variety of "coping aims" or response patterns.

The cognitive system manifests an internal organization consisting of systematically related constructs and sets of constructs. These include ordinary common-sense gestalts of houses, people, and other objects and relations perceived in the environment. They also include various meta-level constructs, such as the entities of science and art (religion, morals, etc.). Constructs evolve in a continuous hierarchical series as the individual interacts with his environment. Piaget, for example, shows that there is a definite evolution of perceptual and logical "schemata" in the process of growth and maturation of the individual. He refers to a process of "equiliberation" which effects a self-regulation of the cognitive field. It involves the assimilation and acquisition of new concepts and categories, as the system maps novelty into its existing cognitive structure.[6] Constructs map the changing environment through continuous processes of matching sensory inputs with established concepts and gestalts. (This mapping is a functional one, guaranteeing only that certain environmental events, which occasion sensory inputs, are functionally correlated with patterns of purposive outputs thus, that the resulting input is predictably modified. There is no further guarantee in the process that the gestalts and meta-constructs form pictorially accurate representations of objectively existing environmental realities.)

Our cognitive system model identifies the input and output channels of the mind as *perceptions* and *conations*. The environmental interaction occurring by means of these channels is a normative one in regard to the system: it leads to the latter's adaptation to the perceived environment. Adaptation can occur by means of one (or both) of the two systems-cybernetical processes. The first is exemplified in the human mind in its purposive manipulation of its stream of experience to stabilize itself in the steady state of its actual cognitive organization; the second, in the reorganization of the existing construct sets to fit the actual stream of sensory experience. The latter process involves *learning*, i.e., the acquisition of new constructs and their integration into the existing sets.

Self-stabilization (*psycho-cybernetics I*) involves the use of conations to structure the stream of percepts to progressive correspondence with the set of constructs already evolved in the system. Thus writing the word 'system'

---

[5]Maslow, *Motivation and Personality, op. cit.*, p. 34.

[6]Cf. Jean Piaget, *Les mécanismes perceptifs,* P.U.F., Paris

on a piece of paper means structuring one's stream of percepts to include a configuration of lines which match the already evolved *construct* "system." The pattern of perception is made to correspond to a construct in one's cognitive organization. Examples of such manipulations range from the spontaneous positioning of one's head for stereoscopic vision and stereophonic hearing, to highly calculative environment-modifications, such as creating objects and streams of events in the image of one's conceptions. Eliminating puzzles, not by learning to see them as something new and hitherto unknown, but by rearranging the puzzling percept-pattern, is likewise an instance of perceptual-cognitive self-stabilization. Building and maintaining a milieu in the image of previous conceptions is a self-stabilizing activity, enabling the individual to live in the "assumed form-world" he evolved through his past experiences.[7] It leads to the stabilization of the cognitive organization despite changeable external conditions. New sets of events are reinterpreted in the light of previous conceptions and, if possible, manipulated to conformity with those conceptions.[8]

In these processes conative patterns are produced and are directed toward restructuring a relevant part or aspect of the system's environment to match the existing construct-sets. That is, "behavioral" response is activated purposively, with the objective of bringing about those experiences which correspond to the requirements of the system. The process satisfies the standard concept of negative feedback: part of the output (conations activating behavioral acts) is fed back to control the input (percept-patterns). Consequently a successful cognitive system can stabilize itself by means of this technique: it can bring about conditions in its environment which confirm and correspond to its existing cognitive organization (construct-sets). All people build worlds around them which satisfy, to a greater or lesser degree, their conceptions of the world. When they command technological capacities, their purposively modified environment bears profound marks of their cognitive constructs, and serves to confirm and reinforce them.

$Q_i = f(\gamma)$ (PSYCHO-CYBERNETICS II)

Mind : adaptive self-organization

The purposive mind may be remarkably accomplished in structuring its stream of experience to correspond with its already evolved constructs

[7]See Chapter 11, below.

[8]A particularly striking form of such self-stabilizing interpretation of experience is described by Kuhn under the heading of resistance to paradigm change in science. He shows that even scientists, trained to see things objectively, are under the influence of the self-stabilizatory negative-feedback function controlled by pre-established constructs. (Cf. Chapter 11, "Scientific Cognition.")

and can, for most of its mature life span, maintain itself in an adapted state. However, the constructs themselves need to be evolved sometime, and they are evolved through overcoming as yet uncomprehended facets of experience. Also, there may be experiences which remain anomalous with respect to all existing constructs. In such cases negative-feedback self-stabilization around the steady states of existing cognitions is dysfunctional, since it maintains the system in a state of maladaptation. Another type of process is called into play, which receives the general name of *learning*. It represents the positive-feedback process whereby new constructs and sets of constructs are evolved as a means of coping with the challenge of anomalous experiences. The adaptation of the cognitive system to its environment can only come about through the elaboration of new constructs which match the anomalous experiences and hence endow them with meaning. The constructs which are thus elicited are not limited to meanings of ordinary perceptual objects (i.e., to gestalts) but include abstract scientific and non-discursive aesthetic constructs as well.

Learning effects a modification of the behavioral programs of the system, and when the system is a *natural* system (e.g., a physical organism), the new program functions as a self-stabilizing device, permitting the system to maintain itself under a wider range of environmental conditions.[9] However, when the system in question is a *cognitive* system, learning produces a reorganization of its essential parameters and becomes morpho*genetic* in character. Hence learning is a function of negative feedback self-stabilization in regard to natural systems (where it leaves the organism's genotype unchanged), and it is a function of positive-feedback self-organization with respect to cognitive systems. The same process of learning thus shifts from a *bio-cybernetic I* process to a *psycho-cybernetic II* process. The shift is occasioned by the framework of the analysis, however, and disappears when cognitive systems are reintegrated with natural systems in a general ontology of psychophysical systems.[10]

The process of learning in cognitive systems can be explored with the aid of the reward–punishment model originally advanced by Thorndike. In this model the feedback effects of behavioral responses are "rewards" if they lessen stimulation in some important way, and "punishments" if they increase it. The model was based on a now superseded "equilibrium" concept of mind which conceives of it as seeking states of minimum tension. Consequently every instance of learning was seen to involve behavior undertaken as a trial-and-error exploration leading toward lessened tension. Mowrer suggests a revision of this model to account for "learning" with-

---

[9]See "Biological Systems," above.
[10]See Chapter 8, below.

out "doing" by conceiving, instead of rewards and punishments, of "secondary reinforcements" and "secondary motivations." According to Mowrer,[11] we do not learn overt behavioral responses, but rather attitudes, meanings or expectations. These consist of token decrements in emotional tension ("secondary reinforcements") as well as of token increments ("secondary motivations"). Consequently learning is viewed as the acquisition of rewarding and punishing feedback from stimuli that have accompanied past action and experience. If a given act has been predominantly rewarding in the past, then the associated incidental stimuli will take on the capacity to produce secondary reward and thus perpetuate the stimulus-action pattern. If, on the other hand, the act has brought about predominantly punishing feedback effects, the sequence acquires the capacity to produce a secondary drive increment, and will tend to guide the organism into different behavior. As Mowrer notes, reinforcing cycles operate by negative, and motivational cycles by positive, feedback.

Mowrer's suggestion has the virtue of lessening the mechanistic bias of the original model and elucidating it at the same time in terms of feedback cycles. On the other hand his model still rests, at least implicitly, on the equilibrium theory of mind which, if consistently pursued, would indicate a progressive search for minimal stimulation. But contrary to this, experimental evidence (e.g., in sensory deprivation experiments) shows that a relatively high level of stimulation is essential for the mind, and that if the stimulation lessens, the system will initiate activities designed to *increase* its level. These factors underlie the exploratory activities of animals, involved in latent learning, and are evidenced by the spontaneous (and not reward-seeking) activities of rats, chimpanzees and, of course, men. Taking account of these factors it is more consistent to hold that the system seeks, not lessened stimulation, but a *match* between its input and its constructs—i.e., "intelligible" stimulation. Puzzles are to be avoided or overcome; meaning is to be acquired and preserved.

Negative feedback stabilizing cycles give way to positive-feedback motivated *learning* cycles when the input fails to match the constructs of the system, or matches them insufficiently. The constructs function as the system's representations of relations in the environment. If the representations are non-confirmed by the feedback of stimulus, the system turns in on itself, engages in a degree of self-analysis and, in free-running activities, attempts to derive the potentially adequate constructs. These are its "hypotheses." They may or may not be reinforced by the feedback of stimulation. If they are not reinforced, they give way to a "consequent

[11]O. H. Mowrer, "Ego Psychology, Cybernetics and Learning Theory," Adams, *et al.*, *Learning Theory and Clinical Research,* New York, 1954.

hypothesis." If the latter is eventually reinforced, the "something new which stands out against the highly organized background of the familiar and requires investigation" (Thorpe) becomes built into the cognitive system as something which is now familiar, and can be utilized, or stored in memory for possible use later. The fundamental reward in the feedback effect is intelligibility, attained by means of the input-construct match. Overt behavioral responses may not be produced at all if no immediate reward is expected of them; intelligibility is the goal, and that will nevertheless have been gained. Thus motivation is assessed here as a striving toward meaning, rather than toward behaviorally obtained reinforcements.

The outcome of learning processes is the organization of the cognitive system to an adequate representation of the significant relations in the perceived and inferred states of its environment. When acquired, these cognitive constructs become stabilized as they are incorporated in the operational program of the system.

The element of self-analysis, required to evolve more adequate environment-representations in place of those which break down or prove to be inadequate in the course of the individual's experience, calls for a measure of what philosophers term "reflexive consciousness." However, unlike in many philosophical schemes, nothing mysterious or *sui generis* is meant by this, but the ability, shared also by certain varieties of artificial systems, to form representations of one's representations of the environment. Reflexive consciousness is a high level "program-selector" or "decision center" which, if it obtains information of the malfunctioning of one of its constructs, undertakes to combine, translate, juxtapose, or otherwise elaborate new constructs on the basis of its existing degrees of freedom, and to test the new constructs against its actual input.[12] When the new construct is confirmed, it becomes part of the operational program connecting the system with its environment. The new construct may be put immediately into effect or be merely stored and used later (or perhaps never); in any case it represents, within the circuits of the system, a more adequate representation of the states of its relevant environment than the one it replaces.

The learning process results in the construction of increasingly functional mappings of relevant states of the environment, permitting a greater degree of predictability of future inputs. It occurs by means of a possibly vast array of "reflexive" and "meta-reflexive" levels, which assign meaning and significance to the first-level "empirical" constructs in terms of generalized "abstract" ones. Here the cognitive activity exemplified in both science and art becomes exemplified. When abstract meta-level concepts cohere into rationally constructed theories, we are dealing with an instance of scientific

---

[12]See Chapter 10, below.

construction. Such constructions can pulverize the ordinary empirical constructs which they initially analyzed and function in their place as a mapping of the empirical world, making use of scientific entities, rather than commonsensical gestalts. Likewise the reflexive analyses of ordinary patterns of cognition in terms of aesthetically meaningful meta-constructs may, in time, come to replace the empirical construct-sets in the artist's experience.[13] Ordinarily, the various levels of constructs coalesce in the unity of individual experience. It takes rigorous conceptual analysis to sort them out.

Psycho-cybernetics II (namely the process which results, in cognitive systems, in the reorganization of the existing construct-sets) may proceed by means of parallel analyses of matches and mismatches between existing constructs and environmental signals on several levels. In relatively simple systems (or in the early phases of the development of complex systems) only one level of operation is functional (for example, in infants, or persons with mental defects rendering them incapable of monitoring their own cognitive processes); however, the rule in human cognitive systems is a multi-level operation, where reflexive consciousness is evolved and serves to analyze cognitive processes on the empirical, as well as on one or more reflexive, levels. The human mind is an exceedingly flexible cognitive system, with positive-feedback dynamics operating on a series of interconnected levels. Inasmuch as the processes are adaptive, they result in cognitive organizations which map the relevant states of the environment with increasing precision and range of prediction. More aspects of the perceptual stream are cognized from more points of view, through various combinations of practical, scientific, aesthetic, and intermixed construct-sets. The respective constructs represent "things," "scientific constructs," "works of art," "persons," "minds," and so on, and form "world-views," "assumed form-worlds," "theories," "styles," "theologies," (etc.) in systematically joined sets. They enlarge the horizons of the system and provide it with increasingly wide ranges of progressively more refined meanings. By providing correlations between inputs from and to the environment in more instances and with more precision, the adapted cognitive system is also the more efficiently manipulative one: it can act on its environment in more ways and with more control, than the relatively non-adapted (uninformed or misinformed) system.

But granted that adapted systems map their environment with greater refinement than non-adapted systems, is it also true that the adaptive reorganization of construct-sets involves a *gain* in the system's level of

---

[13]A detailed presentation of these cognitive processes is given in Chapter 11, below.

organization? In other words, does the here sketched *psycho-cybernetics II* process conduce to *increasingly* organized—or merely *differently* organized—systems?

We have seen that, with some exceptions, natural systems tend to gain in level of organization when their basic structural parameters are reorganized, and that this gain can be measured (at least in principle) as negative entropy or information content. Since cognitive systems are information, and not energy-processing systems, entropy measurements may not be applicable to them; however, we can still apply the concept of information, and the two concepts are formally equivalent. Thus the proposed theory of cognitive systems is isomorphic with the theory of natural systems. If both theories can be empirically interpreted for mental and physical systems, respectively, we obtain evidence for the uniformity of these, apparently radically different, kinds of phenomena. The question is whether the reorganization of construct-sets in a cognitive systems involves an overall statistical gain in information content, paralleling the gain in negative entropy in evolving physical (biological, social) systems. Is it the case, in other words, that *learning* and *evolution* are analogous processes in systems?

Any construct-set may be as organized as (or more, or less organized than) the one which it replaces. Theoretically, the relationships of the old and the new cognitive organization depend on the environmental challenge, i.e., on the changed input which occasions the reorganization. If a pattern of perceptual input is totally new and anomalous for the existing constructs, another set of constructs capable of matching it need not be any more complex and organized. Substitution and not evolution is the case. But if the anomalous input is not an entirely new pattern replacing previous patterns but represents some new aspects or subpatterns within an enduring type of input, then the construct-set capable of mapping the new and anomalous patterns together with the matching ones must be more complex and organized than the previous construct-set, for which the emerging aspects or subpatterns remained anomalous. If Input_1 and Input_2 are basically different and disjunctive, with no increase in complexity of pattern in the switch, then Construct-set_1 and Construct-set_2 need not differ in level of organization. The cognitive system may not, in that case, gain in information-content through its reorganization of its construct-components. However, if Input_2 is a more elaborate form of Input_1 (for example, if "the very same" environment signals itself in hitherto unknown ways and reveals unrealized aspects), then Construct-set_2 must be more highly organized than Construct-set_1; the former must code all items coded by the latter, plus the new items, which are anomalous for the old set.

Since cognitive systems exist in environments of enduring order, it is unlikely that their perceptual inputs should change by a replacement of one type of pattern with another of equal or equivalent complexity and degree of potential intelligibility. It is more probable that changes in the percept-pattern will signify a discovery of new aspects or conditions in the same systematic environment, and that the old patterns will not drop out but continue to present themselves together with the new ones. Under such conditions, the new construct-set must be capable of endowing with meaning a wider range of percept-patterns than any previous one and will, in consequence, be more highly organized. That this is not merely theoretically, but actually the case, is shown by the findings of Piaget and his collaborators. Human cognitive structures, Piaget tells us, are the result of a continuous genesis beginning with the earliest experiences of the infant. All structures owe their existence to a functional genesis in the lifetime of the individual. And, Piaget affirms, "it must be definitely admitted, in the light of the facts, that a genesis always constitutes a passage from a more simple to a more complex structure, and this in an infinite regress (in the present state of our knowledge)."[14] But the complexification of the structures is not due exclusively to the challenge of the differentiated, enduring orders of the environment, but results already from the "homeorhetic" realization of potentials in the human mind. Mental development occurs in an ordered, multi-stage sequence, analogously to organic morphogenesis. The assimilation of gestalts, of logical structures, and the many other "schemata" of human cognition, represents an ongoing self-organization of cognitive systems, as the patterns of empirical knowledge not only change, but expand, differentiate and intensify. The observed trend in mental growth is unmistakably toward the acquisition of structures of greater information content—and hence, by analogy, of more negentropy—as cognitive systems map the many systemic factors of their environment into increasingly complex and organized sets of concrete as well as abstract constructs. (Figure 11)

$Q_i = f(\delta)$ (HOLON-PROPERTY)

Natural systems serve as paradigms in the exploration of cognitive systems, and natural systems are embedded in an embracing hierarchical level-structure. They are located at the intersection of what appears to be, from their individual standpoint, two intersecting hierarchies: an *intra*-systemic hierarchy, of which each system is the highest level suprasystem, and an *inter*systemic hierarchy, in which the given system is a subsystem component. Thus we should expect to find these hierarchical orders present in regard to cognitive systems as well.

---

[14]Jean Piaget, *Le Structuralisme,* P.U.F., Paris, 1968, Ch. 4 (my translation).

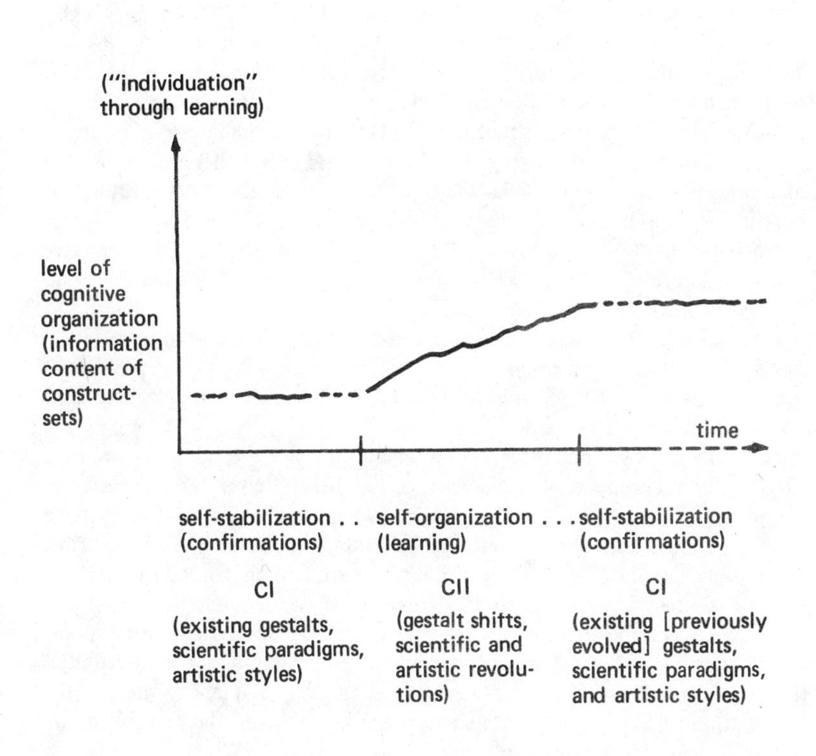

Figure 11: patterns of development in cognitive systems.

Mind: intra- and intersystemic hierarchies

The intrasystemic hierarchy constituting the human mind can be analyzed in a physicalist interpretation to a level-structure composed of the atoms, molecules, molecular aggregates and crystalline substances, cells and sub- and supra-cellular units, tissues and organs which make up the nervous system. These components, however, are not given *in* and *for* the cognitive system: they are not components of mind and experience. Atoms and cells, for example, are not part of the perceptual input, nor are they part of the conative output. If they are said to be found in the construct-sets, we must be careful to point out that not atoms and cells, but *concepts* of atoms and cells can be present there; i.e., not the natural events themselves, but the mappings of these events. Thus the intrasystemic hierarchy constituting the nervous system is not present in a cognitive system but forms its *precondition*—one that is known only through construction. The cognitive system, reflecting upon itself in terms of its gestalts, scientific and aesthetic constructs, can discern an intrasystemic hierarchy culminating in the neural processes which constitute the physical basis of the realm of mind-events. But in so doing, the cognitive system adopts the physicalist attitude and views its own mind-events as resultants of complex integrated physical events in the organism.

Now, the *inter*systemic hierarchy of which the cognitive system is assumed to be a unit component is likewise not a direct "given" of experience. Mappings of it are undoubtedly present in most cultural cognitive systems in the concepts of superordinate societal units, such as the family, tribe, community, nation or culture group, but these mappings are constructs and not the suprasystems themselves. Moreover they are constructed on the basis of perceptual inputs which themselves resolve into patterns of "sensations" which can be variously interpreted. But that they are interpreted (in part) as hierarchical systems encompassing the percipient testifies to the fact that, given the cognitive aims and presuppositions of the percipient, his sensations are well orderable into such systems. When so ordered, regularities recur and are predictable (within limits), and afford a coherent model of some aspect of the environing world. Consequently, construct-sets emerge which map the perceptual input as representing a series of encompassing unities of which the perceiving subject is a part. The sum of these socio-ecological constructs coheres into representations of an *inter*systemic hierarchy; whereas other constructs (those especially of the natural and medical sciences) cohere into representations of an *intra*systemic one. Neither hierarchy is "given" in the experience of cognitive systems, but the experience of such systems is such that these hierarchies are constructable as physicalist concepts and principles which explain and predict the "givens" of experience.

We may conclude that the evidence for the holon-property of cognitive systems is indirect, but nevertheless compelling. Evolved cognitive systems dispose over an organization which forms representations of hierarchical orders in the environment. The organization includes constructs of hierarchy, level, relationship and order, as well as constructs for the units or modules at the levels. Thereby the construct-sets both delineate hierarchies and show hierarchical organization themselves. The organization of evolved cognitive systems provides a fairly detailed mapping of hierarchies, extending in both directions from the given system. The constructs belonging to the sphere of the *natural* sciences probe in detail into the units and relations of the human *intrasystemic* hierarchy, dealing with units smaller than the integrated organism itself. These units occur as constituents of organisms as well as elements in other systems and configurations; and, as constituents of the human organism, including that of the cognizing agent, they are seen as constituting the physical substructure of the cognitive system itself. Constructs belonging to the *social* sciences explicate the units and relations of the human *intersystemic* hierarchy: they deal with units greater than the organism itself. At the intersection of the two constellations of fields we find the constructs of the anthropological sciences, linking the physico-chemical intrasystemic hierarchy with the socio-ecological intersystemic one.

. . . .

The foregoing considerations did not attempt to advance a new or a critical theory of mind and cognition. Rather, they constitute a second-order mapping of mental cognitive phenomena, by means of a conceptual integration of various first-order models, in psychology, personality theory, and epistemology. The mapping undertaken here integrates, in a single and consistent systems framework, generally acknowledged findings in the first-order disciplines. For example, order and wholeness are characteristic suppositions of contemporary theories in the area of cognition (with the exception of extreme forms of existentialism and nihilism). The "open system" postulate characterizes not only the presently much discussed holistic or systems theories of Maslow, Allport and others, but is the basis for all forms of behaviorism. The latter take stimulus and response (input and output) as methodologically decisive in the study of mind-systems, holding that these are commensurate, notwithstanding the black-box machinery in between, of summation, storage and delay.

Adaptation through self-stabilization around a steady state is a widely shared postulate. Some authors (e.g., Mowrer and Stagner) consider the steady state an extension of Cannon's biological homeostasis concept. Others emphasize a *sui generis* steady state or equilibrium. These theories range from the Freudian model of *id, superego,* and environment balance in the *ego,* to Piaget's dynamic equilibrium concept of mind. Most current theories of personality recognize the tendency of the mind (or cognitive component of the organism) to maintain an orderly arrangement of elements in a steady state. But, although such theories dominate much of current theorizing and emphasize stability at the expense of growth and development, other models, as well as further postulates of steady state models, recognize the developmental patterns of cognitive systems, which take the latter toward progressively more ordered states. Goldstein's conception of self-actualization as leading to the enhancement of order in personality, Maslow's theory of growth motives, Kohlberg's stage-concept of moral development, Piaget's epigenesis of cognitive functions, and Jung's principle of individuation, are examples of theories which recognize the inherent tendency in cognitive systems to move toward increasingly ordered states.

The hierarchy concept is likewise well represented. The double aspect of every personality is clearly recognized, for example, by Allport, who advocates that the individual should be viewed both in the content of wider, sociological and cultural interaction (i.e., as a component in a suprasystem) and as the focus of the patterning of all individual acts (as a system composed of its own subsystems). It is also seen that high-level "gating" or control imposes downward-radiating constraints on lower-level processes in all forms of cognition. The higher levels are often identified as values (Parsons), or motivations and meta-motivations (Maslow *et al.*), or ideational processes of diverse kinds (e.g., Freud's *superego*).

The theory here advanced and interpreted conceives of a cognitive system as a dynamic ordered whole, maintaining itself in steady states as well as evolving toward increasingly "informed" states (in both the psychological and information-theoretical senses of the word). Its significance is not that it innovates (which it does not), but that it exhibits the ismorphy of an ontologically presuppositionless theory of "mind-events" with an ontologically realist theory of "physical events." The hereby disclosed isomorphy of the two theories yields an invariance in the *general theory of systems* advanced in this Part, and forms one of the bases of the general *systems philosophy* proposed in Part Two.

GENERAL THEORY OF SYSTEMS: CONCLUSIONS

Atoms and organisms, molecules, cells, minds and societies—units of investigation which appear ordered in their own special way when approached within the framework of the specialized empirical sciences—reappear as units of the same general type in a generally ordered realm of nature and experience when the special theories are integrated within a second-order general theory of systems. We may never know whether the "real" world, that ultimate reality which surely underlies our observations and constitutes our very existence, is truly ordered, and if so, whether it is divided into distinct types of special orders or manifests one overarching type of systematic order. What we do know is that our rational mind seeks order and that when it finds it, it validates its own presuppositions. But not all kinds of worlds can validate all kinds of order and there must be something about the world we experience which permits the validation of certain types of envisaged order. This is the hope and the basic assumption of the empirical sciences. It is also the hope and the basic assumption of general systems theory: a field wherein second-order "meta-models" are constructed and explored on the premise that the "real" world may well be generally ordered.

If the present attempt at outlining a general theory of systems, imperfect as it is, is any indication of what may be possible in more rigorous and informed theories, there are no grounds on which this kind of endeavor could be dismissed or refuted as unwarranted or invalid. The real world may well be generally ordered, and its ultimate modules or units may well be systems maintaining themselves in particular states of organization and evolving to states of higher organization. Moreover, such systems may be ubiquitous within that portion of the universe of which man is a part, and their joint presence could build up the intricate structures which we have come to know as the complex and supremely balanced order of nature.

Part Two

STUDIES IN SYSTEMS PHILOSOPHY

Chapter 8

SYSTEM: FRAMEWORK FOR AN ONTOLOGY

Philosophers have always sought principles of knowledge as a means of disclosing meaning in existence, understanding the environing universe, and guiding human purposes. That such principles are not available on an absolute, unfalsifiable basis does not make the philosophical enterprise futile and inconsequential. Contemporary philosophy can strive to obtain and make use of the best available information, and can attempt to integrate its many facets into consistent general theories, disclosing meaning, eliciting understanding, and offering guidance. It is to these ideals that the here following studies in systems philosophy are dedicated.

The first of these studies considers the problems of ontology. Traditionally defined as the "science of being *qua* being," ontology in systems philosophy becomes the general theory of system *qua* system. Since both physical (energy-processing) and mental (information-processing) systems can be distinguished, systems ontology must first of all come to grips with the famed problem of the relation of *mind* to *matter*, or the *psychic* to the *physical*.

The psycho-physical problem

Our method for considering this problem will be the following. We shall take the general theory of systems advanced in Part One, and draw our conclusions from the isomorphy of the theories of natural and cognitive systems, as they apply to physical, biological and social systems on the one hand, and to the human mind on the other. Inasmuch as a general theory of systems finds empirical interpretation in these domains, we shall conclude that high-level uniformities may exist in the phenomena themselves. But such uniformities do not entail the disappearance of specific differences. Here we shall explore one area of seemingly irreducible differences : those separating "mental" and "physical" events. We shall proceed by emphasizing the differences, and then proposing a framework for integrating the respective events without either disregarding their differences, or concluding to the radical disjunctiveness of the events themselves in reality.

Physical events and mental events

The empirical-experiential immediacy of mind-events contrasts with the inferential nature of physical events and provides the classes of events with different sets of evidence. The systems formed by such events are the cognitive and the natural systems, respectively. Their formulation rests on the evidence of their constituent events. Thus *the evidence for cognitive systems* rests on introspection. Cognitive systems are constructions of "my" experience, since the components of cognitive systems are mind-events. If physical-events are said to figure in "my" experience, I can always point out that such events are not physical as given, but are merely mappings of presumed events outside my mind. And the constructs themselves are mind-events, even if their referents are (presumably) not. A physical event could only be located in the *environment* of a cognitive system, yet the latter merely constructs, and does not include, such environment. What cognitive systems do include are percepts, constructs of many different kinds, and conations.[1] The cognitive organization of the system is the sum of its sets of constructs, elicited by its stream of percepts, or recalled and remembered in its memory banks. By means of a gradual process of learning, it builds up a representation of a non-mental environment: the physical world of seemingly objective reality, outside.

The evidence for physical systems relies on the evidence for the existence of physical events themselves. These events are accepted as occurring outside the mind in a public domain, available for interpersonal observation. No presuppositions of the classical variety need to be made in regard to this domain: it need not be material or substantial in any dogmatic or absolute sense. It need not even be characterized by space and time as eternal "constants" or "absolutes." Space, time, matter, and other physical events may or may not be postulated in an unconditional sense. They may also be looked upon as so many transformations of some unnamed matrix, appearing now as energy, now as matter (rest-mass), and "creating" space and time in their various transforms. Not any specific characteristic functions as the premiss of the natural universe of physical events, but the simple but basic assumption that observations are had *of* events which are *not* mind-events. When this is allowed, theories of the physical universe of natural science can be built up by combining observations with theoretical explanations. But such theories fail, by their very nature, to give a consistent account of mind-events. The latter cannot be observed *qua* events

[1]See Chapter 6, above.

outside the cognizing mind. When one observes someone else's mind-events, he in fact observes that person's physical behavior and attributes analogues of his own mind-events to him. This is unwarranted by the premises of theories of the physical universe: physical systems are not characterized by felt sensations, but by energy transfers issuing in observable modifications of state.

The evidence dictates the separation of physical and psychical phenomena. It raises the time-honored specter of the ghost in the machine. The problem has been stated in the clearest possible form by Descartes and constitutes the foundations of his dualism. Kemp-Smith, writing of the Cartesian view of perception, summarized this argument as follows. "The processes, physical and physiological, above numerated, must have as their ultimate function the bringing into existence, or at least the occasioning so to exist, of certain entities, *viz.* those which we are now accustomed to entitle *sensations* of light and colour. These entities, Descartes further argues, differ in quite radical fashion from the antecedents which generate them. For whereas these antecedents are mechanical processes, occurring in public space, the resulting sensations are, he contends, not so describable, and occur in what may be entitled the field of consciousness . . . In ordinary consciousness the self seems to itself to look out through the eye at X^1; what alone it directly experiences is X^2; and X^2 is a copy, image or repre-

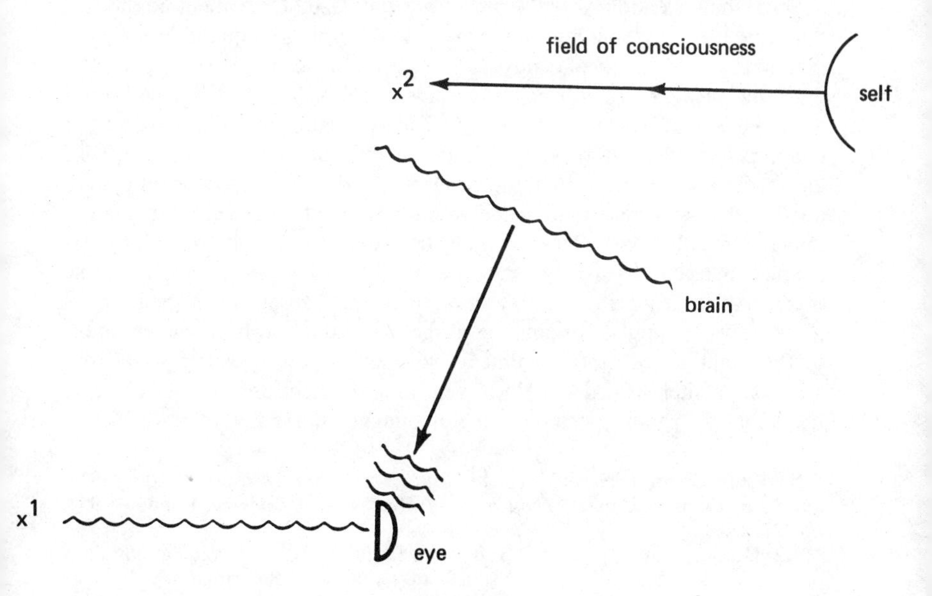

Figure 12

sentation of X^1, constructed by the self, in the light of past experience, out of the sensations that X^1 arouses by acting on the eye, and through the eye, on the brain. X^1 is invisible. What alone can be seen is X^2; and it is not a material body, but a mental image in the field of consciousness."[2]

In the above diagram, proposed by Kemp-Smith, the physical and the mental are clearly disconnected. In it physical events include X^1 "acting upon" (or conducing information to) the eye, and the eye "acting upon" the brain. Separated from these we find a *cognitive* system, encompassed within the Cartesian cogito ("field of consciousness"). Here the self is separated from the non-self ("world"), but both are mental representations: not physical, but *mind*-events. X^2 is said to be *constructed* by the self, out of the sensations that X^1 (the physical event) arouses by acting on the brain, via the eye. Now this scheme, while admirably clear and consistent on the epistemological issue, obviously faces the problem of how the arousal of the brain (a component in a chain of physical-events) can cause or influence a construction in the field of consciousness (a sphere of mind-events). The two spheres must somehow be interconnected; and Descartes ascribed the link to the pineal gland. Unfortunately, neither that gland, nor any other organ within the brain or body, appears to be capable of acting like the requisite physical-epistemic "transducer," transforming physical nerve impulses into sensations of color, sound and smell.

Some thinkers admit that the problem is unsolved.[3] Or it might be thought that the Lockean solution of primary and secondary qualities must be resorted to. On that theory, physical events consist of motions in space and time, and mental events of colors, sounds and other sensations. The former are attributes of matter or "reality," while the latter are additions by our mind, perceived *as if* they were material attributes. The theory explains knowledge by assuming that the primary qualities of nature so affect the mind that they arouse therein the secondary qualities of sensation. Causal interaction still takes place, as in Cartesian dualism, but now it is explained in terms of distinct properties, one set of which is acting on the other. Apart from the consideration that the causal transformation of physical wave impulses originating in the external world into felt sensations in the mind is still not accounted for, the theory is most unsatisfactory since, as Whitehead said, it does seem an extremely unfortunate arrangement that we should perceive a lot of things that are not there—which is

---

[2]N. Kemp-Smith, *Prolegomena to an Idealist Theory of Knowledge,* London, 1924, quoted by Dorothy Emmet, *The Nature of Metaphysical Thinking,* London, 1957, pp. 21–22.

[3]Cf. Dorothy Emmet, *op. cit.*: "The problem of the relation of mind and brain is one for which we have as yet no satisfactory solution." Sherrington and others would agree. (Cf. below.)

precisely what the theory of secondary qualities comes to.[4] Of the two kinds of realities posited by the theory, one consists of entities such as atoms, electrons and photons, and the other of "sense-data" such as colors, sounds, textures, tastes and smells. The first kind is the cause of knowledge but is never known; the second is known but is not the objective reality. Thus there are two natures: "one is the conjecture and the other is the dream."[5]

The "bifurcated" theory of nature is indeed unsatisfactory, whether it is in a Cartesian or Lockean formulation. Yet the difficulty which these theories attempt to overcome is real: it is to exhibit, within one system of consistent relations, mental events such as the redness and warmth of the fire, and physical events, such as the kinetic motions of molecules of carbon and oxygen and the radiant thermal energy acting on the sensory receptors of the body. It is only in the last decade or so that experimental findings have encouraged an approach which sometimes takes the form of an affirmation of body-mind *identity*, or, in a somewhat weaker sense, of body-mind *correlation*. Such research attempts to overcome the problem posed by the disjunctiveness of mental and physical events without bifurcating reality into two externally related layers. It might lead to an internally consistent non-dualist philosophy; it is thus worthy of closer scrutiny.

Brain-mind identity

Let us consider the stronger thesis of *identity* between "mental" and "bodily" (or physical) events. According to the school of thought sometimes referred to as the Australian National Thesis (Smart, Armstrong, Place, *et al.*),[6] physical events such as neuronal discharges in the brain and mental events such as images, pains and tweaks, are identical events. Their identity is not logical (for their properties differ) but empirical ("contingent identity"). The identity theorists make the standard logical distinction between intensional and extensional meaning, i.e., intrinsic meaning, significance, or connotation, and extrinsic reference. or denotation. They agree that the intensional meaning of mental and physical terms is different but hold that it is possible to verify that their extensional meaning (or reference) is the same. Mental and physical terms turn out to refer to the same events, much as "morning star" and "evening star," and "President of the United States" and "Commander in Chief of U.S. Armed Forces," refer to the same entities and persons. The theory thus rests on an empirical hypothesis,

[4] A. N. Whitehead, *The Concept of Nature,* Cambridge, 1920, Chapter 11.

[5] *Ibid.*

[6] As stated, for example, in J. J. C. Smart, *Philosophy and Scientific Realism,* London, 1963; Armstrong, below; Place, below; also Smart, "Sensations and Brain Processes," *Philosophical Review,* LXVIII (1959).

open to confirmation or disconfirmation. But in its Australian versions, the theory is usually made to support a stronger claim: not only that mental and physical events (or the terms by which they are stated) have the same *referent*, but that they are *identical*. Whereas it is logically possible to conclude from the fact that empirical investigation discovers (if indeed it does) that mental and brain events have the safe referent, that they are *correlative*, i.e., that one is the sufficient and the necessary condition of the other, the identity theorists conclude that the shared referent proves the *identity* of the events themselves. And this stronger thesis is not convincing, as I shall show.

Let us take the case of the morning and evening star, or that of lightning and the motion of charged particles. These sets of events are said to be identical in virtue of sharing a common referent, ascertained in different sets of observations and theoretical interpretations. The differences between the events in the set can be ascribed to the conditions of observation (morning or evening) or to frameworks of interpretation (common-sense notions of lightning or physical theory). Further it may be the case that the differences are due to mutually exclusive conditions, so that when one makes one type of observation he is by that token prevented from making the other kind. An example of this would be observing a cloud and observing masses of tiny particles in suspension.[7] When the observer is at a sufficient distance from the phenomenon, he sees it as a cloud—a fleecy substance constantly changing its shape; and when he is closer what he sees are tiny droplets or particles in suspension, i.e., mist or fog. These two sets of observations are mutually exclusive: when the cloud is seen the particles are not visible; and when the particles are visible the cloud cannot be seen.

It would appear that the relationship of mental and physical events is of this variety. When we see a red patch, or feel a tweak or an exalted emotion, we cannot simultaneously observe the discharge of neurons in our brain. And whenever we are in a position to observe nervous processes in the brain we are incapable of feeling the sensations which could be said to be "the same as" those nervous processes. But suppose that someone were to invent a brain-monitoring apparatus which could be not only hooked up to the brain of the subject, but the displays of which could be fed back for his observation. In that event the subject could both have the sensations and observe his own neural processes. To build such an apparatus is technically problematic, but is not impossible in principle. The point is, then, that if such an apparatus were constructed, the identity theorists

[7]U. T. Place, "Is Consciousness a Brain Process?," *British Journal of Psychology*, XLVII (February, 1956).

would regard it as confirmation of their thesis: they would conclude from the identity of the referents, to the identity of redness, pain and exaltation with the activity of certain complex loops in the cortex and related brain regions. And it is precisely this stronger assumption which is questionable.

Consider again the above examples. In every case we have two sets of observations of presumably the same phenomenon. The intensional difference is due, it is claimed, to the conditions of observation, and/or the theoretical framework within which they are interpreted. However, these differences can be reduced and all but eliminated; the two observations, or two theories, can be shown to have a common object with identical properties. We can even do this in the case of observations which are mutually exclusive, such as observing clouds and masses of suspended particles. For we can argue that there is a set of observations forming a continuum within which the two actual observations are subsets at different points, and that, if we could make all observations along the continuum, we would see the phenomenon of one observation gradually transform into the phenomenon of the other (e.g., as we move closer to a cloud we could see it dissolve into masses of particles making up a fog or mist).[8] This assumption will, in general, hold true for any two sets of observations of physical events where there is warrant to assert that they have a common object. If it were possible to observe the evening star continuously, we would see it "transform" into the morning star, and we would likewise see the U.S. President "transform" into the Commander in Chief of the Armed Forces. But this is not the case when the observations are of "mixed" objects, i.e., when one set of observations is of *mental,* and the other of *physical,* events. Even when equipped with a brain-process feedback apparatus we could not see our nervous transmissions transforming into images of red patches and sharp twinges, nor these sensations transforming into discharges of cell assemblies in the cortex. Physical events transform into one another if and when they are truly identical, forming a continuum of transformations elicited by different modes and vantage points of observation; but physical events do not transform into mental events, nor *vice versa,* no matter how the conditions of observation are varied. There will always be a qualitative-quantitative gap, even if the privacy-publicity gap is removed (as in the display of public brain events for private viewing and the still more fanciful—but theoretically possible—display of private mind-events for public viewing through some electrochemical hook-up of the brains of subject and observers). It is simply not the case that a green, odorous, screechy and silky phenomenon ever transforms into corresponding undulatory and molecular structures and processes. The perceived sensation is irreducible to the physical events by which it is scientifically

[8]Cf. Place, *op. cit.*

explained, whether these events are physical objects in the environment or neuronal processes in the brain. Following Malcolm,[9] the irreducibility in question may be summarized under these points : (i) brain phenomena have spatial locations (even if the brain is conceived as a holistic system without exact localizability of functions) whereas it is meaningless to assign spatial locations to at least some kinds of mental phenomena, e.g., thoughts and feelings; (ii) thoughts and other kinds of mental phenomena require a background of circumstances (practices, agreements, assumptions) which cannot be described in terms of the entities and laws of physics, but brain phenomena do not always require such background circumstances, and those it does require (energy and information inputs) are describable in terms of physical theory. Thus, (iii) the contingent identity of brain and mental events is not empirically verifiable; e.g., a thought cannot be shown to be identical with an event in the cortex by means of empirical observation and experiment. The situation is very different from that obtaining in regard to *all*-physical (and possibly *all*-mental) sets of events, in regard to some members of which an identity *may* be empirically verifiable (through a series of observations or theoretical shifts by means of which, e.g., the President "transforms" into the Commander in Chief, and the morning star "transforms" into the evening star). Hence we conclude that it is legitimate to assume an identity between members of all-physical sets of events, if and when their common referents can be empirically verified, but it is logically improper to assume such identity in the case where one of the objects is a physical and the other a mental event; these conserve irreducible differences in their properties, no matter how the conditions of observation are varied. Regardless of how closely we observe a green patch, and how meticulously we examine our feeling of aesthetic significance, the former will not begin to look the least like a physical wave propagation stimulating the optic nerve in the eye, and the latter will show nothing in common with the sound waves generated by a vocalist singing a Schubert *Lied*. Conversely no matter how long and variedly we observe what a physicist takes to be a photon stream, it will not begin to look the least bit green, nor will the wave frequencies produced in the vocal chords of the singer appear moving and beautiful. But it may well be maintained, on the other hand, that the stream of photons, interacting with the optical sensors of the eye, sets up patterns of neural energy transmission which we can introspectively identify as a "green patch," and that the same applies to sound frequencies and our feelings of aesthetic significance. And if so, then "photons" and "green patches," and "sound frequencies" and "aesthetic

---

[9]Norman Malcolm, "Scientific Materialism and the Identity Theory," *The Journal of Philosophy*, LX, 22 (1963).

significance" neither connote nor denote identical events. But they do denote, I suggest, *correlative* brain and mind events.

Brain-mind correlation

We thus come to the weaker, but more reasonable thesis of mind-event —brain-event correlation. The correlation thesis is popular among neurophysiologists who, under the impact of Sherrington's dualism, tend to be wary of collapsing mental into physical phenomena. Rosenblueth states a correlation thesis in his book, *Mind and Brain.* I quote the relevant "scientific postulates."[10]

"Our sensations are causally related to events that occur in a material universe of which our own bodies are a part."

"The physical correlates of our mental events are neurophysiological phenomena that develop in our brains."

"Each specific mental event has as a correlate a specific spatiotemporal pattern of neuronal activity."

Rosenblueth postulates a uni-directional sequence of events *material processes* ⟶ *selective activation of some receptors* ⟶ *coded afferent messages* ⟶ *central neuronal and correlated mental events.* This chain interlinks physical events and mind events without affirming causal relations between them. ("The causality involved in our sensations [he points out] is that which connects the nervous activities and the spatiotemporal events which occur in the rest of the universe."[11]) The relation of brain-events and mind-events is one of non-causal *correlation,* as specified in the above-quoted postulates. Thus Rosenblueth recognizes the non-interreducibility of mind- and brain-phenomena without either affirming a Cartesian and Lockean causal interaction between them (a course taken by another neurophysiologist, J. C. Eccles), or committing the fallacy of the identity thesis.

It appears, then, that the non-causal brain-mind correlation thesis is the more strongly indicated middle road between the Scylla of dualism and the Charybdis of identity. It is this position which is chosen here for explication as a tenet of systems philosophy. Before undertaking this task, it should be pointed out that the correlation of mental and physical events is misleadingly used as a merely heuristic device, as Rosenblueth sometimes appears to use it, for example, when he says that "we need to postulate conscious processes in some entities of the physical universe in order to understand many aspects of their behavior."[12] Thus used, the correlation thesis would have to be affirmed as long as it helps to explain behavior, and

---

[10]A. Rosenblueth, *Mind and Brain, op. cit.*, Chapters 7 and 10.

[11]*Ibid.,* p. 115.

[12]*Ibid.,* p. 67.

discarded beyond that limit. However, there are no independent reasons for restricting the range of mental phenomena to those entities of the physical universe whose behavior shows such complexity that without the assumption of mental events it is insufficiently understandable: this arbitrary cut-off point would make mental events come to a sudden stop somewhere between primitive vertebrates, and plants and insects (the behavior of mice might require the assumption of mental correlates but not that of birches and ants). There is nothing in theory of evolution, however, which would indicate the sudden emergence of mentality from a non-mental ancestry. *Forms* of mentality may, of course, evolve, and evolve by leaps and bounds, but the assumption that at a particular point in evolution purely physicochemical organic systems transform into *psycho*physical systems appears arbitrary and unfounded. The crux of the difficulty lies in treating the hypothesis of mind as a purely explanatory device; if we do so, then we must dispense with mentality for all organisms whose behavior is simple enough to be sufficiently explained without appealing to their awareness of themselves and of their surroundings. But the postulate of mental experiences in other beings is not merely an explanatory device, but a basic alternative to psychological solipsism. If that is to be avoided (which in itself is the one logically ampliative step, since we do not experience anyone's mind-events but our own), then the decision, how far to go in extending the range of mental phenomena, is not merely a decision of heuristic value, but one of logical consistency. There must be good reasons to draw the line above which there are mental experiences in nature and below which there are none, and these reasons must be concerned not only with the explanatory power of the concept of mind but also with general ontological principles. I am not suggesting that we intuit these principles by some mysterious revelation or special insight; only that the criteria for assuming mental experiences are not merely heuristic ones concerning powers of explanation.

Biperspectivism

The above discussions have served to focus the key issues preparatory to their systematic treatment. We can attempt such treatment now in reference to the foregoing assessment of the mind-body problem. We have found that the physical-mental identity thesis is too strong a statement of monism, and that the Cartesian or Lockean bifurcation is unsatisfactory in its dualist modes of explanation. The indicated middle-road appears to be the brain-mind correlation thesis, which brings into meaningful relation physical events in the organism and mental events in the sphere of consciousness. The question we must consider now is, how a brain-mind correlation thesis can function as an ontology which is capable of providing a uniform and consistent view of physical and mental phenomena in nature.

Mind-events and physical-events, we assume, are distinct but correlative types of events. Each type of event forms systems of isomorphic properties, and the question is, how these systems relate to one another. Corresponding to the relationship of cognitive and natural systems, we can assign specific correlations to the constitutive sets of events themselves.

How, then, do natural and cognitive systems relate to one another? Cognitive systems constitute the *mind*; and natural systems range over the full extent of the microhierarchy on the surface of the earth: they include *homo sapiens*, i.e., the *body*. Now, mind is not a public observable; rather than being observed, it is "lived." But that a system is "lived" from one point of view does not mean that it cannot be "observed" from another. In other words, an introspectively lived system of mind-events (i.e., a cognitive system) may well be an externally observed system of physical-events, if the viewpoint of the observer filing the report is correspondingly shifted. If we make this assumption, we get a passport to travel from the realm of mind and consciousness into the realm of physical nature. Consider now the means whereby we could make the trip from physical-events to mind-events. Some physical-events cohere into organized entities: these are the *natural systems*. They are characterized by spatial and temporal organization: structure and function. Observations on them refer to states of the system, quantified and interconnected in theories. What goes on inside the systems is likewise referred to variables yielding transformations of states in information-processing and energy-transfer. No description in terms of sensations "felt" by the systems is warranted—these could not be observed. But there is nothing to prevent us from assuming that natural systems may be "lived" (i.e., observed) from an immanent viewpoint; that one of the possible points of observation is not external to, but coincides with them. Consequently, if and when the observer describes the system, he in fact describes himself.[13]

Now let us see what conclusions are warranted by these assumptions. *First*, we allow that systems of mind-events can be externally viewed. *Second*, we allow that systems of physical events can be internally viewed. We then note that the theories applying to natural and to cognitive systems are isomorphic, with the consequence that when switching from the one to the other system, we do not necessitate changes in theory. The *content*

[13]The identification of the observer with the system is crucial: he is not merely inspecting the interior of the system—he *is* the system. (In the former case he would be an external observer operating within the space enclosed or occupied by the system, and his data would be physical events.)

or *referent* of the theory changes; its *form* remains invariant. We obtain a parallel series of independent variables applying to both varieties of systems. This provides our ontology with a fundamental concept: that of the *natural-cognitive* (i.e., psychophysical) *system*. Such systems are not "dual" but "biperspectival": they are single, self-consistent systems of events, observable from two points of view. When "lived," such a system is a system of mind-events, *viz*., a "cognitive system." When looked at from any other viewpoint, the system is a system of physical-events, i.e., a "natural system." The physical and mental sets of events in the systems are correlates; the systems within which they are found, or which they constitute, are identical. Here identity is predicated on the ground of the invariance of the two theories: if systems of events, however different the latter may be in themselves, are isomorphically structured, so that their respective theories do not change form when passing from the one to the other, then these systems are identical *qua* systems. There are no grounds on which we could differentiate between them.

$$\frac{\begin{array}{l} R = f(\alpha, \beta, \gamma, \delta), \\ Q = f(\alpha, \beta, \gamma, \delta) \end{array}}{R = Q}$$

The psychophysical thesis of this framework for an ontology can be stated as follows. *Sets of irreducibly different mental and physical events constitute an identical psychophysical system, disclosed through the invariance of the respective theories.* The basic entities of systems philosophy are non-dualistic psychophysical systems, termed "biperspectival natural-cognitive systems."

Biperspectivism and evidence

The psychophysical thesis outlined above suggests that mental experience is correlated with neurophysiological process in human (as well as other) organisms. If so, then one is entitled to ask for descriptions, in terms of thoughts, feelings and sensations, of each and every organic event, and *vice versa*, for descriptions in terms of organic processes of each and every mental event. However, even simple reflection will tell us that we cannot give such a description in every case. First of all, some organic processes are not mapped as mind-events under normal circumstances. Those, for example, which are controlled by the autonomic nervous system need not penetrate with signals into cortical regions and thus may not appear as events in the mind. (It must be admitted, however, that the human mind may have as yet unperceived capacities for penetrating autonomic organic processes. Current medical and traditional Yoga experience encourages such expectation.) Secondly, finding a description for even those physical events which are habitually mapped into mental events is a most difficult task. However, the difficulty does not appear to be one of principle, but of observation and experiment. In fact, medical science has long used the correlation of physical and mental events in the organism in diagnosing physical and mental health. (The difficulties of psychiatric medicine lie in the fact that malfunctions in the nervous system are not always, and not necessarily correctly, identified by the introspecting patient; and those of veterinary medicine consist in not at all being able to draw on reports of mind-events [pains, sensations] by the organism under examination.) In simple instances the introspective observer can identify his own sensations as a physical event in his body, such as a toothache signifying a cavity. But in many others he cannot. Not only is a detailed knowledge of medicine needed for this, but also diagnostic instruments, to test the proper correlation of the sensations. In every case, however, the implicit assumption is operative that there is *some* correlation between mind-events (pains, feelings, etc.) reported by the subject and the physical-events (organic functions) taking place in his body.

Matters become considerably more complicated when we ask for the physical correlates of mind-events in the brain itself. The psychophysical thesis might be read to assert that there is a neurological counterpart to every state of the mind, constituting a one-one correlation between mind-events and particular states of associative activity in the neurons of the brain. (The correlation in question does not involve a projection of neural circuits inside the mind, or *vice versa;* these are phenomenologically entirely dissimilar.) But neurophysiology and psychology have not progressed far enough to confirm or disconfirm this assumption by giving us a detailed mapping of both brain and mind-events, capable of comparison and integration. Nevertheless, workers in both fields are aware of the need for cooperation with some such goal in mind,[14] and some beginnings have already been made. The area of research has been galvanized into action by the pessimistic assessment of the relation of bodily and mental functions by Sherrington in his Gifford lectures. Speaking of brain and mental functions, he said that "the two for all I can do remain refractorily apart."[15] Among psychologists, Hebb showed concern over this problem, and remarked that it is not simplified when a physiologist such as Sherrington finds no possible way of reducing mind to neural action.[16] Also

[14]E.g. Hebb, Penfield, Lashley, Hilgard, Huxley, *et al.*
[15]C. S. Sherrington, *Man On His Nature,* New York, 1949, p. 312.
[16]D. O. Hebb, *The Organization of Behavior,* New York, 1949, p. 79.

Lashley expressed regret that "the greatest living neurophysiologist despairs of finding a common ground between the sciences of the brain and the mind."[17] In his view, mind should be seen as the name for an indefinite number of complex structures or relations operating in the organism.

The implicitly Cartesian position of "two substances," into which Sherrington felt himself to be forced, gradually gave way to a more optimistic position concerning the possibility of interrelating mind- and brain-events. The neurosurgeon Penfield performed experiments in which he confirmed the relationship between a mild electrical stimulation of nerve tissue in the temporal lobes and memories evoked in the mind of the subject. He concluded that he was dealing with a nerve-tissue record of consciousness. In his view there is a definite, stable record of the successive experiences in the individual's past in his brain. The key to its activation lies, Penfield says, in the temporal cortex. The record is the product of central integrations of nerve impulses which incorporate in themselves the essentials of consciousness. Past brain-events prove to be capable of being re-activated as conscious mind-events through artificial stimulation.[18]

The electroencephalograph record testifies to other aspects of mind-event–brain-event correlations. Working with a toposcope or one of its modern offshoots, the experimenter has three sets of data to correlate: the physical event serving as stimulus, the electrical configuration derived from this in the brain, and the mental state evoked in the consciousness of the subject.[19] Some significant correlations are found between brain-event responses to the stimulus and mind-event reports by subjects. By comparing the EEG record's statistics of distribution in frequency, phase and space, with the random distribution to be had of "noise," definite "message" factors are disclosed in correlation with introspective reports. For example, visual imagination and the so-called alpha-rhythms appear mutually exclusive. Alpha-rhythms of a certain level correspond to sensations of pleasure (Kamiya's experiments[20]); pleasant sensations also correlate with theta-rhythms. Beta-rhythms are known to be common in tension. These significant patterns appear superposed on the series of regular rhythms which, it is suggested, represent a process of hunting for information. The closest empirically disclosed correlations between EEG recorded rhythmic brain-events and the mind-events reported by subjects occur when the regular rhythmic activity is minimal. Concluding from these findings, Walter holds

---

[17]K. S. Lashley, "Coalescence of Neurology and Psychology," *Proc. Amer. Phil. Soc.*, 1941.

[18]W. Penfield, "The Permanent Record of the Stream of Consciousness," *Proc. Inter. Congr. Psychology*, Montreal, June, 1954.

[19]W. Grey Walter, *The Living Brain*, London, 1953, Ch. 6.

[20]J. Kamiya, "Conscious Control of Brain Waves," *Psychology Today*, 1968, 1.

that "in these observations there is an encouraging correspondence between the extent and complexity of the evoked responses and the subjective sensations."[21]

Reviewing the arguments in support of the theory that brain- and mind-events are correlated (arguments which led him to postulate that "each specific mental event has as a correlate a specific spatiotemporal pattern of neuronal activity"), Rosenblueth classifies the evidence under the following headings:[22]

Phylogenetic evolution: species with evolved brains and nervous systems exhibit behavior that is comprehensible only if mental experiences are assumed. The types of behavior in question are characterized by not being inborn or conditioned reflexes and by being adaptive, i.e., having survival value for the organism performing them. Such "conscious" behavior appears only when the organization of the higher nervous systems has reached a certain level in the evolutionary series.

Ontogenetic evolution: behavior manifesting patterns comprehensible only by assuming that mentation emerges when the anatomical and neurophysiological development of the infant reaches a sufficiently high level. Moreover, abnormal development is accompanied by mental deficiencies.

Deviations from homeostatic equilibria: deviations from homeostatic norms are registered in the higher nervous system and cause important modification, and even the temporary or permanent disappearance, of mental events. (Unconsciousness is occasioned even by moderate and transient hypoxia of the brain, coma by hypoglycemia and high fever, etc.)

The influence of hormones and action of drugs: it is likely that the important influence exerted by hormones on mental processes is due to their action on the central nervous system. Drugs can modify mental processes for the same reason.

Lesions of the cerebral cortex: lesions of the cortex may lead to marked mental deficiencies including deficits in the corresponding sensations and perceptions, perturbations of speech, motor disturbances, ideational or interpretive perturbations (Penfield), and amnesia.

Electrical stimulation of the brain: electrical stimulation of certain areas of the cortex leads to mental processes, although the areas are few which, when stimulated, yield conscious events. Electrically stimulated conscious events include altered meanings of present experiences and hallucinations related to past ones.

---

[21]Walter, *op. cit.*

[22]Rosenblueth, *op. cit.*, Chapter 3, pp. 25–37.

Electrograms: physiological-mental correlations are evidenced in experiments (as previously summarized).

Rosenblueth concludes that the evidence "justifies the inference that the development of mental processes is correlated with the appearance of specific physiological events or activities in some part of the central nervous system."[23]

The evidence here quoted is relatively uncontroversial; however, speculative extensions also deserve mention. Perhaps foremost among these is Hebb's hypothesis concerning neuronal organization, which extends the brain-mind correlation much further. Hebb suggests that, if a stimulus is presented to the individual a sufficient number of times, a "cell assembly" gradually develops in the brain. The assembly is a unit, capable of exchanging innervations with other units of this kind, and they together possess the capacity to evoke particular responses. The series of such units is a "phase sequence"—and this may be identified with what occurs in thinking. Particular assemblies of cells may be activated by external stimulation through the senses, by prior functions within the cortex, or from both sources. If the activation comes from the senses, the stimulus enters a field of ongoing neural activity possessing varying degrees of intensity. If the field is well patterned and forms a constant background for the environmental stimuli, the cell assemblies respond by modification, according to the nature of the stimulus. Thus perception and learning enter causally into the formation of cell assemblies.

Hebb illustrates his theory with the example of the learned perception of a triangle. Not being born with the concept of the triangular, we learn it through the development of sequential cell assemblies. For example, when the eye fixates one angle of the triangle, (A), a corresponding assembly unit forms in the brain. When the eye shifts to angles (B) and (C), the corresponding cell assemblies likewise form. There is a correlation between the visual patterns formed by the angles (A, B, C) and the cortical organization of corresponding cell assemblies (a, b, c). As the individual in his experience repeatedly encounters triangles, the cell assemblies stabilize. Eventually, the assemblies may be maintained and activated through "reverberatory circuits" in the absence of the visual pattern.[24] Similar, as well as considerably more complex, relations are established between perceptual patterns and cortical organizations during the lifetime of the individual. The more numerous the cortical organizations, the greater the depth and range of the responses. All learning, Hebb suggests, consists in the building of systems of cell assemblies. At the level of conceptual thought elaborate phase cycles evolve, constituted as a series of series of cell assemblies.

[23]*Ibid.*, p. 38.

[24]Hebb, *op. cit.*, pp. 73, 142.

These are relatively independent of external stimulation. Consequently, while learning depends in part on impulses introduced from the senses, intracerebral processes may give rise to new associations on their own within the cortex.

Monistic alternatives to biperspectivism

Economy of thought is an ever-present motivation of systematic theories, and systems philosophy is no exception. Viewed in the light of Occam's razor, Berkeley's elimination of material substance is one way a philosophy of reality can be built; it is a way, however, which turns out to be uneconomical because of its entailment of a universal mind or Spirit. The opposite kind of ontology (although not so named by their proponents) is implicit in the modern behavioristic and materialistic elimination of mental substance, leaving a purely physical world as the principle and substance of all phenomena. Our approach, however, recognizes irreducible differences between sets of mental and physical phenomena, and attempts to integrate them through the discovery of invariances. It thus flirts with dualism. But is this flirtation, and the complex solution delivering us from the embrace of a bifurcated nature, truly necessary?

The answer is: it is necessary if it is the *simplest* theory which is adequate to the facts. Here 'facts' is taken to mean properties of observed events. Basically, we are dealing with two sets of events: the events of immediate experience, and the events of theory based on reason. The latter proposes to explain interconnections of the former, but does so with the help of "constructs" which are entirely dissimilar from the explicanda. Now, it may be that a non-idealistic form of Berkeleyanism would furnish a way out of the dilemma: we could assume that science merely provides entities of calculation (*Rechengrössen* [Mach]) and that the "real" world is composed of the qualities we perceive. Such is essentially the position of naïve realism, of which Russell said that it leads to physics, and is falsified by physics.[25] But empirical science makes no claim to absolute truth, and it is always possible that modern physics is valid but unsound, i.e., that its laws give confirmable predictions and map actual processes, but that the predicted events and mapped processes are entirely different as they are in themselves. They could be composed, for example, not of electronically bonded atomic systems in further mathematically calculable weak interactions, but of colors conjoined with sounds, tastes, odors and textures. The world would indeed be a comfortingly knowable and reliable place to live in

[25]"Naïve realism leads to physics, and physics, if true, shows that naïve realism is false. Therefore naïve realism, if true, is false; therefore it is false." Bertrand Russell, *An Inquiry Into Meaning and Truth,* London, 1940, *Introduction.*

under these circumstances: we could say, with good conscience, "'the greenness' is *there*." We would not be expressing a mental event by falsely attributing it to a physical one—we would be pointing to a chunk of objective reality.

On close scrutiny, there are no valid theoretical reasons why such a reaffirmation of naïve realism could not be upheld. Colors, sounds, etc., could be furnishings of the real universe, without having to be treated as perceptions in some mind. Minds themselves could be clusters of sensations which are explained, merely for purposes of calculation and prediction, in terms of electrochemical discharges of neurons. The world would be as we see it; not as we explain it.[26] And is not "seeing" more reliable as an instrument of empirical knowledge than "explanation"? Why, then, should we believe that the unseeable is the real, and the seen the illusion?

When we press the issue in this way, the only cogent answer that remains in favor of the scientific universe of physical, chemical and biological processes is that their laws *explain* and *work*. They furnish the means whereby the world around us, whatever its ultimate nature may be, can be known and manipulated. There is no science of a colored, noisy, tasty, textured and odorous universe. It is always possible that not having such a science is due to a bias of the Western intellect, and not to the intrinsic impossibility of evolving it from the givens of sense experience. The Western intellect attempted to seek release from the chains in Plato's cave and to strive toward the "sun" ever since the dawn of Hellenic civilization, with no other assurance of finding reality beyond the senses than a trust in the power of clear and exact thinking to deliver a true picture of the ultimate nature of the universe. The comfort of living in a perceived world was readily surrendered for greater quantitative precision, although unsophisticated minds continued to live in the cozy "cave" of sensory images and shadows all through the centuries. But what if beyond the opening of the cave there was but darkness—if the light existed truly *in* the cave? A cave is a snug place and it is a monistic universe. Therein only sensory events flit hither and thither. And like Plato's prisoners, we might develop a fair ability to predict them and reason about them.

Contrast, now, this rather wistful picture of a monistic perceived world with the monistic world-picture based on the assumption of the reality of

[26]Russell concluded his dictum on naïve realism and physics (previous footnote): "And therefore the behavourist, when he thinks he is recording observations about the outer world, is really recording observations about what is happening in him." And Feigl presented a version of the brain-mind identity theory wherein sensations are the realities, and neurophysiological theories but their theoretical descriptions. (H. Feigl, "The 'Mental' and the 'Physical,'" *Minnesota Studies in the Philosophy of Science, Vol. II.*)

the objects of scientific explanation. This is a view of the universe as a realm of physical events, with the possible addition of certain "emergent" laws and principles, such as those of biology and psychology, when physical laws appear inadequate for describing certain phenomena. In this view the constructs of empirical science do not merely explain observations: they tell us what the objects of observations actually *are*. They tell us that the observed green is really a wave propagation of specific frequency, originating in an energetic source of quanta of photons. The photons and the waves are not even visible, let alone green. Nevertheless, they are said to be the true actualities of this world.

The monistic scientific theorist goes even further : he describes *mental* events themselves as physical events. The differences between the observed properties of sensations and the nature of scientific entities are not considered a deterrent: if the flash of lightning in a thunderstorm can be ascribed to the motion of charged particles in the air, then the perceptual event whereby we know the lightning can also be ascribed to the activity of neurons in the projection areas of the cortex. According to one proponent of such a monistic-physicalist theory, "the claims of psychical research are the small black cloud on the horizon of a Materialist theory of mind. If there were no questions about paranormal phenomena to consider, there would seem to be little serious obstacle to the *complete* identification of mental states with physico-chemical states of the central nervous system."[27] The assumption of non-physical *mental* states is regarded as superfluous, and a violation of the principle of economy. Mental states could only be inserted into the causal chain of physical processes as additional, unverifiable links. Although a hypothesis inserting mental states is a logically possible one, and is logically compatible with the observed facts, it is said to have nothing to recommend it.[28] Hence we end up with a world of physicochemical events in which mental states are uneasy illusions, to be dispelled by their contingent identification with states of the central nervous system.

The internal consistency, explanatory power, and overall neatness of the contemporary scientific edifice is what primarily recommends the above position. The world reduces to optimally elegant, rational and calculable mathematical proportions; sets of relationships in which entities are nodal points, whose independent nature or substance can be called "material" or "physical," or may be simply left unspecified. The world of sensations constitutes merely the set of codes whereby information is received of the

---

[27]D. M. Armstrong, *A Materialist Theory of the Mind,* New York, 1968, p. 364 (italics in original).

[28]*Ibid.,* p. 365.

objective physical world; and the latter is not colored, or odorous, or tasty, but predominantly mathematical and relational in character. Man is one such entity—a physical system in a physical world. The principle of economy reigns supreme; this time by espousing the reality of the objects of scientific explanation, rather than by reifying the objects of immediate sensory experience.

These are alternative monistic philosophies of reality, each excelling in its own way by its internal neatness and consistency. However, such merits have been won by a rather arbitrary disregard for the evidence supporting the case for the opposite kind of ontology, i.e., the disregard by the physicalist theory for the irreducibility of mental events, and the disregard by the sensationalist philosophy for the cogency of rationally postulated physical events. But rather than reducing mental events into physical ones through some stratagem such as declaring them to be contingently identical, or doing away with the world of scientific entities by means of an essentially Berkeleyan type of argument, we can recognize the fundamental character of both types of events and still produce an internally neat and consistent ontology. The way to do this, I suggest, is through biperspectivism. We can assert the identity, not of the component events themselves, but of the total systems constituted by sets of events; i.e., the natural system as constituting a model of the externally observed human being (or other concrete entity), and the cognitive system, modeling the human (or other) being in the introspective analysis. By this procedure the human mind is not inserted as an unverifiable link in a causal chain of physicochemical causation. Rather, it constitutes another chain of events; one where epistemic, rather than physical, processes furnish the causal links. Thus I agree that inserting a "mental" event somewhere into the loop of physical stimulus, neural energy transmission and processing, and responsive behavior, is uneconomical and, moreover, unverifiable on physicalist premises. I also agree that it is redundant to assert that there are mental events which consist simply in being there, and have no connections with other events, perceptual or behavioral. I admit, with Armstrong, that "the concept of a mental state is primarily the concept of *a state of the person apt for bringing about a certain sort of behavior*."[29] Although there may be mental states which analyze other mental states rather than issue directly in actual response, ultimately all mental states are interconnected with the environment in a continuous loop with feedback properties. Where I disagree with the identity theorists is in holding that there are no better alternatives than placing the mental state *within* a physical chain of causation. There, it becomes a superfluous and unverifiable "ghost in the machine."

---

[29] *Ibid.*, p. 82.

In the present view, mental states are states ultimately issuing in responses, but these responses are not *physical behaviors,* but *conations,* i.e., acts of will. Just as a felt sensation of greenness is not a physical stimulus describable in terms of neural transfers, but a perceptual quality, or "sense datum," so a volitional response is not physical behavior describable in terms of the efferent discharge of neural impulses into the motor centers. Rather, perceptual sensations and volitional responses are mental acts constituting, jointly with other mental acts, information-processing loops with self-corrective feedback properties. Thus we do not put a ghost into the machine, and do not build a machine around the ghost. We have a ghost, and we have a machine, and when we examine them in detail, we find that they constitute identical systems. Such identity is due not to their "material" or "substance"—which remains "ghostly" in the one case and "physicochemical" in the other—but to the isomorphy of the theories mapping the systems.

Conclusion

The question concerning the basic building blocks of reality, a typically ontological question, is answered here in regard to that segment of nature which constitutes negentropic systems in integral hierarchies. The basic units here are systems, more exactly, natural-cognitive systems. This position is reached by (i) recognizing the mutual irreducibility of physical and mental events, (ii) constructing separate models for sets of physical and mental events, and (iii) exploring the models for isomorphies. The uniformities disclosed by the applicability of isomorphic models are integrated, as an ontological principle, in the concept of a *biperspectival* (internally and externally viewable) *natural-cognitive system.* Mind-events thus disclose themselves, under introspective analysis, as the correlates of some species of physical-events in the organism. The evidence for such correlation is cogent in regard to human beings, and it may be generalized on theoretical grounds to include all negentropic systems in the microhierarchy.[30]

With the postulation of biperspectival natural-cognitive systems we have laid one cornerstone of the complex edifice of systems philosophy. Next we consider the hierarchical level-structure constituted by the different species of natural-cognitive systems, and outline a more specific framework for a "philosophy of nature."

[30]See below, pp. 168ff.

Chapter 9

HIERARCHY:
FRAMEWORK FOR A PHILOSOPHY OF NATURE

Organization : the basic criterion

Human sense organs and forms of thought evolved for a rigorously functional purpose : to enable man to survive by conveying signals from the relevant segments of his environment and endowing these signals with operationally significant meaning. Consequently human experience, as conveyed by the senses and coded in practical, commonsensical modes of thought, is not necessarily a reliable instrument of objective empirical knowledge. In fact, biological and cultural factors combine to form a kind of existential "bubble" for each person, which permits him to get along remarkably well in his ambient, but can be misleading when used as a guide to objective knowledge. The "bubble" includes concepts of *things* and of *relationships,* generally restricting concreteness and reality to the former. By limiting the class of things to entities of the human, as well as the subhuman, level, it enables the human being to effectively manipulate them, conceiving of the most complex things as his equals, and of all lesser things as either components in complex things, or simpler environmental background entities. Above the level of the human being, however, thinghood ceases and spheres of "relationships" commence. These are conceptualized as constituted by the communal endeavors of concrete human beings, thus giving license for the manipulation of superordinate (social and ecological) structures.[1] *Things* closest to the human level are most sharply distinguished; with simpler things, of smaller magnitudes, fading into the periphery of conscious recognition. The smaller entities in the hierarchy are known but in the form of aggregations, including relatively large numbers of components (such as crystals, molecules and atoms). *Relationships*

[1]As a matter of expediency, human organizations (teams, corporations, etc.) are often conceptualized as "group persons." They are not, however, attributed concrete thinghood; the latter ceases at the level of the human individuals.

immediately above the level of individual human beings are likewise those which stand out in sharpest relief: these are the personal relationships. Beyond them lie the progressively vaguer areas of more complex and extensive relationships, extending in the direction of larger social and ecological systems–ultimately, of the world system. The great divide, or "common-sense Rubicon," is the human subject himself, and those he recognizes as such. This is the level of "persons." From here on down, "things" are manipulated; and from here on up, "relationships" are ordered. Hence the common-sense existential bubble of the human being is highly efficient: it permits control-behavior to extend in both directions in the micro-hierarchy. (Figure 13) But that this bubble is also true, in the sense of representing an objective mapping of the environing world, must be questioned.

Regretfully, functional efficacy and objective truth-value are often confused. Merely because something is a functional coding of an objective state signalled to a system, does not mean that it is also an exact reproduction of that state. A series of programs, operated by feeding punched tapes into computerized hardware, can control the manufacture of complex products, such as automobiles; but this is not equivalent to having a true representation of the finished vehicle in the programs. In the program, certain sets of relationships are functionally matched, and these control the process; but examining the codes or programs is to examine something very different from the product the manufacture of which it controls. There is no "little car" inside the production-line control system at the automobile plant. But it is precisely such an image that we expect of human beings, when they claim to know their environing world. And it is precisely a series of images of things, and of relationships constituted by things, that constitutes the existential world-picture. But these images are more like functional codes for producing efficient responses than objectively true mappings of existent entities. When they are taken for the latter, the functional bubble is expanded into an anthropomorphic and subjective world-picture.

The options for achieving empirical knowledge are clear-cut. One either chooses to argue out of one's functional, but subjective, bubble and take the functional code for an objective image; or he can attempt to bypass such subjectivities and search for non-anthropomorphic conceptual categories. The former alternative represents the empirical approach of most philosophers who take human experience as the touchstone of their knowledge of the world. Whether they argue on the basis of sense-data, language, or aesthetic and perhaps religious intuition, they explicate intrinsically subjective structures into seemingly objective world conceptions. But the latter alternative is embraced by the contemporary sciences when they concentrate, not on experienced things and their relationships, but on structures

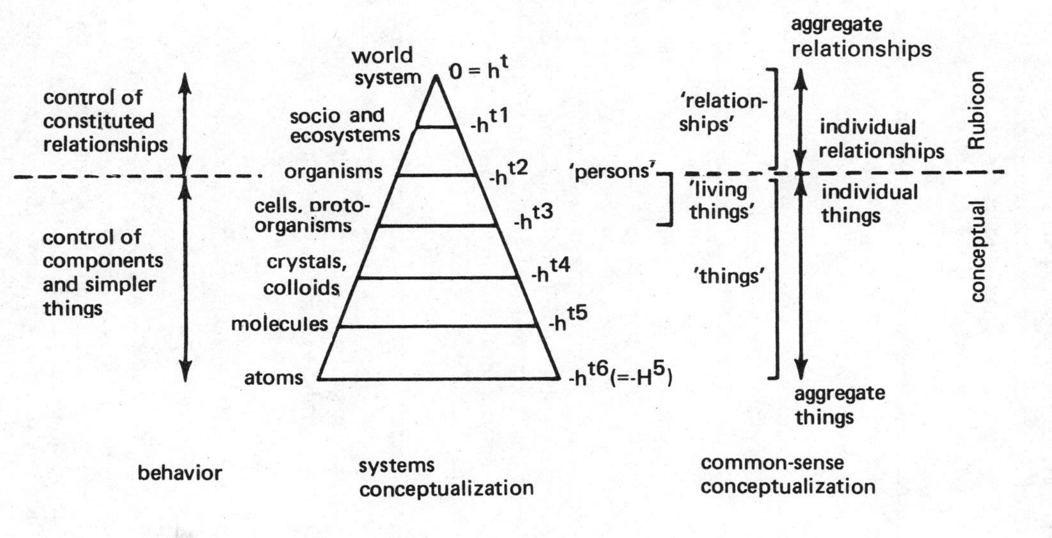

Figure 13

Profile of the anthropocentric "existential bubble" of human beings, mapped against the organizationally defined level-structure of the microhierarchy.

and functions—in one word, on *organization*.[2] By taking the way certain events are structured, in space and in time, as the criterion of their reality ("concreteness"), the anthropomorphic profile of the commonsensical bubble is exploded. Man is removed from the topmost position he occupies in the behaviorally efficient bubble-model of the world; "persons" no longer define the common-sense Rubicon. Man becomes but a system embedded in the systematic interconnectedness of terrestrial nature. There is no longer a categorical gap between "things" (including "animate things") below the level of persons, and of constituted "relationships" above. There are systems, subsystems and suprasystems, and which is which depends on the level of one's conceptual analysis. By achieving alternate foci of investigation to the human, man can be considered now a suprasystem (e.g., of his organs, tissues and cells), now a subsystem (of his group, society, and nation). Likewise any given system can achieve these alternate classifications, by relating it to its parts, or to superordinate wholes.

Using organization as our criterion of concreteness in reality, we conclude that natural systems populate the microhierarchy, being already exemplified in atoms, as well as in cells, organisms, and even in the progressively emerging global social and ecological system. If we observe the previously advanced biperspectival principle, according to which natural systems, when introspectively viewed, are cognitive systems (and cognitive systems, externally observed, are natural systems), we obtain a coextension of the class of natural and cognitive systems. It follows, then, that if any natural system, e.g., an atom, is introspected upon, i.e., if the observer himself coincides with that system, the latter appears to consist of sets of "mind-events." We thus attribute subjectivity to such systems as atoms and molecules. We do the same in regard to *supraorganic* systems, of course; coinciding with a social or ecological system likewise yields mind-events for the introspective analyst. In making these assumptions, however, we

---

[2]As a rule, scientists are not given to outlining general philosophies of nature, but when they do, they may turn out to be like the "metaphysics" proposed by David Bohm: "The basic metaphysics that I am now considering is one in which *process* is fundamental. I therefore suggest that one entertains the notion 'All is process.' That is to say, 'There is *no thing* in the universe.' Things, objects, entities, are abstractions of what is relatively constant from a process of movement and transformation. . . . Rocks, trees, people, electrons, atoms, planets, galaxies, are . . . to be taken as the names of centers or foci of vast processes, extending ultimately over the whole universe. Each such center or focus refers to some aspect of the overall or total process, which is relatively stable (i.e. which has a certain tendency to 'survive')." David Bohm, "Further Remarks on Order," *Towards a Theoretical Biology*, C. H. Waddington, ed., Chicago, 1970. (Italics in original.) Compare with the here proposed "Systems Metaphysics," Chapter 15, below.

appear to be making undue demands on credibility. Does such a hypothesis not contradict empirical evidence?

I believe that it does not. Introspective evidence is limited to the analyst himself: direct first-hand information can only be had of our own mind-events. Indirect (second-hand) information can be had of other people's mind-events, when they verbally report on their introspections. We can also infer to their mind-events by observing their behavior and physical changes of state. The principle of coextension holds, by inference, in regard to other members of our species: we readily admit that if we would coincide with their physical organisms, we would also perceive their mind-events. These would very likely be similar to our mind-events inasmuch as most other people have organisms much like our own.

But natural systems other than members of our own species cannot communicate their mind-events to us verbally. We can only infer them through the observation of their behavior, and their behavior differs from ours to varying degrees. The behavior of a chimpanzee differs relatively little, while that of a molecule does so considerably. Ordinary common sense would assert that the chimpanzee has mind-events somewhat like we do, while the molecule does not have them at all. But what is the basis of this assumption? Clearly, it is the relative similarity of the chimpanzee's behavior to ours, and the dissimilarity of the behavior of the molecule. Yet the similarity in question is always relative, and it shades into dissimilarity in a continuous gradation. Chimpanzee behavior has much in common with the behavior of more primitive animals, and those in turn behave similarly to marine forms of life, and to plants. A now outdated dictum in biology holds that there is nothing in man which is not also present in the amoeba. It is outdated because amoebae are not radically different from still simpler organisms, such as bacteria and viruses, and these show striking similarities in behavior to crystals and organic macro-molecules. The boundary of the inorganic is imperceptibly transcended, without encountering a similarity-dissimilarity gap sufficiently wide to warrant the assertion of mind-events on this side of it, and their denial on the other. Common-sense must be challenged to give a better criterion for the assertion of mind-events than overt similarity to human behavior; and common-sense may be unable to do so. It may be that a better criterion is *organization*: structure, self-direction, self-maintenance, control, non-summativity and other systemic properties. These could be held up as criteria for the presence or absence of mind-events.

If systemic properties, such as have been advanced in this study, serve as criteria for the attribution of "interiority" to the modules of the micro-hierarchy, then any module physically describable in these terms is also describable in terms of mind-events from the *introspective* viewpoint. Thus if the observer could coincide with an atom, he would describe electron bombardment in terms resembling minor pain and distress, and nuclear fission in terms approximating sensations of crisis. Of course, there is no warrant on this theory to attribute mental events of the *human* variety to atoms—indeed, to any natural system other than the human. *Mind*-events are generalized in the microhierarchy, and not *human* mind-events: specific types of mind-events can be assumed to correlate with specific types of physical-events in each system. Thus if we remember that some 10^{11} neurons form the complex cerebral nets (physical-events) which correlate with human mind-events, we must beware of attributing anything resembling human mind-events to lesser systems, such as atoms and molecules. Their mind-events must be entirely different in "feel" from ours, yet they can be mind-events nevertheless, i.e., types of sensations, correlated with, but different from, physical processes.

We must beware of the fallacy of attributing levels of mentality to systems proportionately to their position in the level-structure of the microhierarchy. If we did make such an assumption, we would have to attribute a higher grade of mentation to social systems than to biological ones, and this assumption is strongly counterindicated in many instances (for example, in the case of relatively loosely integrated groups of individuals, such as the faculty of a college or university). Whereas it is generally true that the full biological system is likely to possess a higher grade of mentality than its component cellular and organic systems, and that these in turn have interiorities which yield a set of more evolved mind-events than their own molecular and atomic subsystems, it is patently false that the suprasystems formed by biological systems are more conscious, or more "mental," than their members. There would be no empirical warrant for making this assumption.

However, there is no *conceptual* warrant for such an assumption either. Mind-events define the integrated processes of a total system, viewed introspectively, and are proportional in level of development to the systemic functions performed on the basis of the integration and differentiation of the subsystems. Thus, where the functional integration of highly differentiated subsystems is pronounced, there the mind properties of the system are expected to be evolved; in the opposite case mentality remains embryonic. For example, the functional integration of the differentiated organs of the human physiology in an organic whole of tremendous cybernetic function-capacities correlates with a high level of mentality. If either the differentiation, or the level of integration, of the subsystems were reduced, the mind-phenomena would be correspondingly impoverished. Now, social systems in our experience never achieve a comparable degree of integration of their own, greatly differentiated subsystems, even when they are highly regulated. An army, for example, or the Church, comprises a much smaller number of differentiated subsystems in a much less integrated whole than the nervous system of even relatively simple forms of life. Hence, despite the fact that social systems are multi-organic suprasystems, their own level of mentality is incomparably simpler than that of their most modest member. The principle in question can be stated in the proposition: *The level of mind-events of a system is proportional to the level of differentiation and functional integration of its subsystems.*

With the proviso, that it is the level of differentiation and functional integration of the subsystems which correlates with the level of the system's mind-events (rather than the hierarchical level of the system) the postulated coextension of cognitive and natural systems becomes a general principle of actuality within the microhierarchy. Often in more cosmic contexts, many such principles have been offered in the history of thought. Spinoza, for example, affirmed one such principle in suggesting that the physical aspect of the correlate constituting the human mind is the human body. And he added (in the *Scholium*) that this applies "not more to men than to other individual things, all of which, though in different degrees, are animated." The subjective psychical correlate of objective physical constitution is "intuition" for Bergson, "enjoyment" for Alexander, and "feeling" for Whitehead. Morgan espouses such a principle explicitly in saying that "there are no physical systems of integral status, that are not also psychical systems; and no psychical systems that are not also physical systems. All systems of events are in their degree psychophysical. Both attributes, inseparable in essence, are pervasive throughout the universe of natural entities."[3]

Recent formulations of this principle include Teilhard de Chardin's theory of evolution in terms of the parallel emergence of mind (the "interiorization of matter"),[4] and the Soviet Marxist thesis of mind-emergence from mind-potentials advanced by Tjuxtin and based on Lenin's thesis that the ability to "reflect reality" is the property of all matter.[5] There is no dearth of further parallel formulations in contemporary philosophic and meta-scientific literature, either.

In the present theory, the psycho-physicality of concrete systems is operationally analyzed: whether a system is physical or mental depends on the viewpoint of observation. The operation of passing from the one to the other viewpoint permits the alternate inspection of the complementary (but not simultaneously appearing) aspects, even if a given system can only be

---

[3]C. Lloyd Morgan, *Emergent Evolution,* London, 1923, Lecture I.
[4]P. Teilhard de Chardin, *The Phenomenon of Man,* New York, 1959.
[5]V. S. Tjuxtin, in *Voprosy filosofii,* Moscow, 1964, 2.

observed from one viewpoint at any one time and can thus be described *either* as a natural *or* as a cognitive system (and not both.) Thus for me, I am a cognitive system and everyone else is a physical system. But I allow that if I *could* view myself from an outside standpoint, I, too, would appear as a physical system, and if I *could* project myself as an observer into anyone else, I would see him as a cognitive system. And this biperspectival theorem applies to the full range of the microhierarchy, from atoms to societies.

If objections are still raised to the attribution of psychical qualities to natural systems beyond the range of organic or life phenomena, the critic should remember that he does not have certain warrant for asserting such qualities even in regard to his own wife. Mind-events are limited to the person reporting on them, and are speculatively asserted of everyone else. If nature can be systematically and consistently explained in a framework which does not recognize fundamental categorial schisms between diverse natural (physical, biological, social) systems, denying mind-events to any such system is no more justified than denying it to any other. Thus the critic has a choice; he can be a skeptic and allow for mind-events only in himself, or he must specify criteria for deciding why his wife, and not a carbon atom, should be attributed mind-events. If under "mind-event" not human sensations are meant, but sensations appropriate to the organization of the system, he will be hard put to find criteria which coincide, within any reasonable margin of error, with common-sense restrictions on the prevalence of mentality in nature.

Building a philosophy of nature : (i) Systems and reality

Nature appears to bring forth ordered dynamic systems which exhibit definite forms of invariance. They are non-summative ordered wholes, irreducible to the functions and properties of their parts. They are self-maintaining by means of a variety of dynamic processes, subsumable under the general heading of negative-feedback steady-state homeostatis, and positive-feedback structural evolution. Diverse populations of such structured systems evolve at different rates and provide environments for one another. Thus the less evolved system may become a subsystem in a more evolved one, and we get hierarchical structuration. Consequently a combination of key state and function descriptions defines *physical* events which constitute *natural systems*. Introspective analysis likewise discloses structured sets of qualitatively distinct *mental* events, and our biperspectival theorem combines these with natural systems in the concept of *natural-cognitive* systems.

Since everything we can identify in the microhierarchy is either a self-maintaining dynamic structure or an aggregate of many such structures, the fundamental existents are definable in reference to organizational invariances. We thus get the equation,

specified organizational invariance = real microhierarchical entity = "natural cognitive system,"

or,

class of natural-cognitive systems = class of real microhierarchical entities.

Now, one of the great questions of traditional philosophy is answered in reference to the above concepts. *What is real*? or, *What are the principal furnishings of reality*? receives the simple unequivocal answer: *natural-cognitive systems.* If any set of events in the microhierarchy constitutes a non-summative self-maintaining and self-evolving dynamic structure, it is "real." The level of organization of the system, its spatial extent or temporal endurance, do not qualify this assertion. Systems may be cybernetically or structurally stable: they are then likely to endure for a relatively long time in their environment. Other systems may have a low level of stability, and these could thus have an ephemeral existence, disorganizing to their stable subsystems when perturbed. Stable neutral atoms exemplify the former, many social institutions the latter, type of system. Yet if the criterion is not endurance, but systemic organization, both types of system qualify as furnishings of reality. Highly organized systems tend to be structurally less stable than relatively primitive systems, but compensate for their structural instability through a wide range of cybernetic control functions. If these break down, the complex system is prone to disorganization into its primary stable components. If the latter are considerably more stable than the system itself, the system appears to an observer as an ephemeral epiphenomenon, consisting of the periodic grouping of the stable subsystems into higher-order associations. Yet *when* the system is exemplified, and *as long* as it is exemplified, there are no grounds on which the predicate "real" could be denied in regard to it: if its subsystems are organizations of subsystems and are hence real, the system itself, being such an organization, must be equally real.

The origin of such systems is as immaterial to their reality as is their relative persistence. Systems of the kind we have explicitly considered arise in nature, i.e. independently of conscious human purposes. But it is entirely consistent to assert that human conscious purposes could cause systems of this kind to form. They would then be called "artificial" as contrasted with "natural" systems. Yet the difference between them would amount to a difference between a consciously planned, and a naturally arising, template; and the amount of conscious planning which goes into a template is immaterial in regard to the structure which arises on it. Hence artificial systems, *qua* organizations with certain specifiable invariant properties of state and function, qualify for furnishings of reality as well as any naturally originating system. However, not *every* artificial system can so qualify, only those which possess the systemic characteristics applicable to natural systems. Hence when self-repairing and self-evolving machines with a high level of integration of heterogeneous parts are devised, these machines will represent forms of existence on a par with natural systems.[6] Populations of such machines could evolve selectively, and adapt to the hierarchy of other, naturally originating, systems populations. That the machines have come into being due to a *conscious* blueprinting by existing natural systems, rather than through a *genetic* blueprinting by them, makes no difference to their concrete actuality.

Classical conceptions of real things, as naturally originating solid particulars, must be surrendered as inconsistent when the more fertile perspective of organizational invariance is adopted as the criterion of real entities. And when we do adopt organizational invariance as the criterion, then relative persistence, origin, substance, level of integration, manifest functions and properties, are so many specifications of *characteristics* of systems, and not touchstones of their *reality*. According to the here advanced theory, any organization of events that satisfies the state and function postulates of systems is real (concrete, veridical), and all such actualities are biperspectival, analyzable to physical as well as to mental sets of events. The consequent proposition, that transient social organizations, as well as artificially created machines, have mental events, must be accepted: using the differentiation and functional level of integration of subsystems as the criterion of the mentality of the systems, we do not attribute anything like human minds to less organized systems. And if each of us has mind-events and is systemically organized, then other systemic organizations have mind-events in the analogously oriented introspective analysis. When organization is the criterion of existence, then it is also the criterion of mentality: the alternates are either an arbitrary cut-off point for mind, or the logically consistent but unfruitful tenet of solipsism.

(ii) *Hierarchy and emergence*

The hierarchy concept in regard to terrestrial phenomena is becoming

---

[6]Self-evolving machines do not exist yet, but they are perfectly feasible in principle (cf. Wiener, Ashby, von Neumann *et al*).

generally accepted today as a key ordering principle. Reviving modes of thought which have first been developed in classical philosophy,[7] contemporary investigators are wont to offer such sweeping statements as "the universe contains a hierarchy of systems, each higher level of system being composed of systems of lower levels,"[8] and "reality (=the world) is a level-structure such that every existent belongs to at least one level of that structure."[9] Statements such as these represent, as Bunge himself points out, sweeping generalizations of evolution.

The empirical world-view entailed by such generalizations conflicts with monism, in asserting the irreducibility of the levels, and with dualism and pluralism in assigning generative relationships between the levels (the higher levels are formed by coactions of systems on lower levels). Consequently the indicated position of a philosophy of natural hierarchy is *integrated pluralism*: "an ontology that proclaims both the diversity and the unity of the world."[10]

Originality or novelty is conceived in such a philosophy as emerging in successive levels. Not a "metaphysical" emergence is meant by this, i.e., the appearance of something out of nothing, given the proper conditions (e.g., the emergentism of Lloyd Morgan and Samuel Alexander), but the appearance of new transformations of invariant properties. General systemic properties appear on each of the different levels of the hierarchy in transformations appropriate to that level. Each level is characterized by systems of which the subsystems are systems on all lower levels, and the coaction of the subsystems in their diversity and complexity is what determines the kind of transformation we find at the systems-level. Hence ordered wholeness arises on the basis of nuclear and electronic coactions on the level of the atom, and it arises on that of economic, legal, political and civil coactions on the level of human society. The forms of adaptation likewise differ, without connoting thereby the emergence of new qualities out of something entirely different. The emergents may appear phenomenally new (as valence at the molecular level and abstract thinking at one point in the evolution of the human biocultural level), but on closer examination they are found to be *new* transformations of invariant properties characterizing all natural-cognitive systems.

The word 'new' in the foregoing sentence is italicized in order to emphasize that, notwithstanding the analysis of the "newness" as novel

---

[7]Cf. L. L. Whyte's historical introduction to the concept in *Hierarchical Structures op. cit.*

[8]James G. Miller, "Living Systems: Basic Concepts," *op. cit.*

[9]Mario Bunge, "The Metaphysics, Epistemology and Methodology of Levels," *Hierarchical Structures, op. cit.*, p. 22.

[10]*Ibid.*

transformations of already existing (i.e., exemplified) invariant relationships and their correlated properties in systems, what emerges is indeed new for the system in question. Emergentists, both modern and classical, often confused three forms of novelty emerging in regard to a particular phenomenon, of which only one is truly new, i.e., an *emergence,* although not a metaphysical or "inexplicable" one. There is, first, novelty introduced by a purposive external agency, for example, the assembly of a watch by a watchmaker. Here the information necessary to bring about the novelty is supplied from the outside where it pre-existed fully (i.e., in the watchmaker's head). Secondly, there is novelty based on information already present in the system where it appears in the form of a set of codes (information-carrying structures), as, for example, in chromosomes, which carry the information necessary for the development of the mature organism. But only the third form of emergence constitutes significant and "true" emergence: here the information itself is generated in the process whereby the reorganized structure comes about. Evolution in systems at all levels of nature's hierarchy is of this form: it is not *teleology,* the realization of an existing purpose, but *telenomy,* the emergence of purpose itself or, more exactly, of the dynamic organization which manifests it. Indeed, unless a divine creator is postulated, who carries in himself the information necessary for building all systems capable of realization in nature, the watchmaker form of emergence breaks down; and unless all information was somehow encoded in the original hydrogen nuclei which marked the beginnings of the present cosmic epoch (or in the primeval atom, if the "big bang" cosmology is preferred), the chromosomal coding type of emergence also has to be discarded. Both of these alternatives lack the very possibility of confirmation via the existing methods of scientific inquiry. Thus true emergence appears to be the case, mitigated, however, by the consideration that what emerges is not an unforeseeable novelty, but a new and more complexly organized dynamic-structural variant of existing types of organization. Thus the initial set of information ("codes" or "instructions") brings about the appearance of new *organizational variants,* as information is generated in the process of creative adaptation and natural selection. The initial set of information is the set of constraints which causes ordered wholes with non-equilibrium steady states to appear and maintain themselves in a changing environment. The emergence consists, then, in the generation of new forms of organization in which these mechanically and thermodynamically improbable structures appear over a long series of developmental transformations. Hence there is emergence of organizational *form* but not of organizational *type;* where "form" is read as a concrete transformation of a theoretically discovered invariance specified as the "type." "Natural-cognitive systems" as *type,* emerge in increasingly organized irreducible *forms,* as evolution progresses from level to level in a developing hierarchy of systems.

(iii) *Levels and evolution*

The levels of nature's hierarchy can be but tentatively identified, due to the tremendous richness and complexity of empirical phenomena. Nevertheless, several schemes have been advanced in recent years: one was proposed by Gerard (cf. p. 52); another was given by Miller which we quote as follows: "*Atoms* are composed of *particles*; *molecules,* of atoms; *crystals* and *organelles,* of molecules. . . . *Cells* are composed of atoms, molecules and multimolecular organelles; *organs* are composed of cells aggregated into *tissues*; *organisms,* of organs, *groups* (e.g., herds, flocks, families, teams, tribes), of organisms; *organizations,* of groups (and sometimes single individual organisms); *societies,* of organizations, groups, and individuals; and *supranational systems,* of societies and organizations."[11] A third scheme, offered by Bunge, gives the following categories:

Elementary particles—atomic nuclei—atoms—molecules—bodies

Physical systems—chemical systems—organisms—ecosystems

Physical processes—chemical processes—biological processes—psychical processes—social processes (= *human histories*)

Material production—social life—intellectual culture

Physics—chemistry—biology—psychology—sociology—history.[12]

Bunge's categories are convenient tools to define those indicated for systems philosophy. *First,* we make a distinction between reality *in toto* (= the world), and that portion or aspect of reality which tends toward hierarchical order and pronounced structuration. In the latter, two types of hierarchical structures are to be distinguished: the *macrohierarchy,* encompassing astronomical units from the space-time field to the astronomical universe as a whole, and the *microhierarchy* (or microhierarchies), including organizations from atoms to planetary socio-ecosystems.

Second, the terrestrial microhierarchy is defined neither by the substance nor by the identity of its component systems, but by their *organization.* Of Bunge's categories the second in the order listed applies: *physical systems —chemical systems—organisms—ecosystems.* The other categories either

[11]Miller, *op. cit.*
[12]Bunge, *op. cit.*, p. 21.

focus on the individual substance or characteristics of the unit systems, or on the fields investigating them (with the exception of the third listing, which focusses on processes rather than on systems—a conceptualization which runs the risk of losing sight of uniformities disclosed by systemic state and function invariances).

Third, in the microhierarchy all systems are adaptive, ordered wholes, partially surrounded by other such systems in their environment. Consequently, such systems tend to form higher level suprasystems as a function of their mutual adaptations. For example, when in a solution atoms mutually "adapt" through the sharing of their outer electrons, they form molecular or crystalline suprasystems. These systems, when mutually adapted, may form a multicellular organism, and organisms mutually adapted form ecologies and societies. The key property is adaptive self-maintenance of differentiated parts in interdependent, coacting relationships.

Fourth, if a physical ground is sought for the existence of such systems, it must be sought, I submit, in the nature of space-time as a multi-dimensional field generating sets of constraints which produce relatively stable, non-probabilistic sets of energy-flows. Gravity, electromagnetism, and perhaps as yet undetected force-fields, are required for configurations of flows with internal, non-equilibrium constraints, expressed already in the structure of hydrogen nuclei. Given such "partially ordered chaos," continued process will be in the direction of higher levels of order and complexity. Inputs ("disturbances") in these embryonic systems will lead to the differentiation of some dynamic properties, the merging of others, and the selective evolution of dynamic systems fitted to their environment. Thus atomic nuclei may acquire a complement of filled shells in suitable environments, and these manifest irreducible properties (chemical valence) which, under suitable conditions, lead to the formation of multi-atomic associations (molecules). These in turn can evolve through continued environmental stimulation toward crystalline forms on the one hand, and polymeric associations on the other. The more stable among these evolve under constant environmental pressure and selective adaptation toward still more differentiated and integrated forms, adding weaker interactions to stronger ones, and building up the Chinese-box hierarchy of natural organizations. Thus at a certain point life appears, and later multi-organic affiliations in ecosystems and societies. (cf. Figure 2)

The process is determined as to overall level of organization, but not as to local form or rate of progress. Its key characteristic is an overall increase in level of organization, which, on the one hand, entails the cybernetic stability of the systems and the diversification of their properties, and, on the other, their structural instability and decreasing numbers. (Figures 2′ and

2″) Thus structural stability is proportional to primitiveness in level of organization, whereas cybernetic stability, manifested as an increasing variety of self-stabilizing functions and properties, is proportional to complexity. For example, complex organisms, such as mammals, are more vulnerable than protozoa, but are able to cope with environmental changes which could be lethal to the latter; and social systems of the modern technocratic-bureaucratic variety are considerably less stable than primitive tribal societies, but they, too, are more efficient in coping with pernicious influences in their particular surroundings.

Great functional efficiency goes hand in hand with the diversification of phenomenal properties, and with the branching of the specialized sub-assemblies performing the requisite functions. There are but 82 stable forms of atoms but some one million species of plants and animals. However, the quantitative abundance of atoms is considerably greater than the abundance of living organisms. Hence there is a pronounced tendency toward *individuation,* as greater variety is introduced into populations of smaller size. Ultimately a single, highly diversified and qualitatively rich organization tends to emerge, lending systemic order and unity to all subsidiary levels of systems in the microhierarchy. On earth, this highest-level organization is the currently forming global sociocultural ecosystem.

An unknown number of microhierarchies may be superimposed on the levels of the cosmic macrohierarchy on suitable planetary surfaces, possibly yielding an immensely rich array of forms of dynamic self-maintaining and self-evolving systems, driving toward increasing individuation and functional autonomy, and paying for it by inherent structural instability. This giant process of organization could have started with no more than a partial ordering of randomness in some cosmic regions, as local complexities arose due to the combined action of known, and perhaps as yet unknown, forces, establishing non-random relations between packets of energies in structurally cohering forms. Thereafter the action of physical constants on these primitive systems could branch into the complex panorama *before* us and *in* us. Without mystery and special acts of creation, it could lead to the emergence of new forms of order and their correlated qualitatively fresh and irreducible properties. It could bring forth man, mind, and culture. It brought us to where we are, and takes us irrevocably along, toward new levels of differentiation and integration, if not in our physiological, then in our cultural, social, and ecological structures.

The particular forms of future development are unpredictable: they are partially dependent on us. In that sense we are the masters of our destiny. But the general trend of the developmental pattern is forseeable: it leads toward increasing levels of order and organization, as differentiated sub-systems integrate into larger functional wholes, and system is superimposed upon system in an ongoing hierarchical sequence. In this sense, then, our future is determined, for we are part of nature, and subject to its general laws of development.[13]

[13]But this developmental sequence need not reduce the effective degrees of freedom of the human individual: cf. Chapter 12.

Chapter 10

CONSCIOUSNESS: FRAMEWORK FOR A PHILOSOPHY OF MIND

In the history of thought, the human mind was often conceived as a transcendent "spirit" or "soul," somehow inhabiting the body without being identical with or reducible to it. The scientist's working assumption, on the other hand, has always tended to be the reductionist doctrine which attempts to explain mental phenomena in terms of neurophysiological processes. The concept of biperspectival natural-cognitive systems overcomes these alternatives and presents us with a concept of man as a psychophysical yet non-dualistic entity. It affords a conceptual basis from which a constructive, fresh look can be taken at the basic, as well as the highly sophisticated, functions of the human mind. In this chapter we shall consider the problems associated with the *philosophy of mind,* taking man as a natural entity, embedded in the microhierarchy, and exemplifying at his own level the invariant general characteristics of natural-cognitive systems.

Brain-mind isomorphy (biperspectivism restated)

The evidence for conceiving of man as a biperspectival natural-cognitive system can be restated by comparing the systemic properties applicable to natural and to cognitive systems. The properties will prove to be isomorphic, but the language in which they are described will differ for the two kinds of systems: there will be a "mind-language" as well as a "brain-language."

We can use Hebb's theory to furnish us with the vocabulary of our "brain-language." Cell-assemblies are said to form on the basis of the neural interaction of sensory stimuli and intracortical activity. The organism responds on the basis of these assemblies. Its response affects its stream of sense signals, which in turn reacts on the formation of cell-assemblies. Cell-assemblies can also form on the basis of intracortical activity alone, in autonomous processes. Now, the cell-assemblies (or neural pathways, engrams, reverberating loops, or however we may call such cerebral organizations) represent the connecting link between stimulus and response.

"Cell-assemblies" are equivalent, within the brain-language, to "cognitive constructs" within the mind-language. The isomorphy of the physical brain-system with the cognitive mind-system can be summarized as follows. Responses effected on the basis of existing constructs (=cortical organizations[1]) react on the system's environment and structure its input to match its cognitive (=cortical) organization. This is a process of negative-feedback self-stabilization, and it applies to man as a natural system (organism) as well as a cognitive system (mind). Persisting mismatches in the input-organization relation trigger cognitive construct (=cortical cell-assembly) activity, leading to the establishment of more adequate constructs (=cortical organizations). Here we encounter self-organization, likewise applying to man both as a natural and as a cognitive system. Since his construct-sets are hierarchically organized and also map external hierarchies, and since we know that as a natural system man is embedded in the microhierarchy, we can take the above parallelisms as signifying complementary perspectives of a single system. On the biperspectival theorem, the qualitative discrepancy between what I experience of myself, as well as of the world as given to my mind, and what I understand of the processes of my body, brain, and of the external world as affecting my body and brain, is explained as the difference between two perspectives of myself as a natural-cognitive system. One of these perspectives is mapped in the theory of natural systems; the other, in that of cognitive systems. Since the theories represent special cases of a general theory of systems, their pertinent conclusions can be integrated in the biperspectival theorem: "sets of irreducibly different mental and physical events constitute an identical psychophysical system (termed 'natural-cognitive system')."[2]

The functional origins of brain-mind

Brain and mind are different perspectives of the same system. In man, brain (and hence mind), evolved in the context of adaptations to the environment, assuring the survival, and later the dominance, of the species.

It is now believed that a relatively rapid growth of the brain of *homo erectus,* a species ancestral to modern man, took place about half a million years ago. Although *erectus'* brain was smaller than that of *sapiens,* it was much larger than the brain of *Australopithecus.* The increase in brain size meant more neurons and possibilities of more numerous associations between them. The increased capacity for forming cell-assemblies (Hebb) also meant (on the biperspectival principle) an improved capacity

---

[1]The sign "=" stands for biperspectival correlation, not identity.
[2]Cf. Chapter 8, above.

for construct-formation. Indeed, anthropologists believe that *homo erectus* was most likely able to think in a quasi-human manner, and could speak some rudimentary language. He was also a tool-maker and user. Then, with the emergence of *sapiens*, at least 35,000 years ago, we find a brain size ranging from 1,200 to 1,500 cubic centimeters, and contemporaneously there is more evidence of the ability to think, speak, create artifacts and, in some instances, to produce a culture.

The functional use of a larger brain, with its richer potentials of organization, enabled man to outwit his physically stronger rivals and adversaries in hunting, food-gathering, building of habitats and organizing communal activities. Improved cognitive capacities correlated with increased brain size paid off in the great Neolithic revolution: agriculture and the domestication of animals. As man relied extensively on his cerebral-mental powers, selective pressure in evolution favored the survival of mutants with particularly highly organized brains (=individuals with evolved construct-systems). Brain-mind on this level was clearly a tool of adaptation to the environment.[3] It enabled man to locate those states in his surroundings which gave him an advantage over other species, and to utilize them for his own advantage.

Now, higher cortical organizations could evolve only when the basic homeostatic regulations of the organism were integrated in an autonomic nervous system. Evolution may have taken many thousands of years to perfect in *homo sapiens* a mechanism which could operate with automatic controls and thus, despite its complexity, not require the services of all the neurons capable of accommodation in the cranium. The stability of homeostatic self-regulation in maintaining the constancies of the "internal environment," Claude Bernard said, is the condition of free life. And Walter referred to Sir Barcroft's asking in his classes at Cambridge, "What has the organism gained by the constancy of temperature, constancy of hydrogen-ion concentration, constancy of water, constancy of sugar, constancy of oxygen, constancy of calcium, and the rest?" The answer was illustrated by the following example: "How often have I watched the ripples on the surface of a still lake made by a passing boat, noted their regularity and admired the patterns formed when two such ripple-systems meet . . . *but the lake must be perfectly calm*. . . To look for high intellectual development in a *milieu* whose properties have not become

---

[3]The adaptation-value of larger brain size and higher association-activity in the circuits derives from the accrued flexibility of that organ: there is a speed-up of over six-hundred times in neural transmission in *homo's* brain, as compared with his more primitive ancestors. (Cf. Julian Huxley, *Evolution in Action*, New York, 1953, p. 73.)

stabilized, is to seek . . . ripple-patterns on the surface of the stormy Atlantic."[4] Homeostasis becalms the lake; it stabilizes the internal environment. It thus opens the possibility for further evolutionary development: for developing parts of the nervous system (such as the frontal lobes of the brain) to interpret more refined "ripples"—those namely, which are signals conduced from the senses. Functions can evolve to fulfill the complex operations required for the amplification and interpretation of rather weak and fluctuating sense signals, and the correlation of the adequate motor innervations. Cognition can develop when autonomic homeostasis has paved the way.

The kind of cortical organization which is likely to have appeared first in the history of our species must have been that which enabled *homo* to respond purposefully to a recurring variety of objects in his immediate surroundings. In "mind-language," such organization afforded the meaningful *gestalt*: a perceptual pattern seen as a significant object, with available patterns of purposeful responses. Now, such a gestalt-endowed, perceptually cognitive human being has a wider repertory of responses than a genetically programmed homeostatic organism. He incorporates more of his environment into his perception-response loops, and can maintain himself *vis-à-vis* conditions which may exceed the range of error-tolerance of less evolved organisms. He compensates for mismatching environmental conditions through a manipulative modification of the environment. The threshold of error-tolerance has been increased: man can now exist in—since he can compensate for—a wider range of environmental conditions.

The origin and autonomization of cognitive capacities

The above argument may have been read to assert that higher cortical brain-events, and their correlated cognitively constructed mind-events, are "*nothing but*" survival functions of man. This is a reductionist position which, however, does not follow by entailment from what has been said.

The survival function of perceptual cognition is affirmed only with respect to the *origin* of the pertinent brain-mind capacities. These capacities evolved as a more refined means of orienting the human being in his surroundings and ensuring the satisfaction of his open-system transactional needs. But to interpret the present position to suggest that human perceptual cognition is "nothing but" a refined means of ensuring one's survival is to commit the reductionist's "genetic fallacy": to confuse the origin of something with its actual function and use. Perceptual cognition may have evolved in such a survival-need context but may, in time, be put to uses which have little immediate connection with biological survival.

---

[4]Quoted by W. Grey Walter, *The Living Brain, op. cit.*, Ch. 1.

In fact, perceptual cognitions may even come to *frustrate* survival needs: explorations leading to clarification of some aspect of the perceptible environment may impair actual biological survival-functions. Curiosity killed the cat, and curiosity is a typically cognition-oriented activity. Its presence in all species of animals equipped with evolved nervous systems suggests that the information-extracting capacity of such systems, once developed, acquires autonomy, and is not entirely subject to the imperatives of biological need-fulfillment.

But let us get the issues clear: the progressive autonomy of perceptual cognition does not change its functional mechanism, which is a direct consequence of its origin. Although it may be put to purely cognitive uses, perception (in the "brain-language") is nevertheless the interaction of organism and relevant environment; a process whereby the states of the latter are mapped by the higher nervous centers of the former, and his correlated motor responses impose states on the environment which correspond with the pre-established mappings. Thus when the individual perceives his environment as an intelligible sequence of meaningful gestalts he is adapted to his environment: he knows how to respond to it so as to maintain himself under its conditions. But *knowing* how to do this and *restricting* oneself to doing it are two different things. When the natural-cognitive system evolves the brain-mind structures which map the environment within its organization, it can elaborate patterns of responses which exceed the range of that which is of direct survival use. Perceptual cognition crystallizes as a functional element in the progressive adaptation of system to environment, but when crystallized, tends to gain partial autonomy and serve ends of its own.

The here outlined thesis, of the genesis of a function in the context of progressive adaptation and the subsequent partial autonomization of the generated function, overcomes the specious alternatives of "*dualism or reductionism.*" It starts with the position espoused by the reductionist and ends with that held by the dualist. But this thesis itself is neither reductionist nor dualist; it maintains that there is genetic continuity from below, and logical irreducibility from above. Genetic continuity avoids the trap of dualism, and logical irreducibility eliminates reductionism. For example, cortical organizations in the brain evolve in a continuous sweep from homeostatic self-regulations in the autonomic sections of the central nervous system but, when evolved, such organizations serve cognitive purposes of their own and are logically irreducible to homeostatic functions.

The perennial problem, how higher cultural cognitions (such as those incorporated in science and art) relate to the biological basis of the human organism, can be likewise clarified through the genetic continuity–logical irreducibility principle. The cognitive activity of science can be seen as a mapping, with increasing degrees of refinement, of various carefully selected states of the environment into cortical organizations in the brain. These organizations function as scientific *constructs,* coding the perceptual information and lending meaning to observation and experiment. Stated in the language of brain-events, observation in science can be described as the conduction of sense-signals to cerebral centers, where complex phase-sequences of cell assemblies interpret the signals and produce responses on the basis of the signal–cortical-organization match. In mind-event language, scientific observation is the "seeing *that*" certain predicted and epistemically correlated constructs of the theory are instantiated. Since the content of observation in either case is information reaching the system from its environment, the match of perceptual signals and cognitive organization decodes data concerning external events within the system. Scientific observation, in the context of a theory, represents a mapping of a certain state of the environment, via the system's actual cognitive equipment. In this regard, then, scientific cognition contributes to the orientation of the system in its environment, by constructing the latter with greater refinement than the concrete gestalt-cognitions of everyday life.

The above remarks apply to "normal" (as opposed to "extraordinary" or "crisis") science. But I do not wish them to be read to suggest that normal science is *nothing but* a more refined mode of ensuring the survival of the human being in his relevant environment. The genetic "nothing but" fallacy need not be committed, and it is not what is meant here. (Indeed, it would be difficult to account for the construction of, say, astrophysical models as a means of ensuring the continued survival of the scientist, or of the scientific community.) Rather, what I claim is that science is a logical *outgrowth* of the environment-cognizing functions of man; i.e., of those functions which furnish important information concerning the states of the human environment. (Prediction and control remain basic and essential components of scientific endeavor.) But, just as pure cognitive capacities intervene in ordinary perception, notwithstanding its origin in the context of biological need-fulfillment, pure cognitive aims intervene, and in fact often supervene, in the context of scientific investigation.[5] The genesis of

---

[5]These notions are exposed by a number of thinkers in the contemporary literature. Cf. Craik: "It is clear that, in fact, the power to explain involves the power of insight and anticipation, and that this is very valuable as a kind of distance-receptor in time, which enables organisms to adapt themselves to situations which are about to arise. Apart from this utilitarian value it is likely that our thought processes are frustrated by the unique, the unexplained and the contradictory and that we have an impulse to resolve this state of frustration, whether or not there is any practical application." Kenneth Craik, *The Nature of Explanation,* Cambridge, 1943, p. 7.

science as a more refined means of extracting relevant information from the environment does not prevent the means from becoming an end in itself. Pure cognitive science, "science for science's sake," or better "science for the sake of understanding alone" is always a possibility. It guides scientific effort, and in some cases even conflicts with the more pragmatic aims of humanly useful prediction and control. The origin of science no more determines its present uses than the origin of concrete modes of cognition determines *its* uses. In fact, many a scientist has given his life, or a good part of it, for the acquisition of knowledge regardless of the concrete benefits he (or anyone else) could reap from it.

Genetic continuity from below, due to the functional origins of the modes of thought which later crystallized into science, and logical irreducibility from above, once science as such has developed, applies to the phenomenon of scientific cognition. But does a similar process apply also to art? What, in sum, is the relevance of *aesthetic cognition* to the self-maintenance of the human being?

To affirm a relation between aesthetic experiences and existential needs is usually considered to entail a particularly rampant kind of reductionism—the reduction of art to biological survival function. But a relevant relationship may be discerned in the genetic context, without involving the reduction of art in its crystallized and autonomous form to the self-maintenance of the organism. Art, we hold, clarifies a large segment of human experience: the segment which is infused with and dominated by feeling. The human being relates to his surroundings on many levels, and each of his relationships requires control brought about by comprehension. But neither abstract scientific nor concrete perceptual cognition can tell us about the objects and nature of our *felt* experience—the former uses reason to compute relations between the observed states of our environment, and the latter brings us the intelligible objects and events of the perceptible segment of that environment. But not only do we *see* the various aspects of the environment *that* they signify lawlike events, and not merely do we see them *as* intelligible gestalts; we also *feel* many of the states of our surroundings, and we are not fully oriented in our ambient until our feeling-mediated apprehensions are clarified and comprehended. Herein lies the relation of art to the total pattern of self-maintenance of the human being. It brings him an elucidation of his emotive apprehensions—of himself in the world—and thus orients him in his ambient in a way in which no other type of cognition can. Aesthetic cognition may be described as a *"feeling as"*: with the 'as' referring to some aesthetically elucidated aspect of the relation of self and world. A single song can bring about such elucidation, and so can a gesture, a line of trees against the horizon, a happily expressive phrase, or the smell of wood smoke on a winter day. The publicly sanctioned masterpieces are universal statements of objects and events which can be "felt *as*" thus significant by human beings (at least by those acquainted with the given style and medium). They help people feel at home in, and know, their worlds—they render experience meaningful and significant.

Thus aesthetic cognition embodies a kind of a-rational, primarily emotive, mapping of states of the environment within the system. The art experience, often described as "sublime," "beautiful," etc., is in fact a way of feeling the experient's existential relations. Here, a match between exteroceptive stimuli and cortical organization (in mind-language: *percept-patterns* and *aesthetic constructs*) brings about intelligibility for aspects of experience not elucidated either by science or by concrete modes of cognition. How it feels to live in this world—the existence of the artist in his time—is analyzed in terms of the aesthetic constructs which, for the introspective artist, represent his spontaneous cognitions of his environmental relations. The cognitive power of an accepted style in art resides in its capacity to elucidate the relations of the artist and his surroundings. It is primarily for this reason (discounting for the moment fads, sensationalism, snobbism and similar factors) that schools of art crystallize around accepted styles, and also the main reason why, if the style in question fails thus to function, new styles are evolved with new schools and adherents. It thus appears that works of art (as all aesthetic cognitions, including informal ones in daily experience) have a basic survival function: they clarify the person's existential experience and enable him to better orient himself in his ambient.

This assessment may give a fair picture of the historical origins of art, but is inadequate to explain its subsequent development. It can hardly be claimed that works of art, as understood and admired in contemporary culture, contribute to the orientation of the individual in his world. They no doubt endow the lived worlds of artist and art-lover with greater meaning and significance. But they add nothing to the facility with which the biological organism meets its survival requirements in the environment. Thus in art, too, we must beware of the "nothing but" fallacy. Art may have originated as an emotive means of locating oneself in relation to one's *milieu*; and in primitive art, intertwined with myth and magic, aesthetic cognition could have served to direct behavior. But in its subsequent development the type of cognition brought about by expressive statements in colors, sounds, words, movements and acts, could, and in fact did, gain autonomy. Modern men no longer paint pictures to influence the outcome of a hunt or to cast spells, beneficial or otherwise. They create and view pictures, and other works of art, because of the intrinsic satisfaction it affords them. The satisfaction is still in the form of a comprehension, intuitive and fundamentally a-rational in nature, of the relation of self and world. But now the comprehension itself is the main motivation of the activity, and not its practical use. Just as concrete modes of cognition gain autonomy in pursuing purely cognitive goals (sometimes called "explorations"), and as abstract scientific cognition gains autonomy in pursuing likewise cognitive goals (there called "basic research"), so does art gain autonomy in pursuing its own cognitive goals—called "enjoyment" and "creation." The search for significance is many sided, and all its varieties derive from basic activities designed to maintain the organism in its ambient. But when evolved, the many cognitive modes achieve a purpose of their own: their goals become relatively autonomous. From instruments of survival, they transform into ends in themselves. As a result, man is an autonomous, self-directed agent, although he must still perform his existential transactions. As the development of the central nervous system followed upon the autonomization of the control of homeostatic functions in the lower reticular brain regions, so the evolution of autonomous cognition followed upon the fulfillment of survival functions by readily mastered cognitive processes. The accomplished cognitive capacities are there, and they are free much of the time to pursue their independent goals.

The history of the higher cognitive modes of man is the history of the functional development and gradual autonomization of the ways he orients himself in his ambient. As a result, man's world has been enlarged from the narrow limits of that which is of immediate relevance to his biotic needs, to the width and depth of a world which encompasses the reaches of the observable and measurable universe on the one hand, and the realm of all that is aesthetically intuitable on the other. The enlarged environment is functionally present in the biperspectival organization of man, as cerebral organizations mapping the environmental states in his *brain*, and as constructs carrying the meaning of his experience of the external world in his *mind*.

The origin and autonomization of consciousness

The principle of genetic continuity from below and logical irreducibility from above can overcome the specter of the false alternative of dualism *or* reductionism when dealing with human cognitive capacities. But there is still another major area of inquiry to which it must be applied if the alternative is to be eliminated, and that is the field of events denoted by the term *consciousness*. Man not only knows; he also knows that he knows. Does this not, finally, connote the existence of something *more* than a functionally evolved organism-environment transactional capacity? Is not consciousness evidence for the presence of something transcendent—a soul or spirit—in man?

If the genetic continuity–logical irreducibility principle is to apply to consciousness, we must show how consciousness evolved in human beings as a functional capacity of their environmental transactions. Because, if consciousness should turn out to be a non-functional phenomenon, we may have to assume a transcendent dualist principle. But if it can be shown that it contributes to the performance of the self-stabilizing and self-organizing existential functions of the human being, its functional origins can be elucidated, and the large sphere of "pure" consciousness can be exhibited, similarly to "pure" cognitions, as the outcome of the progressive autonomization of an originally functional capacity.

I suggest that the role of consciousness in the syndrome of functions of human beings resides in their capacity to *learn,* i.e., to modify their cognitive and behavioral patterns on the basis of experience. In the "mind-language" appropriate to cognitive systems, learning means the substitution of new sets of constructs, more adequately matching actual percept-patterns than preexisting ones. To do this, the cognitive system must command a remarkable capacity: it must be capable not only of *using* its constructs, but also of *evaluating* them in relation to their capacity to determine the perceptual input. The cognitive system must measure its actual constructs against their success in analyzing the encountered percept-patterns, and evolve them by selecting those of its imaginatively hypothesized constructs which have the best chances of success. The same process, stated in the "brain-language," concerns the ability of the central nervous system to produce new cell-assemblies (engrams, nervous pathways, associations, reverberating loops); to lock on to those to which enduring or repeating inputs correspond, and to discontinue the others, not reinforced by the existing pattern of stimulation from the exteroceptors. This calls for an intracerebral process, evolving new phase-sequences, and then selecting those which are the most adequate to the stimuli. Such processes are not dependent on mysterious pyschic forces; they function on relatively well-understood principles. Our understanding of them comes from dealing with automata of the learning variety. Although the circuits of these are not collodial but electronic, their information-flows are simplified analogs of the learning brain. The machines choose their subroutines on the basis of norms incorporated in a high-level program-selector. The latter has access to information concerning the relative success or failure of the subroutines in accomplishing the programmed goals. It evaluates these routines by measuring them against their chances of success, and reorganizes them to optimize the chances. Thus such machines perform a process which is analogous to consciousness: they not only *have* operational programs, but also *know* of them. They are "aware" of their own subroutines, and reorganize them if needed. As Minsky said, in 1968, "it is not very hard, technically, to put in [machines] some features that you would have to call self-awareness of a kind." [6]

Artificial system language seems unduly mechanical when applied to living systems, and psychological language appears too vitalistic when used to describe artificial systems. Yet the great lesson to be learned is that despite the traditionally separate terminologies, "learning" is a process which applies to any system which adopts new modes of operation which are more certainly conducive to its goal than its previous modes. (In fact, some writers, such as Minsky and Johannesson,[7] now believe that machines can learn to be superior to human beings in intelligence as well as in various technical and moral respects.) Since all learning involves a phase of self-evaluation, it involves, for machines as well as for men, what philosophers traditionally called "reflexion" or "reflexive consciousness." In the contemporary view, these terms stand for the capacity to receive information from, and to evaluate, one's own operational programs, such as codes, gestalts, principles and values.

Reflexive consciousness–the seemingly uniquely human capacity to know one's own states of mind—thus turns out to have a definite functional role in the evolution of learned behavior. We may assume that the largely pre-programmed insects have little of it, and that higher primates, insofar as they are capable of acquiring new behavior patterns, manifest it to the corresponding extent. It also follows that the greater the problems to be overcome by learning, the more need there is for self-evaluation. When the evaluative process is intricate, involving decisions of high orders of prediction, the level of consciousness must be correspondingly evolved, to high orders of clarity and rationality.

The functional role attributed here to reflexive consciousness derives from our cybernetical model of natural and cognitive systems. It is likely that, whenever cybernetics serves as the framework for the explanation of mental phenomena, a similar role will be assigned to consciousness, regardless of the further specifics of the theory. A good example of this is Deutsch's concept of consciousness as a collection of internal feedbacks of "secondary messages." The latter are messages *about* changes in the state of parts of the system; "primary messages" are messages moving through the system in consequence of its interaction with the outside world.[8] Secon-

---

[6]Marvin Minsky, "Can Computers Evolve to Super-Human Levels Before Man?" Paper presented at the 1968 conference of the Institute of Religion in an Age of Science.

[7]Olaf Johannesson, *The Tale of the Big Computer*, New York, 1968.

[8]Karl W. Deutsch, "Some Notes on Research on the Role of Models in the Natural and Social Sciences," *Synthèse*, 7 (1948–49).

dary messages function as "symbols" or "internal labels" within the net itself. They are fed back into it as additional information, thereby influencing, together with the primary messages, the net's subsequent behavior. Consciousness, Deutsch tells us, does not consist in these labels, but in the processes by which they are derived from and fed back into, the net.

Deutsch's concept of consciousness is that of a functional process whereby some of the interactive processes of system and environment become available for analysis to the system itself. He points out that such processes occur in electronic calculators, where they serve important functions in recall.[9] They also occur in the division of labor of large human teams which collectively process information and perform certain functions of thought, e.g., industrial research laboratories and political or military organizations. In such groups the "secondary messages" are the guide cards and index tabs added to the information moving through, or stored within, libraries, card catalogs, or document control centers. Being unable to take in the vast amounts of primary messages, the heads of the organizations deal mostly with the titles, description sheets, routing slips, etc., which become further attached to the messages. Thus it is only the "secondary messages" which are picked up and processed by the internal network, the functions of which represent "consciousness" in the organization. They influence the processing of the data moving through the organization and, if secondary messages attach to some of the operating rules themselves, the organization influences itself by its own internal data processing (by modifying and replacing defective rules).

The cybernetical analysis of information processing nets assigns a functional role to self-analysis, regardless of whether the nets are human, social, or artificial systems. The efficient processing of vast amounts of information in a complex self-regulating system involves accurate checks and rectifications through circuits which evaluate the processing functions themselves, and not only the data carried by them. In the present analysis, the human being *qua* cognitive system analyzes his own cognitions of his world; and *qua* natural system, commands cerebral organizations which analyze his own cerebral networks. When there is less than adequate performance in the construct-sets representing the cognitive system's empirical cognitions, and in the cerebral networks which, in the natural system, process the information from the exteroceptors, their analysis within the system is a precondition of self-regulation and hence of efficient operation.

The functional origins of reflexive consciousness do not mean that,

---

[9]Avizienis, the designer of STAR ("Self-Testing and Repairing" computer) describes his machine's self-analyzing functions as the monitoring of the "inter-office memos" passed between the computer's subassemblies. (*Time,* Dec. 7, 1970)

when it is evolved, it must be restricted to the development of new constructs and cerebral organizations when existing ones break down. To assume this is to commit the "nothing but" fallacy. Reflexive consciousness emerges in the context of its functional role, but it may then persist in contexts where it has no functional role to fulfill. In fact, reflexive consciousness, like cognition itself, may at times be at odds with efficient functioning. For example, it is cognitively functional to explore the repertory of one's known gestalts when faced with puzzles such as are presented by Gestalt psychologists, where an odd pattern can take on sudden meaning if one looks at it the right way (i.e., finds the applicable gestalt). But it can be a source of trouble and embarrassment if one reflects on his gestalts while engaged in the manipulation of their external counterparts, for example, in driving a car. The wheel, pedals and shifts must be handled spontaneously; a consciousness of experiencing these perceptual objects only interferes with their manipulation. Similarly, consciousness of one's mental processes is useful and necessary in learning situations, but it prevents one from sleeping if undertaken when going to bed. Yet, *homo sapiens* can always make use of his ability to reflect on his constructs (i.e. call into activity organizations analyzing other cerebral organizations); and he often does so outside biologically relevant contexts. It is due to this fact that man is a creature that is not only adapted to his actual environment, but also knows factors of change and events which are unrelated to his existential needs of the moment. Man, by being self-reflexive in a biologically excessive measure, can become "objectively" cognizant—taking an interest in, and effectively mapping, states and events which do not directly influence his existence.

The socio-cultural phenomenon that corresponds to, and results from, this disinterested cognitive capacity of individuals may be conveniently termed "*philosophy* (*theory*) *of* . . ." an empirical field. In this view, "philosophies of . . ." are assessed as reflective endeavors, having as objects of cognition other items of cognition, gathered on different fields of inquiry. For example, philosophy of science may be defined, following Collingwood, as differing from science in degree of reflexiveness: while the scientist investigates objects directly, the philosopher investigates the scientist's investigation of objects. Such "second degree reflexion" (or "building models of models") is a generic characteristic of philosophy of science; it is generated when the existing paradigm in a science is challenged by persistent anomalies in observation and experimentation. In these periods, the leading men of the field turn to basic theory and engage, not in direct empirical inquiry, but in speculation *about* their science; i.e., in *philosophy* of science. But, although philosophy of science is effectively called into life at such times (e.g., at the time Galileo and Newton laid the foundations of modern science and challenged the Aristotelian paradigm, and again when Einstein laid the foundation of relativity physics, challenging the Newtonian paradigm and, still more recently, when Heisenberg, Dirac, De Broglie and Schrödinger founded quantum mechanics, replacing the anomalous classical theories in microphysics), philosophy of science does not retire into the womb of normal science at all other times—it continues to be practiced even when the new paradigm is evolved and well functioning. When the new paradigm *is* evolved and functioning, the practitioners of the resulting normal science divest themselves of their extraordinary role of philosopher and return to science proper. But not so the philosophers of science generated in the interim: they profess interest in the analysis of scientific methods and concepts rather than in research and discovery, and continue to practice philosophical reflection on science. Thus the analog of reflexive consciousness obtains on the sociocultural field of science: reflexive analysis is generated in a functional context when the reorganization of cognitions is a necessity, and then survives on its own as a "disinterested" or "purely cognitive" field of inquiry. And the same goes, *mutatis mutandis*, for the other "philosophies of . . . ," such as aesthetics, legal and political philosophy, philosophy of religion, etc.[10] In the same sense in which reflexive consciousness is not "nothing but" a functional way of overcoming a problem in the commerce of man with his world, so "philosophies of . . ." are not merely means of resolving problems arising in other areas of inquiry. Both evolve into permanent and autonomous cognitive activities, pursuing their own specific objectives rather than subserving those of more directly empirical fields.

There is some suggestive evidence to the effect that reflexive consciousness can and does occur in an entirely "pure" form, with all connections to empirical cognitions fully severed. The evidence concerns the nature of the states of consciousness which are variously described as the "oceanic condition" or "state of ecstasy" of the mystic. In these states and conditions the empirical world is purposively neutralized; the relevance of "objects" in the environment is almost fully ignored or suppressed. Shands, who made detailed studies of such experiences, points out that the person

---

[10]In a more sweeping generalization, philosophy as a whole could be viewed as a problem-oriented genesis of reflexion with subsequent autonomization in a discipline. For example, while Greece was a victorious power in the ancient world, her thinkers produced daring feats of speculation on the nature of reality without engaging in searching analyses of the meaning and nature of their concepts and presuppositions. The classical age of philosophy was founded with the reflective methods of Socrates precisely when Greek civilization burned on the pyre of the Peloponnesian War. (Hegel said, "the Owl of Minerva spreads its wings with the falling of dusk.")

embarking upon meditation or contemplation seeks out a setting in which there is either a comprehensive similarity in isolation or a comprehensive similarity in the performed routine. "The anchorite seeks freedom from change by retiring to a limited and isolated area; the religious seeks to achieve uniformity and to avoid novelty by establishing a fixed pattern of the day and night which varies little or not at all. The Yogi occupies himself with reflexively oriented exercises, and the Zen Buddhist concerns himself with repetitive rumination over nonsensical ideas."[11] Shands concludes that the state of ecstasy "appears to be a state in which there is an intense feeling of *knowing,* but a total denial of *object.* When the mystic finds himself knowing *something,* the moment of intensity has passed. The ecstasy is in knowing—but in this sense knowing becomes an intransitive verb, since it does not take an object—or, if one wants to put it the other way around, it is knowing knowing."[12]

A scientific explanation of ecstatic and meditative states has been suggested by Fischer.[13] On his theory these states are at opposite ends of a continuum of mental states which includes normal states of awareness at the center. Ecstatic states are at the apex of a perception-hallucination continuum of increasing ergotropic arousal, whereas the meditative states of Yoga samadhi constitute the peak of a perception-meditation continuum of progressive trophotropic arousal. During the ecstatic state there is neither the capacity nor the necessity for motor verification of the subject's intense sensations. Likewise at the peak of trophotropic arousal in samadhi, the meditating subject experiences nothing but his own self-referential nature, emptied of compelling contents. In both types of mental states the separateness of the external world gradually vanishes. This "Oneness" with the universe is explained on Fischer's theory by the integration of cortical and subcortical activity in the subject's nervous system. Thus the entirely "pure" forms of knowledge obtained in transcendental states of consciousness now receive a scientific explanation which does not reduce them to epiphenomena but link them along a continuum with normal states of brain functioning and awareness.

The principle of genetic continuity from below and logical irreducibility from above takes us in stepwise progression from homeostatic survival functions to the transcendental forms of knowledge and experience. It illustrates Piaget's dictum, "Life is essentially autoregulation" and his principal

---

[11]Harley C. Shands, *The War with Words: Structure and Transcendence,* New York, 1971.

[12]*Ibid.*

[13]Cf. Roland Fischer, "A Cartography of the Ecstatic and Meditative States," *Science* **174,** No. 4012 (1971), 897–904.

hypothesis, "Cognitive processes seem . . . to be at one and the same time the outcome of organic autoregulation, reflecting its essential mechanisms, and the most highly differentiated organs of this regulation at the core of interactions with the environment, so much so that, in the case of man, these processes are being extended to the universe itself."[14] In a systems philosophy of mind we postulate a biperspectival correlation of cognitive constructs and brain processes and affirm the functional origin of cognitive development. At the same time we insist, perhaps more than Piaget does, on the progressive autonomization of each of the attained stages. Thus we trace the gradual stepwise progression from survival to knowing, to knowing knowing, and ultimately to knowing the knowing knowing itself.

In this account, the remarkable varieties of human psychological capacities receive a non-reductionist and yet non-dualist explanation. The noblest fruits of our brain-mind are seen as functional products of a self-stabilizing and self-organizing system, attaining autonomy and launching the individual on a wide variety of cognitive and behavioral pathways, over and above those required for his biological existence. The typically psychological characteristics of man are not slighted in this account, nor are they reduced to biotic functions. Rather, these capacities are shown to derive from basic systemic properties of natural-cognitive systems as special cases. The functional origin of the capacities does not entail their reducibility to the conditions which gave birth to them; the human psyche crystallizes in a process of gradual autonomization as an irreducible, yet by no means supernatural, phenomenon.

[14]Jean Piaget, *Biology and Knowledge,* Chicago and London, 1971, 26.

Chapter 11

COGNITION: FRAMEWORK FOR AN EPISTEMOLOGY

The structure of cognition

The new theories of perception and cognition recognize the active role of the knowing subject in the processes of knowledge. Human beings are environmentally transacting open systems; and they do not perceive and cognize something just because it *is* there, or necessarily the *way* it is there. But the Cartesian and Newtonian conceptions of man and universe held man to be a spectator of an objectively existing universe of matter and substance. The role of cognition was reduced to one of directing one's attention to the objects; from these scrutinies (Whitehead called them "attitudes of strained attention") we were to receive sensations ("impressions" for Hume, "ideas" for Locke) and construct our knowledge of the external world. These conceptions survive in contemporary philosophy in as diverse forms as Russell's logical atomism, in which the basic building blocks are "sense-data," and the ontological school in phenomenology which, using Husserl's *Konstitutionslehre,* seeks to derive a knowledge of the world from the "givens" of consciousness.

However, the entire thrust of contemporary scientific findings in the psychology and physiology of perception is contrary to such optimistically inductive procedures. Even when we strip the content of experience of emotional overtones and intellectual bias, the residue is still likely to be a mirroring of the nature of the percipient rather than the nature of the percipiendum; the neat division of the cognitive processes into a spectator here and a spectacle there does not apply. In recent decades, systems theorists, including this writer, have remarked: "a new concept of man has evolved: he is now thought of as an active system built or 'thrown' into the environing world. His knowledge of the world is no longer conceived as that of a disinterested spectator who sees what encounters his eye; the existential behavior of the human being in his surrounding medium has emerged as a vital determinant of his cognition. The latter is shaped, and in turn shapes, an ongoing transactional relationship between man and environment. In this process each man evolves his own model of the world."[1] I have called these individually evolved world-models "world-perspectives,"[2] and "effective environments."[3] They constitute diverse cognitions of the environing universe, as the latter relates to, and is significant for, the needs, motivations and concerns of the individual.

Since about the 1940s, experimental psychologists and engineers have come to realize the remarkable *under*determination of the physical stimuli which human beings obtain from their environment. It was originally thought, for example, that a machine functioning like a voice-typewriter could be easily built, analyzing the frequency and intensity of the sound-waves produced by a speaker and translating them into a print-out of the words. Even a phonetic print-out proved to be impossible, however. The reason is that the actual sound-waves are remarkably underdetermined by the speaker, and also show great variation depending on context. (For example, the 's' sound in 'slip' is quite different from the 's' sound in 'sloop.') Thus when one hears someone speaking and understands what he is saying, he is doing considerably more than identifying determinate sounds by preestablished rules with determinate meanings (e.g., 't'-sound means "letter 't' of the alphabet" followed by the next sound, and so on). The hearer sharpens the sounds and assimilates them to meanings in the light of his own linguistic categories. It appears that to proceed thus requires familiarity with the entire linguistic structure of which the heard phrase is a part, i.e., with the entire grammar. The same goes for other cognitive perceptions. The full structure ("grammar" or "set of rules") of the pattern of which the perceived situation is a part must be known before the situation can be understood, or even fully perceived.

A striking example is provided by de Groot's experiments with chess-players.[4] In these well-known experiments, de Groot placed positions on the chess board before subjects for about five seconds. He then removed the positions and asked his subjects to reconstruct them. Chess grand-masters and masters reconstructed the position almost without error, while non-players were unable to correctly locate most of the some 20-24 pieces on the board. However, the chess masters and grandmasters were not able to reconstruct chess positions which were due to a random shuffling of the pieces; their visual memory functioned no better in such cases than that of the non-players. Thus they remembered only what they knew from pre-

[1]Ervin Laszlo, "Cognition, Communication and Value," in Lee Thayer, ed., *Communication: The Ethical and Moral Issues,* New York, 1973.

[2]In "Cognition, Communication and Value," *op. cit.*

[3]In *System, Structure and Experience: Toward a Scientific Theory of Mind, op. cit.*

[4]Adriaan D. de Groot, "Perception and Memory versus Thought: Some Old Ideas and Recent Findings," B. Kleinmuntz, ed., *Problem Solving,* New York, 1966.

vious experience: positions likely to come about in expert games of chess. Their cognitive and recall abilities turned out to be dependent on their actual knowledge of the rules of the game and the likely moves. The case is similar to someone hearing a complex sentence in (say) French: he will understand it if he knows that language and also knows the "grammar" (or conceptual categories or rules) of the universe of discourse to which the phrase belongs. To a person not knowing French the sounds appear remarkably undifferentiated; and to a person knowing French but not knowing the area of discourse, familiar words stand out against an undifferentiated background of unfamiliar ones. (Orientals look alike to Westerners, and Westerners to Orientals.) These examples underscore the fact that the physical stimulus itself does not impose its meaning; it is merely a general scheme into which the meaning can be "read" by assimilating it to known patterns. With some exaggeration one can say that we perceive what we know, rather than know what we perceive. (This, of course, is a simplification, since ordinarily we can also *learn* to perceive—to know—the initially unknown.) The examples suffice, however, to establish a general assertion concerning the importance, in cognition, of prior cognitive categories (somewhat like Kant's transcendental schemata, though they are not logically necessary, but empirically evolved). One and the same physical stimulus can mean different things to different people, depending on the cognitive categories by which they interpret it. And if they do not have the appropriate cognitive equipment, the stimulus itself may escape attention—it may be experienced as no more than a rather undifferentiated and unremarkable pattern of color, sound, smell, taste, or touch.

It is but a short step to conclude from this complete personal or cultural relativism. Individuals know the world as they happen to conceive of it, and every type of knowledge is as good as every other. There are no criteria of "truth" and "reality," no ways to decide between the varieties of human cognitive experience. But this conclusion, although based on powerful arguments, is overly hasty. It is not to be denied that, as Dorothy Lee puts it, "culture is a symbolic system which transforms the physical reality, what is *there*, into experienced reality."[5] And it also follows that "the universe as I know it or imagine it in the Western world is different from the universe of the Tikopia, in Polynesia."[6] Cultures provide different codifications of reality, and different responses in terms of these codifications. One's own culture, as Lee affirms, offers a strongly bounded and precategorized view of reality. For one thing, it is based on a language which Whorf has shown to be a major element in the formation of

---

[5]Dorothy Lee, *Freedom and Culture*, New York, 1959, Introduction.
[6]*Ibid.*

thought. "Man's very perception of the world about him is programmed by the language he speaks, just as a computer is programmed. Like the computer, man's mind will register and structure external reality only in accordance with the program. Since two languages often program the same class of events quite differently, no belief or philosophical system should be considered apart from language."[7] In Hall's extension of the Whorfian thesis, people not only have different "programs" for interpreting reality, but they "*inhabit different sensory worlds.* Selective screening of sensory data admits some things while filtering out others, so that *experience as it is perceived* through one set of culturally patterned sensory screens is quite different from experience perceived through another."[8] All this is undeniably true, and must be taken into account when comparing and evaluating "world-perspectives" (or empirical models) constructed by people in different cultures. But it is also becoming evident that all men, regardless of the culture they happen to belong to, have basically similar nervous systems, are equipped with analogous sense receptors, command like patterns of response, and use patterns of thought (whether rationally or emotively motivated) which obey very similar laws or regularities. In other words, there appear to be some "universal" traits underlying cultural-cognitive relativities: Chomsky could locate "linguistic universals,"[9] and Kluckhohn discovered a number of "universal categories of culture."[10]

Finding such universals is rendered difficult, if not impossible, by arguing out of one's own culturally or individually relativistic categories. In that light every other world-model becomes but a special case of one's own, and is forced into the latter's structural scheme. But, in using the neutral framework of a cybernetical model, one is no more arguing out of his own culture-categories than out of that of a thermostat. Conceptualizing the cognitive process with such categories we can reach universal structures, for we are not dealing with particular contents. Regardless of whether a person conceives a sensory pattern as *trees,* meaning "standing peoples, in whom winged ones built their lodges and reared their families," or interprets the (presumably) very same pattern as obstructions to be cut down and burnt, he is using a construct (or gestalt) which endows his perceptual input with meaning. And the development of constructs and gestalts obeys some general regularities, already manifest in biological evolution and set forth in cultural development.

---

[7]E. T. Hall, *The Hidden Dimension,* New York, 1966, pp. 1–2.

[8]*Ibid,* p. 2.

[9]Noam Chomsky, *Cartesian Linguistics,* New York, 1966.

[10]Clyde Kluckhohn, "Universal Categories of Culture," *Anthropology Today,* A. L. Kroeber, ed., Chicago, 1953.

There are "quantum leaps" in cognitive capacities as one goes from lower to higher phyla. Behaviorally, the differences can be characterized as increasing complexity of adaptive capabilities and increasing breadth of transfer and generalization of learning. Every animal, at least above the level of worms, has a capacity for learning by means of forming stimulus-response associations, conditioned and operationally reinforced reflexes, and adaptive response patterns. As we move along the phylogenetic scale toward the higher (more differentiated) organisms, the simpler learning capacities persist, but are found in combination with more complex forms of learning. The new capacities are emergent functions, superimposed on those which are characteristic of the less evolved species.[11] Human levels of learning are reached when adaptive behavior includes the cognitive capacities required in the use of tools and weapons to maintain the evolving patterns of survival in social groups. "Culture" is the emergent quality at that level, and it is centered on the individual's ability to learn (as well as to learn-to-learn) the environmental factors which are pertinent to his being and well-being both as an individual, and as a member of a collectivity. Such cognitive abilities can attain subsequent autonomy, as cultural development focusses on goals and objectives which have little to do with well-being, and nothing with being. But the cognitive capacities acquired in the gradual processes of adaptive behavior (learning) enable the physiologically relatively weak and clumsy human to maintain his complex patterns of existence, and fit him into an ecological niche which he could not occupy but for his superior cognitive (tool- and weapon-using) powers. In ontogeny, certain invariant universally human mental capacities are likewise found to evolve. Child psychologists[12] have found that cognitive development proceeds by distinct and qualitatively different stages in the child at different ages. Each stage of development forms a structured whole, rather than being a link in an additive chain of stimulus-response behavior patterns. Evidence has been gathered for two general stages of mental development, termed "concrete-associative" and "abstract-conceptual" thinking. The transition from the one to the other occurs between the ages of five and seven. Similar to the evolution of adaptive patterns and capacities in animals, these stages do not displace one another, but form a hierarchically integrated structure. Individual differences among the cognitive abilities of mature individuals can be wide, and are accounted for by

---

[11]Cf. M. E. Bitterman, "The Evolution of Intelligence," *Scientific American,* **212** (1965).

[12]Cf. the works of Piaget, and such recent reviews as J. L. Phillips, Jr., *The Origins of Intellect: Piaget's Theory* (San Francisco, 1969), and L. Kohlberg, "Early Education: a Cognitive-developmental View," *Child Development,* **39** (1968).

genetic as well as socio-environmental factors.[13] The lower abilities (associative learning and memory-span) are found to be the necessary, but not the sufficient conditions of the development of the higher (abstract reasoning) capacities. Regardless of the genetically and empirically induced differences, however, basic modes of thinking characterize all human beings, and indeed all higher biological species. These are rooted in, and explained by, the fact that all such organisms are self-maintaining open systems using a specific mode of reproduction, and forming part of some similarly specific social structure. The mental capacities needed to maintain such systems in their environments are adaptive functions; they crystallize as cognition in the more evolved species, and culminate in man.

Without reducing the higher capacities, and their possibly autonomous goals and motivations, to biotic need fulfillments and simple stimulus-response conditioned responses, we can glimpse a genetically continuous scale of adaptive (and eventually consciously cognitive) behavior, with each of its phases operating by means of certain invariant control-codes (=constructs). Those most immediately pertinent to human cognitions make up an ascending order of categories, universally human in principle but variously evolved in different real individuals. These categories may be listed as follows:

(i) *Gestalts* (invariant patterns with established meanings to which the input patterns are assimilated); (ii) *rational constructs* (theoretic entities postulated through abstract reasoning and connected to the input patterns by means of some established rule of correspondence); and (iii) *aesthetic constructs* (non-discursive meanings discovered in the input and illuminating some part of the knower's "felt experience"). These are types of constructs which represent the limits of human cognition, given the kind of perceptions, cognitive organizations and effective output channels at our disposal. I argue that the many forms of human experience do not constitute disjunctive culture-conditioned categories, but a set of universal structures which transcends individual and cultural differences and relativities, and accommodates, as sub-classes, the many varieties of cognitive patterns as environment-mappings and constructions of natural-cognitive systems on the specifically human level of nature's inclusive hierarchy.

Ordinary perceptual cognition

The classical theories of perception—sometimes called "theories of immaculate perception"—have been replaced by the theories constituting

---

[13]Cf. B. K. Eckland, "Genetics and Sociology: a Reconsideration," *Amer. Soc. Rev.* **32** (1967); A. R. Jensen, "Patterns of Mental Ability and Socioeconomic Status," *Proc. National Academy of Sciences,* **60** (1968).

the "new look in perception." Cantril summed up the difference between the old and the new theories succinctly in the phrase, "the world we experience is the product of perception, not the cause of it." Classical empiricists held that the world (the domain of the primary qualities) is the cause of our perceptions (the secondary qualities), and struggled with the problem of how non-perceivable properties—motion or extension, *per se*—can give rise to perceptual sensations—colors, sounds, and so on. The problem is not solved, but it is now reconsidered: it is stated within the framework of an interaction between world and perceiver. In this interaction, the need-patterns of the perceiver are co-determinant, with the properties of the perceived object, of the content of perception. As Murphy said, "the perceived world pattern mirrors the organized need pattern within."

The general conclusion emerging from recent experimental work on the psychology of perception is that seeing (hearing, etc.) is not a simple registering of externally induced stimuli, but a complex interactive process whereby relatively underdetermined stimuli are assimilated to cognitive sets derived from the past interactions of the organism and its environment, and are thereby endowed with meaning.[14] How the brain performs this feat of extracting the relevant features of a complex and often confusing stimulus pattern and "sees" the derived pattern as an instance of a known thing or relationship, is not known with certainty.[15] Fodor likens the problem to that of building a model for a "black box" automobile factory in which cars consisting of some eight million parts are manufactured, and of which all that is known is that the raw materials go in on the one end and the finished cars come out on the other.[16] Some basic data on the "machines" of the factory have come to light recently (e.g., in the work of Adian and Hartline on the eye-mechanism which exaggerates boundaries and edges and thus makes us perceive "things" with sharply defined bounds against background patterns). But the more complex operations in perception, whereby the features of familiar things and events are distinguished and endowed with perceptual constancies, may elude experimental workers for some time to come: the complexity of the operations is enormous. Fortunately, complexity does not call for spiritualistic principles but is explainable by the availability of some 10^{11} neurons in the nervous system, with their possibly still more staggering number of interconnections.

---

[14]Cf. *Explorations in Transactional Psychology,* F. P. Kilpatrick, ed., New York, 1961.

[15]Cf. Ervin Laszlo, *System, Structure and Experience,* New York, 1969, esp. "Further Notes on the Perception of Invariant Intelligible Gestalten," pp. 101–108.

[16]In a talk given at the State University of New York at Binghamton, on March 21, 1970.

Thus perception, as we said, is a highly interpretive process. It presupposes a knowledge of the entire "grammar" of the situation in which the perceived pattern is a part, or of which it is a representation. Thus seeing (to use the convenient visual metaphor for the entire range of perceptual cognition) is never a mere registering of what meets the eyeball; it is always, in Hanson's words, "a theory-laden understanding."[17] Hanson asks, "How should we regard a man's report that he sees *x* if we know him to be ignorant of all *x*-ish things? Precisely as we would regard a four-year-old's report that he sees a meson shower."[18] But we do know many "*x*-ish things"–ordinarily we see our physical stimuli *as* something exemplifying our already acquired categories of knowledge. We see intelligible objects, known from previous encounters either with them, or with their kind. The "mind-language" name for such cognitively constructed entities, to which stimuli underlying our perceptions are assimilated, is *gestalt.* Gestalts are organized wholes wherein parts exemplify relationships which have meaning relative to the total pattern. A jumble of lines perceived as such is not a gestalt, but when in it we suddenly perceive a tree, the lines become organized, and the pattern exemplifies a gestalt. The appearance of every single item changes compared to that it had when viewed in isolation; the total pattern gains articulation, unity, closure and a tendency to symmetry. In recall, these gestalt elements are further emphasized to the detriment of irregularities, as specified in Koffka's "law of good organization."

Gestalt perception emerges in the course of the ongoing adjustment of the organism to its environment; a process which requires the differentiated perception of the factors of significance in the stream of perceptual stimulation. A successful high-grade organism perceives its environment in the form of significant and meaningful gestalts. These it derives from its perceptual patterns by the operations discussed above. But human beings tend to *reify* their gestalts; to assume that they truly exist in their environment. Indeed, we ordinarily do our best to make them exist in this way. We surround ourselves with things that we can perceive as the familiar things we "know" to exist in our world, and tend to ignore that contingent of our environment which is puzzling and unfamiliar. We make and manufacture things in the image of our gestalts. We also develop new gestalts if and when a puzzling set of perceptions cannot be dispelled, and thus forces itself on our attention. Thus we learn of new kinds of things in our perceptual world.[19] By means of a combination of manipulating our everyday

[17]N. R. Hanson, *Patterns of Discovery,* Cambridge, 1958, p. 19.

[18]*Ibid.,* p. 21.

[19]See the learning theories discussed in Chapter 5 ("Biological Systems") and Chapter 7 ("The Mind").

world and of adapting to it, we build our environment into our patterns of perceptual cognitive activity, and build our activity patterns into our environment. The outcome is a stream of experience in which deliberate choices can be made, based on one's established patterns of knowledge, satisfying the basic goals of survival and understanding, as well as the many instrumental tasks and objectives of human existence.

Scientific cognition

Natural-cognitive systems construct their environment in their internal structure: in diverse neural loops and associations (in the physicalist view appropriate to natural systems), and in many sets of empirical and reflexive constructs (in the introspective analyses of cognitive systems). In ordinary perceptual cognition, a configuration of signals from the environment (neural energy transmissions in the physicalist view, percept patterns in the mentalist one) matches, or fails to match, pre-established cognitive organizations in the system, and is accordingly meaningful, or puzzling. These organizations are *gestalts* in the cognitive system.

Gestalt-perceiving capabilities result from the progressive differentiation of the manifold components of the stream of sensory stimulation; a process in the course of which the system builds up the common-sense world of self and world. The first step is drawing an ego-boundary: the differentiation of "self" from "world" (non-self). Even when the ego-boundary is thoroughly established, it tends to remain flexible and hazy—many of the "self" qualities are projected into the "world" in anthropomorphizing cognitive modes. Only in the mature stage of ordinary perceptual cognition is there a fixed boundary between "objects" *out there* and "self" *in here*. The scene is then set for the commonsensical spectator-spectacle view of the knowing process, with the subject gazing at the world of objects and knowing it as a result. The gestalt-system hardens into relatively inflexible, constant patterns, and the knower locates himself within this fixed and largely predictable perceptual world.

The process of differentiation continues beyond this point, however, for individuals who evolve abstract-conceptual modes of thinking. Pattern perception and interpretation acquires a new level of refinement: it is more analytic of the sensory input than concrete-associative gestalt-cognition. The global state of "undifferentiated absolute of self and environment" (Berlyne, Piaget) is broken up not only into "self" and "world," but also into subdivisions of the latter: a $world_1$ of science, $world_2$ of art, and so on. A hierarchic order is then established between the various sets of constructed "worlds." Werner pointed out that "Wherever development occurs it proceeds from a state of relative globality and lack of differentiation to a state of increasing differentiation, articulation, and hierarchic order,"[20] and the truth of his observation is evident when we consider the empirical world-pictures of a person who is not only perceptually, but also scientifically (and perhaps also aesthetically and religiously) cognizant. He constructs and lives in an *Umwelt* which could include the natural universe of the sciences, the meaningful felt surroundings of the arts, and the transcendental world of religion and mysticism. The worlds are hierarchical in that the more differentiated ones tend to control the lesser ones, and encompass the latter without assimilating them. For example, the scientist conceives of his surroundings as the immediately accessible portions of an abstract-conceptually known natural universe, which extends in space and time beyond his limits of sense perception, and which, moreover, is composed of elements (atoms and particles) which similarly elude his direct acquaintance. He thus reinterprets his concrete gestalt world in the light of his scientific conception of the universe. Nevertheless he does not fully assimilate the former to the latter: he continues to eat on tables (and not aggregates of molecules), turn on lights (and not streams of photons), and speak to people (and not complex homeostatic biosystems), etc. Similarly, the artist's *Umwelt* is controlled by his aesthetic conceptions, yet his everyday world, though it takes on the coloring of his aesthetic cognitions, remains distinct in everyday life and for everyday purposes.

We are concerned at present with the type of cognition exemplified in science. The fundamental difference between science and ordinary perceptual cognition is that whereas in the latter percepts are assimilated to the confirmed components of the cognitive organization spontaneously and immediately (in being "seen *as*" gestalts), in science the analysis of percepts to constructs is conscious and deliberate. Contemporary science can only be understood if a clear distinction is drawn between spontaneously cognized sense-objects (percepts seen *as* intelligible gestalts), and epistemically percept-correlated scientific entities, non-perceived, and in some cases intrinsically non-perceivable. Science does not deal with perceptual gestalts except in a primitive and preliminary way. To illustrate this contention, let us take one of the least theoretic of sciences, botany, which does make use of gestalts in identifying its objects of study. For example, "this is an oaktree" counts as a proposition in that science, and an oaktree is something which can be perceived with the same degree of immediacy as any other sense-object. But consider the meaning of 'oaktree' in botany. It is not the same as in the language of a child or of a farmer. A child sees the

[20]Heinz Werner, "The Concept of Development from a Comparative and Organismic Point of View," *The Concept of Development*, D. Harris, ed., Minneapolis, 1957.

perceptual configuration he calls 'oaktree' as a massive brown-green complex object, swaying in the wind. The farmer sees it in relation to his land and to his plans. Any scientific meaning of 'oaktree' includes, however, further factors as part of it, and these factors (such as rate of growth, photosynthesis in the green leaf, relation to soil and other plants, etc.) are not seen *in* the oaktree-gestalt. They are confirmed by that gestalt, to be sure, but there is no sense in which one could see an oakleaf *as* photosynthesis. Rather, one can see *that* photosynthesis occurs in it. The distinction marks the difference between a concrete-associative cognition of sense-configurations, and an abstract-conceptual one. In perceptual cognition, the particulars projected by our senses to our minds are given the attributes of the universals familiar to us: we see this particular green object as a universal gestalt, e.g., "oakleaf." In scientific interpretation, on the other hand, we formulate an abstract-conceptual scheme which tells us what photosynthesis is, and then in perceiving an oakleaf we see *that* photosynthesis takes place in it. The seeing becomes metaphorical; it is thoroughly charged with abstract-conceptual meanings.

It has been customary to contrast the metaphorical sense of "seeing" in science with the spontaneous act of directly seeing that which meets the eye in ordinary perception. In the light of the findings of the new psychology, however, the contrast reduces to one of degrees on a continuous scale. In ordinary sense-perception, what is seen is known by one's previous acquaintance with it; thereby the object is categorized and placed into a familiar context. If these processes appear as a direct "seeing that which is there" it is because the brain, in as yet not fully understood functions, almost instantaneously assimilates the new and variable sensory patterns to pre-established nets and loops, which confer intelligibility upon them. If the sensory patterns create puzzles, the subliminal process of analysis reaches consciousness: the subject (e.g., in the experiments of Gestalt psychologists) is aware of trying to fit the perceived pattern into this or that familiar scheme. Such patterns have to be specially devised; ordinary experience yields patterns which lend themselves to immediate and spontaneous assimilation. Science, however, requires familiarity with a set of abstract concepts ("constructs") if observation is to be informed and meaningful. It is not ordinary experience, involved in living in the immediately environing world, that generates these constructs, but prior theoretical endeavor: the scientist's imagination, disciplined by a rigorous method and informed by the constructs already evolved in the history of his science. When his constructs have been evolved and the method assimilated, the scientist "sees" his sensory patterns in terms of the theoretical entities of his science: he sees *that* certain scientific entities are instantiated and that certain scientific laws or principles are operative. "Seeing" in such informed observation can regain a degree of spontaneity approaching that of regular perceptual experience—the scientist may at times come to believe that he sees "that which is there" much like the common-sense percipient. Yet, as noted earlier, in recent years it has become clear that the physical stimuli of all perceptual experience is underdetermined, and that they are lent meaning and significance through a prior knowledge of the "grammar" of the situation in which the experience occurs. It is not merely the sound of the word 'Chicago' which lets the hearer identify it with the name of the city on Lake Michigan, and it is not merely the physical appearance of chessmen on the board which allows a chessmaster to instantly memorize their position. Likewise it is not the physical stimulus in a scientific experiment or observation which points to the theoretical constructs of that science. In every case, familiarity with the entire conceptual scheme relevant to the stimuli functions as referent, and catalyzes the assignment of meaning and significance—even if the process is subliminal and practically instantaneous.

Interestingly enough, the epistemology of contemporary science and ordinary perception have been reintegrated, not by science turning out to be more like the classical notion of perception, but by perception turning out to be more like the modern concept of scientific cognition. The inductive ideal seems remote in both: neither perception nor scientific cognition proceeds by examining the data and then inducing from them to the indicated conclusions. Rather, both operate by some process of (not necessarily conscious) deduction: from presentationally concrete gestalts, or from abstract conceptual constructs.

The hypothetico-deductive approach has been used in science for some time and is now emerging as its fundamental method. As Northrop emphasized, the primitive concepts of Newton's physics can be put on one page of a book, and the definition of its eight most important defined concepts can be put in sixteen lines.[21] Yet the main assumptions of the theory are nevertheless adequately noted. Such a bare minimum provides the grounds for theories explaining the motions of the planets, the falling of an object to the ground, the pendulum, the lever, the balance, and the gyroscope. This is a typical instance of a high-level theoretical science. In such science there are a small number of undefined or primitive concepts making up a small number of propositions or postulates. From the combinations of the primitive concepts in the postulates, all theorems of the science can be deduced by rules of formal logic. (Nevertheless, rigor of this kind remains an ideal of empirical science, attained only in exceptional

[21]F. S. C. Northrop, "The Problem of Integrating Knowledge and the Method for Its Solution," *The Nature of Concepts, Their Interrelation and Role in Social Structure*, Stillwater, Oklahoma, 1950.

cases in some privileged branches. These branches deal with relatively simple systems, where the number of variables permits the manipulation of mathematical models of manageable complexity.)

Hypothetico-deduction in the construction of experienced reality is one, and perhaps the only way in which the egocentric predicament of epistemology can be overcome. The mind, as cognitive system, is constituted by sensations, memories, conations, etc., i.e., by what we call "mind-events." These are the things we know with immediacy and without possibility of doubt. If they occur, they occur; nothing can falsify them. But these are, at the same time, the events which make the mind a closed system in the absence of imaginative constructions of external reality. There is nothing among mind-events which would stand for, or point to, physical events with absolute certainty. Mind-events can be interpreted as physical events, but such interpretations are hypothetical and falsifiable. It is always possible to argue that mind-events are just that: mind-events, and that physical events are something unknown, or perhaps even non-existent. To know physical objects the mind must undertake a rational *inspectio* of its objects of experience (Descartes), and this involves the prior assumption of objective non-mental space and time ("extension"). Hume, Berkeley, and more recently Husserl explored the main avenues which could lead to an inductive penetration of the external world of physical events from the given world of mind-events, and they ended up either with skepticism or with some species of idealism. It appears that the external world of physical events in space and time is a necessary assumption for the realist interpretation of experience, and constitutes the framework for the postulates of science as well as for the "animal faith" (Santayana) of common sense. Contemporary natural science could progress freely when scientists realized that the physical universe is only known as a construction described by postulates built of primitive terms, and entailing theorems relevant to one's sensory experience. "The belief in an external world independent of the perceiving subject is the basis of all natural science," said Einstein, and added that we can only grasp this "physical reality" by speculative means.[22] Twentieth century science discovered at last the need for what perceptual cognition has been actively engaged in doing all along: integrating rational assumptions with empirical sense-data. One can immediately apprehend only those differentiated things which are conveyed through the senses, but clearly the senses do not convey a three-dimensional grand piano through space to a neutral sense organ, then up one's nerve to his cortex, and then across in some mysterious manner into the consciousness (Northrop). All that the senses convey are colors, sounds, odors, pain and

[22]Albert Einstein, *The World as I See It*, New York, 1934, p. 60.

pleasure, and similar sensations. These are not three-dimensional objects possessing a back side which we do not sense, and right-angle corners constant through varying sensed perspectives. Even a common-sense object requires an imaginative leap beyond the percept-plane, into the realm of "construction" by means of rational assumptions (Margenau). In this regard perceptual cognition and scientific cognition are similar. Their differences enter with the nature of the "leap." In perceptual cognition it is tacit and functions subliminally. We see the particular *as* the universal, without realizing that we imported our prior "constructs" or gestalts into the cognitive act. In science, however, the leap is explicit and conscious. The imported rationalistic constructs must be evolved through purposive conscious effort and then applied to the experience in question. By that token the experience becomes an "observation *of*" the postulated entities, and can serve to confirm or disconfirm the latter.

Confirmation is had when the observation bears out the prediction flowing out of the construct system. More exactly, a theory is confirmed when the perceptual data at time t_n and place s_n can be "seen *that*" the epistemically correlated and predicted construct *y* is the case. The match between the predicted construct and the analysis of a perceptual pattern to its epistemic correlate is determined both by the theory and by "nature." The theory determines how the observation is to be "seen." For example, gazing into a microscope, the biologist "sees" a virus. But no amount of gazing without his prior known theory would enable an observer to see that entity. He would see an odd shape, surrounded by other shapes and moving in relation to them. Even more striking is the example of observation in a Wilson chamber. The informed physicist *sees* the vapor trails and infers *that* they condense on the passage of electrons and trace out their trajectories. The uninformed observer could gaze at the vapor lines for an eternity and would not see that an electron has passed. But this is not to be construed as a self-determination by science of its own observations, i.e., a manufacturing of the evidence. Nature determines the limits within which the perceptual patterns can be seen *that* the predicted constructs of the theory are the case. Not every odd shape appearing in the microscope can be seen that it is a virus, and not every vapor trail corresponds to the passage of electrons. Nature supplies the limits of the interpretable perceptual patterns, and science supplies their interpretation.

Some, but not all, constructs of science have observable ("epistemic") counterparts. In fact, the more theoretical a science, the more of its constructs are indirectly correlated with observations via other constructs. The epistemically correlated constructs are connected to observations by what Margenau calls "rules of correspondence" and Bridgman terms "operational definitions." The indirectly correlated constructs are so chosen that they lead to the predicted epistemically correlated construct with the optimum simplicity, adequacy, fertility and elegance.[23] Confirmation thus involves a match between one out of a possibly large number of constructs, and a sensory pattern which is "seeable that" the construct pertains. The selection of the indirectly correlated constructs and their interrelation within the theory is not guided by purely empirical principles. To illustrate this, take the construct "time." Its operational definition correlates it with the observation of a clock, when a "rule" or "correlation" is specified which states how the observed position of the hands (or other features) of the clock should relate to the concept of time. But, since there are many different types of clocks, from sand-clocks, mechanical, electric, electronic, atomic, astronomical and biological clocks, it is not clear which series of observations should serve to operationally define time. The decision is dictated by the internal consistency of the theory in which time is taken to be an independent variable. For this reason astronomical clocks are often used in modern physics: Newton's laws become true in a large number of cases and provide the simplest equations for the observed phenomena. Non-operational ("constitutive") definitions dictate the selection of operational ones whenever alternatives are available. The choice assures optimum simplicity, consistency, adequacy and elegance in the theory. In view of this dual role of definitions—partly operational, partly constitutive—it is incorrect to affirm in the manner of operationalists that every definition of scientific quantities must be operational. Some definitions are constitutive, and are dictated by the scientist's preference for certain types of explanations over others (Margenau).

Thus the match between the perceptual data and the operationally defined (epistemically correlated) segment of the scientific construct-set (theory) is dictated not only by observations, past and present, but also by the construct-set itself, many parts of which are chosen on non-empirical grounds. Consequently when such a match is held to signify an isomorphy between the theory and the relevant states of the universe (in the sense that the theory is said to be a mapping, or a conceptual representation, of these states), it is clear that not an "absolutely known," "objective" reality, but a rationalistically postulated and deductively confirmed "scientifically constructed" universe is meant. A confirmation of this "discovered" or "invented" universe (as De Broglie points out, the distinction between invention and discovery becomes increasingly hazy in contemporary science) proceeds by means of conceptual operations ("predictions") which, purely by quantitative calculations, forecast the occurrence of an epistemically correlated construct of the theory, at a specified time and place. Behavioral and physical operations ("experiments") performed on, or in relation to, the

[23]Henry Margenau, *The Nature of Physical Reality*, New York, 1950.

scientific object, bring about states or conditions conforming to those predicted. The observations at the specified time and place thus provide a criterion for the theory: the perceptions either correlate with, or fail to correlate with, the predicted constructs.

The operations involved in confirmation function by self-stabilizing negative feedback, reducing, and in the optimum case eliminating, the margin of error between the predicted and the actual observations. This can be rephrased intuitively in the following simple example. Suppose that I see a particular speck of light in the night sky. The light itself is a percept, and its explanation requires a body of systematically interconnected constructs, some of which furnish its operational definition. If I have such a system of constructs at my disposition, I can pass from the seen light to the epistemically correlated construct, via a rule of correspondence stipulated by the theory. That is, I can assume that the light "is" Venus. (More exactly, I then see *that* the light emanated from the planet Venus.) Using the formulas of celestial mechanics, I can calculate the position of Venus two hours from now. The calculations involve constitutively defined constructs, such as gravity fields, inertial forces, masses, light-velocity, and so on. Passing through these indirectly correlated constructs I end up with another epistemically correlated one: the predicted position of Venus relative to my point of observation. I then undertake the necessary operations to confirm or disconfirm the prediction. This involves returning to the specified spot two hours later and directing my eyes toward the specified portion of the night sky. I call my calculations correct if (in the absence of extraneous inhibiting conditions, such as clouds, smog, or blindness) I can see a light of the expected intensity when and where predicted. The observation will again be a "seeing *that*" there is light emanating from Venus, i.e., a complex theory-laden seeing, rather than the mere cognizance of a light-spot on the sky. Negative feedback is involved in this case simply in waiting for two hours, and then positioning oneself for the relevant observation. Thereby the margin of error between the predicted and the actual observation is optimally reduced. The reduction involves a manipulation of the relevant test object: I "manipulated" Venus in relation to my perceptual field so that its light-ray could interact with my eye when and where predicted.

Technical experiments in science are normally suggested by considerations flowing out of an existing theory and are designed to exhibit the agreement of the theory with observation (e.g., Atwood's machine, Foucault's apparatus, the Wilson chamber, etc.). The point of the simple example above is that *all* confirmation involves "experimentation" in some sense, and that all successful experimentation is self-stabilizing through negative feedback.

We can identify the conclusions of this brief assessment of scientific method with the postulates of our theory of natural cognitive systems. The chosen system (e.g., a scientist) commands a cognitive (=cortical) organization which includes, in addition to concrete gestalts, scientific entities termed *constructs*. When the scientist sees a luminous spot on the night sky as light emanating from Venus, or when he sees an oddly contorted moving shape in his microscope as a *paramecium*, he interprets his, in itself underdetermined, physical stimulus by his previously acquired conceptual (=cortical) reference frame. By means of connecting the seen entities with unseen and unseeable ones (such as protons, force-fields, temperature, etc.) he cognitively maps his environment in the form of a scientifically known physical universe. He thereby establishes order and coherence in his experience beyond the range of immediately perceivable actualities.

The scientist matches his hypothesized constructs (theories) to the relevant observations of the physical environment, through the process of experimentation. The construct-set is taken as (hypothetically) valid and the experiment is designed to match the predictions flowing out of the set with percept-patterns in observation through the purposive manipulation of part of the physical environment (i.e., test object). If successful, the manipulated part of the physical environment yields a percept-pattern which matches the predicted epistemically correlated constructs of the theory, as and when predicted. Thereby the construct-set has been validated and the corresponding aspect of the scientist's cognitive organization stabilized. The world has turned out to be such as the scientist expected it to be, and he can go on to explore further aspects of it in the hope that more of it will prove to conform to his conceptions. He has found a "paradigm" and confirmed it by experiment.

"Science"

Systemic control processes can be understood both in relation to an arbitrarily chosen individual system and in regard to populations of systems. In many cases (as in that of evolutionary self-organization), the latter perspective is fruitful; in most other cases, it is at least helpful. Hence now we take a population of scientifically cognizing natural-cognitive systems ("scientists") and examine their collective cognitive structures ("science"). Such a population-study is more feasible for scientific than for ordinary perceptual cognition, since science is a relatively standardized mode of cognition, with its degrees of individual variation contained by its method. What one scientist does in a given context differs from what others would do only by the degree of skill he possesses in manipulating his constructs as well as his equipment. Individual fancy is largely eliminated from science, and so are differences due to temperament and feeling. The "rules of the game" must be adhered to, if the result is to count as science, rather than as fiction or poetry. But the skill here referred to is no menial skill; it is a highly cognitive one. It permits gradations from routine laboratory or theoretical exploration to advancing highly novel ideas, issuing in scientific "revolutions." Nevertheless, no scientist creates his theoretical or experimental innovation out of nothing (or from "pure insight"), but is restricted by the existing situation in his field. Thus to appreciate the dynamics of the formulation and confirmation (as well as of the disconfirmation and reformulation) of scientific construct systems, it is useful to take into account the state of the discipline as a whole, and not only the experience of a given worker on it. The framework of a single natural-cognitive system can be transcended: we can assume that our chosen system is a member in a scientific community in which all members together possess all the constructs and make all observations in the discipline at its present stage. As a measure of linguistic expediency, I shall refer to science itself as a multi-personal natural-cognitive system, meaning by it the patterns of observation, theory-construction, and experimentation of all its members.

The above device permits the exploration of the here given account of cognitive self-stabilization in the experience of a given scientist as the systematic self-stabilization of "science." The kind of science which proceeds in this manner is that which Kuhn calls "normal science."[24] It operates on the basis of an established paradigm, accepted by its members as the valid theory, used in interpreting observations and conducting experiments. The paradigm is a universally accepted model and, like an accepted judicial decision in the common law, is an object for further articulation and specification under new or more stringent conditions. Paradigmatic normal science is not innovation-seeking. In fact, Kuhn emphasizes that it tends to suppress problems ("anomalies") which would require innovations. Rather, normal science actualizes a promise inherent in the paradigm on which it is based: the promise that the predictions generated by the paradigm will be matched by the relevant observations. Matching new observations to the theory is the principal motivation of normal science. It attempts to extend the knowledge of the facts displayed by the paradigm by means of increasing the extent of the match between the facts and the paradigm's predictions, and by further articulating the paradigm itself.[25]

[24]T. S. Kuhn, *The Structure of Scientific Revolutions,* Chicago, 1962, Chapters II, III.

[25]Kuhn, *loc. cit.* In agreement with this, Kluckhohn reported on an instance of "normal" activity in his field, anthropology: when an examination was made of

None of the "normal foci" of paradigmatic scientific investigation aims at anything new or different from the "promise" of the paradigm. Indeed, Kuhn marshals considerable evidence to show that observations which do not fit the predictions generated by the paradigm are simply not seen. If called to the attention of the scientific community, they will be branded irrelevant or "metaphysical."

Paradigmatic normal science is basically self-stabilizing—research is limited to the extension and the refinement of the accepted theoretical framework. Its value lies in the self-imposed limitation of its members: they concentrate on problems that only their own lack of skill can prevent them from solving. Thereby a given theoretical framework becomes rapidly and thoroughly explored, in regard to all its possible experimental applications and theoretical entailments. Normal science does not aim at novelties, and when successful, it finds none. It is deviation-reducing, aiming at the elimination of all differences between the constructs of the theory, from which predictions can be generated, and the epistemically correlated percepts, which furnish the pertinent data of observation. This characterizes a large part of the scientific endeavor but not, of course, all of it. Innovation through basic reform of theory—the emergence of new paradigms due to anomalies forced upon the attention of the practitioners of normal science—is a recurrent feature of the evolution of science. Let us consider its grounds and characteristics.

The principal task of normal science is to extend and refine a given paradigm by piecemeal experimentation and research. In the course of such activity persistent mismatches may occur between the predictions generated by the paradigmatic theory and actual experimental findings. If and when such mismatches occur, we encounter an "anomaly" which strains the credibility of the paradigm. Now, it takes a considerable amount of anomalous findings to induce a paradigm-shift; initially, there is a built-in resistance to taking the mismatches seriously, and even to perceiveing them. Bruner and Postman reported on experiments with healthy adult subjects in which a series of normal cards were presented to them interspersed with some abnormal ones (e.g., a red six of spades and a black four of hearts). They found that the subjects almost always identified the anomalous cards as normal ones: they were fitted to one of the previously known categories without awareness of the misfit. With further exposure to the abnormal cards, the subjects showed hesitancy and confusion until

---

the tables of contents of ninety ethnographic monographs published in English within the last twenty years, it was found that there is a stereotyped scheme with numerous but comparatively minor variations, and that genuine innovation is exceedingly rare. Cf. Kluckhohn, *op. cit.*, p. 509.

finally, and sometimes suddenly, most of them would make the correct identification. Once the subjects did so with a few cards, they had little difficulty with the others. A few did not make the right identifications even at forty times the average exposure, and these experienced acute distress. They seemed to forget what the pertinent categories were: what a given color looked like, or a spade.[26] There is eloquent evidence here of the natural tendency to reinforce one's conceptual categories in experience, selecting those experiences which can do so, and seeing the rest in terms of one's preconceptions. The same difficulty obtrudes itself in perceiving anomalies in normal science. Members of the community of normal science initially experience only the normal and expected phenomena, even under circumstances where later anomalies are observed. Further acquaintance brings about something like hesitancy and confusion: an awareness that the paradigm does not fit properly. A period Kuhn calls "crisis" commences at that time, marked by exploratory research, philosophical analysis of concepts, and the reconsideration of fundamentals.

Reviewing the data provided by Kuhn, several interesting facts come to light. Anomalies require frequent exposure to be observed, and then they tend to be first observed by men who have *not* spent a great deal of time working with the paradigm challenged by the anomalous observations. These are either young men, or men coming into the discipline from elsewhere.[27] Vested interest in the paradigm, as well as reinforced modes of looking at the data through the viewpoint provided by it, are effective checks on the perception of anomalies. Although there are three types of phenomena which can later be seen to have warranted the replacement of the paradigm (Kuhn), only one kind is first perceived as an anomaly. This set constitutes the observations which stubbornly refuse to be assimilated to the paradigm. The other two—observations already explained by the paradigm, and observations which can be understood in detail only through further theory articulation—do not provide motives for the reformulation of the paradigm; they are not apprehended as anomalies, although at a later time, when a new paradigm has been evolved and accepted, they may be seen as such in relation to the former paradigm.

Anomalies of the initially perceptible kind are at first seen as just another form of puzzle to be overcome within the framework of the existing paradigm. On repeated exposure, however, the stubborn persistence of the anomaly brings into question the paradigm itself. More and more attention is devoted to it, and the field's eminent men come to view its resolution as the key issue of their discipline. Early attacks on it closely follow the rules

---

[26]Quoted by Kuhn in *The Structure of Scientific Revolutions, op. cit.*, Chapter VI.
[27]*Ibid.*, pp. 89–90.

of the paradigm, but, given continued resistance, new rules and principles (many of them *ad hoc*) are evolved, so that the paradigm is continually violated. Though there is still a paradigm, few members of the science now quite know what it is, or how far to accept it. This is a distressing period for the discipline. Einstein wrote of the time when the Newtonian paradigm was questioned by various experimental evidence (black-body radiation, the Michelson-Morley experiments, etc.), "it was as if the ground had been pulled out from under one, with no firm foundation to be seen anywhere, upon which one could have built." [28] Copernicus and Wolfgang Pauli left testimony of similar distress at periods of crisis in their disciplines.[29]

Although the breakdown of the established paradigm is not necessarily explicitly recognized by all members of the scientific community, the apparently random, exploratory type of research activity which follows in its wake is a telling effect. Unlike the refinement of an established paradigm, the transition to a *new* paradigm is not a cumulative process, of adding data to data and construct to construct. Rather, the entire foundation upon which the old paradigm rests is called into question; its most basic categories are re-examined. There is a shift in the evaluation of the observations: from seeing that observation A signifies x, it is becoming evident that it does not signify x, but the new entity, y, is not yet articulated.[30] Its articulation constitutes "extraordinary science." The scientists in such periods seem like men searching at random. They first seek to specify the problem and to resolve it within the paradigm, but move away from the latter when they fail to do so. Since no experiment can be conceived without some hypothesis, scientists will be constantly trying out new speculative theories which, if successful, may lead to a new paradigm, and if not, can be easily discarded.

It is in such times that scientists, who do not ordinarily need or want to be philosophers, turn to philosophical analysis. They have no need for philosophy in normal science: to the extent that a paradigm functions without disturbing anomalies, its basic assumptions need not be made explicit. But when the paradigm breaks down, philosophical analysis is called for. It is no accident, Kuhn suggests, that the emergence of Newtonian physics in the 17th Century, and of the new physics in the 20th, should have been

---

[28]Quoted by Kuhn, *op. cit.*, p. 83.

[29]*Ibid.*, pp. 83, 84.

[30]Kuhn remarks on the parallel to a change in visual gestalts but calls it possibly misleading. (p. 85) Scientists, he claims, do not see something *as* something else (e.g., as in visual perception, where lines on paper can first be seen *as* a bird and then *as* an antelope), but simply see it. This is a gross oversimplification of the process of observation and obviously an unintentional error, not justified by Kuhn's own subsequent statements and assessments of the epistemology of science.

preceded as well as accompanied by fundamental philosophical analyses. These periods have also been characterized by the prevalence of "thought experiments," calculated to isolate the problem and show the inadequacy of the old paradigm to an extent and with a clarity not attainable in the laboratory.

Scientific paradigms are not wholly rejected until new ones are established. Science without a paradigm is science without an accepted theory—it is a regression to non-science. It is unacceptable to a scientific community brought up on theory, and having acquired its skills and techniques in relation to it. Thus the decision to reject a paradigm is simultaneous with the decision to accept another, and these decisions involve a comparison of both paradigms in regard to their success in dealing with the phenomena investigated in the science.

The view that basic scientific frameworks are replaced rather than modified, and fall rather than change with the emergence of anomalies, is gaining general acceptance and requires no special discussion here. Instead of belaboring this point, let me restate the historically exemplified processes of scientific progress in the framework of systems philosophy. Normal science is self-stabilizing through negative feedback in the planning and execution of experiments and applications. The percept–construct (observation–theory) match is assured by the circular method of assuming the validity of the paradigm and then demonstrating it in areas, or with a degree of refinement, not previously attained. Science as a collective cognitive system switches from self-stabilizatory *cybernetics I* to self-organizing *cybernetics II* when mismatches occur persistently, in many key areas of investigation. Although at first there is considerable resistance to acknowledging these, as scientists operating within the conceptual framework of the existing paradigm are negatively conditioned for their perception, sooner or later, if and when attempts to overcome the already perceived mismatches by an extension of the existing paradigm fail, or involve ungainly patchworks on the theory, negative-feedback self-*stabilization* yields to positive feedback self-*organization*: science receives its "initial kick" which sends it on deviation-amplifying explorations. A period is initiated which calls for self-analysis: existing basic assumptions, fundamental rules and entities are reflected upon in an effort to locate the roots of the trouble.

The period of crisis, marked by a philosophical awareness of the suppositions of existing theories, comes to an end when, out of a number of hypotheses advanced as candidates for the new paradigm, one set is confirmed and accepted. In order to be accepted, the new paradigm must explain all observations explained by the former paradigm, and explain in addition the observations which present anomalies for the latter. Upon the acceptance of the new paradigm all observations of the given science are reassessed: there is a general construct-shift, and instead of seeing "A *that* x," scientists now see "A *that* y" (where A stands for an item of observation, and x, y, for scientific constructs).[31]

Aesthetic cognition

Experience (to paraphrase Croce) is mute until man makes it speak. The stimuli activating neural signal activity in the sense organs are relatively underdetermined and not necessarily meaningful *per se.* Meaning is assigned by the possession of prior knowledge, and such knowledge means having confirmed sets of cognitive constructs. The construct categories so far discussed include gestalts and scientific entities. We now come to distinguish a third variety: the *aesthetic* construct.

Experience can be meaningfully conceived under various categories. A perceptual situation structured according to the familiar shapes of known entities in familiar relationships gives a meaningful experience; a set of selected observational data occurring as and when predicted likewise yields a meaningful experience, although it belongs to a different category. Ordinary experience is spontaneous and carries with it a belief in the truth and reality of the things experienced; scientific observation is methodologically disciplined and self-critical and does not necessarily lose its theoretical flavor. And there is the third variety of meaningful experience which we shall discuss here—the *aesthetic experience,* which, although as interpretive as the others, can (and usually does) appear to be perfectly spontaneous. An artist or aesthetically sensitive person sees, hears, touches, smells, tastes —in short, *senses*—the world around him in reference to his already evolved aesthetic tastes and constructions, and normally believes that the meanings he derives from his sensory experiences inhere in the very nature of the experienced world. The world is said by such persons to *be* beautiful, sublime, significant and aesthetically meaningful, in much the same way as it is said to contain tables and chairs by the common-sense observer. Yet

[31]Such a shift is operative even when the difference between x and y is unmeasurably small. For example, even if the numerical values entering into the equations of motion for an object on the earth's surface moving with velocities well below that of light are not different in Newtonian and in relativity physics, the observation of moving bodies is interpreted by contemporary scientists within the Einsteinian framework. Although space-time reduces to Euclidean geometry and relativistic effects become unmeasurably small, the phenomena observed in physics are not placed within a three-dimensional absolute space and time and are not related to phenomena in other frames by Galilean transformations. The new paradigm dictates the interpretation of *all* observations in a science.

aesthetic cognition, like any variety of cognition whatsoever, is highly interpretive, and is determined as much by the pre-established construct-sets (=cortical organizations) of the experiencing person as by the things he experiences.

In calling aesthetic experience "meaningful" I am transgressing the limits set by the positivistic theory of meaning, which thus regards only factual statements. Yet experiences which cannot be described by factual statements can sometimes be entirely meaningful. Meaning emerges when a sensory stimulus pattern is cognitively assimilated to a construct-framework, and there are many varieties of constructs and none has a monopoly on meaning. This is not to embrace an unconditional relativism—I am not proposing that any method of interpreting experience is as good as any other. For purposes of reaching practical decisions in daily life, perceptual gestalts are more fruitful and functional than scientific or aesthetic constructs; in setting up precise calculations and predictions, scientific constructs have the edge over all others. And when the objective is to grasp some otherwise elusive factor of significance about the environing world, aesthetic constructs can outperform the rest. Great works of art are embodiments of such constructs.

The history of art extends as far back as the history of our species. *Homo* is not only an intellective *sapiens* but also an emotive one. When the synchretic unity of his primitive experience gave way to more accurate but also more specialized modes of cognition, two great branches of knowledge appeared: philosophy (including the early forms of the sciences), and art. The rivalry between thinkers and poets was a basic fact of life in ancient Greece, and there represented the alternatives of rational and emotive cognition, vying for dominance. Both trace their origin to mythology: in the great myths, rational and emotive factors are interwoven. If they in turn are the surviving expressions of still earlier modes of cognition, we may assume that primitive experience, in its syncretic unity, encompassed the seeds of both science and art. But when one evolved to the extent that it became incompatible with the other, the lacuna it created needed to be filled. Science divested of emotive factors is irrelevant to much of human experience, and so is art disdaining the formulas of reason. Plato's aesthetics is an attempt to unify the dichotomic Greek experience under the banner of reason, allowing emotion the purely instrumental role of helping the mind discipline itself in the appreciation of things of which reason approves. Over the Middle Ages, Western art subserved the syncretic cognitive mode of Christian *Weltanschauung*, and functioned as an instrument of reinforcement of the prescribed beliefs and sentiments. But with the coming of the Modern Age, internal splits in religion weakened its hold on art, and art achieved autonomy: it served no other master but the laws of meaning proper to itself. It thus came to fulfill a role in which it complements science: it renders meaningful large segments of human experience which are non-measurable and thus unavailable for scientific treatment.

The world of art is a meaningful world, as everyone knows who has experienced it. It elucidates man's relation to his surroundings: it makes him feel and know what it is to be alive. These are poetically worded statements themselves, but they have empirical validity. Experience, human as well as animal, is emotively charged. Philosophers like Whitehead did not hesitate to tell us that "primitive experience is emotional feelings, felt in its relevance to a world beyond."[32] The "feeling-cognition" of the "world-beyond" is, to judge from recent experiments in animal psychology, shared by all species with relatively evolved nervous systems. Olds and his collaborators could stimulate the septal area and other regions of the brain of rats to produce attraction toward, as well as repulsion from, given objects.[33] The presence of "push-pull" emotions in animals was confirmed in many other experiments. It is reasonable to assume that introspectively known *human* emotions can be traced to some primitive functioning in the organism, which has at its base the "pull" of attraction and the "push" of repulsion. It may be that Whitehead was right: "anger, hatred, fear, terror, attraction, love, hunger, eagerness, massive enjoyment, are feelings and emotions closely entwined with the primitive functioning of 'retreat from' and of 'expansion towards.' They arise in the higher organism as states due to a vivid apprehension that some such primitive mode of functioning is dominating the organism."[34]

It certainly does stand out with some degree of clarity that feelings are components of all but the most carefully "purified" types of human experiences. The constant concern of scientists, to eliminate personal feelings from their observations and thus to preserve them from subjective bias, illustrates this contention. It also illustrates the fact that science must do away with feelings, rather than use them as a tool of cognition. I do not mean that science cannot attempt to cognize feeling; I mean that science cannot use feeling *to* cognize. The distinction is important, for it is clear that feelings can be investigated through science, both by the abstract-conceptual scheme erected to explain and control mind-events in psychology, and by the correlation of mind-events with brain-events in empirical neuro-

---

[32]A. N. Whitehead, *Process and Reality, op. cit.*

[33]J. Olds and P. Milner, "Positive Reinforcement Produced by Electrical Stimulation of Septal Area and Other Regions of Rat Brain," *Journal Comp. Psych.*, XLVII (1954).

[34]A. N. Whitehead, *Symbolism, Its Meaning and Effect*, New York, 1927, II, 4.

physiology. But in so doing the scientist does not cognize his data in virtue of his feeling—he eliminates them as much as he can. What he cognizes is the event denoted "feeling," and that event is not *felt* but *observed*. Due to this condition, science can only explain feelings arising in the course of experience, not use them to *understand* experience. Feeling is not a means of cognition in science; but it is that, I suggest, in art. To understand the large segments of human experience which are infused with, and sometimes dominated by, feelings, is the *raison d'être* of art, and of aesthetic activity in general.

Art (and henceforth I shall use "art" as an umbrella term to cover all emotive-expressive cognitions, not excluding the common or garden variety) differs from science not only in the emotive nature of the construct which lends meaning to experience, but also in the immediacy by which the construct relates to percepts. In Northrop's words, "the distinguishing mark [of art] is that its *primary* concern is with immediately experienced materials. The painter does not talk about colors, after the manner of the mathematical physicist; he puts before one the directly sensed particular blue, red, yellow, etc., each instance of any one being unique, as are their similar seen shapes. Conversely, the primary concern of the mathematical physicist is to use the impressionistic painter's emotively moving immediately apprehended colors, not in and for themselves, but merely instrumentally as the means for inferring, and the data for confirming or disconfirming, the quite different invariant laws of electromagnetic propagations. . . ."[35] The difference in this regard between science and art can be concisely stated as that between "concrete" meaning and significance discovered *in* experience (art) and "abstract" meaning found *in virtue of* experience (science).

But what makes the art-experience "intrinsically" meaningful? A major part of aesthetics is devoted to the analysis of this question. We have already indicated an answer and shall now enlarge on it in more detail.

Traditionally, studies in aesthetics and art criticism have tended to proceed on the ground of introspection, examining the writer's own experience and drawing generalized conclusions from it. We thus find descriptions heavily charged with adjectives such as *enlightening, sublime, satisfying, transcendent,* and many more. But these adjectives describe an introspected sensation; they do not explain it. The explanation can be provided, however, within the here proposed epistemic framework.

Aesthetic meaning is carried by one variety of construct which decodes certain patterns and sequences in the sensory input. The aesthetic construct

---

[35]F. S. C. Northrop, "Toward a General Theory of the Arts," *The Journal of Value Inquiry,* I, 2 (1967).

is not a gestalt: the objects of our experience are not just "seen as" familiar things. It is also not a rational construct: perceptual data are not "seen that" some theoretic entities are exemplified in experience. If all aspects of human experience could be satisfactorily coded in terms of common-sense perceptual gestalts, and the rational constructs which elucidate relationships and structures in our environment, there would be no need for aesthetic constructs; art may never have arisen. But many aspects of experience are (or appear to be) too unique to permit cognition in terms of concrete gestalts and abstract scientific symbols. Something more adequate is needed : something that will grasp just how it feels to experience *this* thing *now*. Susanne Langer put this well when she said, "there is a great deal of experience that is knowable, not only as immediate, formless, meaningless impact, but as one aspect of the intricate web of life, yet defies discursive formulation, and therefore verbal expression : that is what we sometimes call the *subjective aspect* of experience, the direct feeling of it—what it is like to be waking and moving, to be drowsy, slowing down, or to be sociable, or to feel self-sufficient but alone; what it feels like to pursue an elusive thought or to have a big idea. All such directly felt experiences usually have no names—they are named, if at all, for the outward conditions that normally accompany their occurrence. Only the most striking ones have names like 'anger,' 'hate,' 'love,' 'fear,' and are collectively called 'emotion.' But we feel many things that never develop into any designable emotion. The ways we are moved are as various as the lights in a forest; and they may intersect, sometimes without cancelling each other, take shape and dissolve, conflict, explode into passion, or be transfigured. All these inseparable elements of subjective reality compose what we call the 'inward life' of human beings."[36]

Knowledge, Langer affirms, is considerably wider than our discourse. In the present conceptual frame this means that meanings and cognitions transcend the limits of ordinary language—and even the languages of abstract thought. The residue, left over when everything that can be grasped by common-sense cognition and by theory construction has been deducted, is still very large. It is the ineffable sphere of intimate human experience, unique but no less meaningful. Artists seek to give it meaning by creating works of art and preserving for all eternity, as it were, the fleeting impressions of the moment. The creation of such works helps clarify meaning; perfects and elucidates the but dimly perceived aesthetic construct. Art can *express* vague meanings and render them conscious. This is a function of artistic creativity which makes it one of the many kinds of human *cognitive* activities. Collingwood said, "When a man is said to

[36]S. K. Langer, "Art As Expressive Form," in *Problems of Art*, New York, 1957.

express emotion, what is being said about him comes to this. At first, he is conscious of having an emotion, but not conscious of what this emotion is. All he is conscious of is a perturbation or excitement, which he feels going on within him, but of whose nature he is ignorant. While in this state, all he can say about his emotion is: 'I feel . . . I don't know what I feel.' From this helpless and oppressed condition he extricates himself by doing something which we call expressing himself. This is an activity which has something to do with the thing we call language: he expresses himself by speaking. It has also something to do with consciousness: the emotion expressed is an emotion of whose nature the person who feels it is no longer unconscious. It has also something to do with the way in which he feels the emotion. As unexpressed, he feels it in what we have called a helpless and oppressed way; as expressed, he feels it in a way from which this sense of oppression has vanished. His mind is somehow lightened and eased."[37]

Since for Collingwood art *is* expression, the above passage suggests that art is one of the basic forms of human cognitive activities. Activities of this kind are information-directed: they seek out meaning in experience. When they are frustrated, or simply prevented from being exercised, the subject feels oppressed and helpless. When meaning has been discovered, his mind is eased; he feels satisfied. Meanings are sought and found in all spheres of experience. But constructs adequate to the patterns of experience must be had. Gestalts code the meaning of most of the recurrent configurations of percepts in our everyday surroundings; scientific constructs code the meaning of many events beyond the range of the directly perceivable world. But the large segment of experience for which these constructs are inadequate must also be coded to acquire meaning—and this is the function of aesthetic constructs. Here the fragrance of a rose *means* something more adequate to our perception of it than the scientific physicochemical explanation; and the sight of the morning sun rising over Venice *means* something far more appropriate to its "feel" than an explanation in terms of geography and celestial mechanics. Here are meanings which are non-verbalizable and non-symbolizable even in artificial languages. The language adequate to them is not a conventional one of symbols, but an expressive one, where meanings are immediate and spontaneous. While it takes a knowledge of the English language to understand the word 'sunrise,' and a knowledge of geography and history to understand the meaning of 'Venice,' it takes but sensitivity to perceptual experience to *understand* the meaningfulness of a sunrise over Venice when seen, or when represented in the impressions of a great painter. The aesthetic construct is *concrete*: it is a representation of "felt experience"

[37]R. G. Collingwood, *The Principles of Art*, Oxford, 1938, Ch. VI.

in its intimacy and immediacy. It is what we see "in" a picture, hear "in" a symphony and understand "in" a poem, over and above the sounds and the sights registered by our senses. It is that which is either matched, or mismatched by perceptions. If it is matched, we say that the perceptions are *of* an object of art, or instance of natural beauty. If they are not matched, we normally refuse to call what we perceive "art" or "beautiful."

It will be noted that the aesthetic construct functions as the code for the constructive representation of the relevant object or event in the physical environment. It has isomorphy in some functional respect with the perceived environmental state—i.e., with the artifact or natural object which is experienced. The isomorphy is similar to that which obtains between a program and the object manipulated by the programmed machine, and between the genetic code and the molecule for the synthesis of which it carries the instructions. It is a functional relationship, and not a pictorial projection of the qualities of the perceived object inside the mind of the appreciating subject. Thus the entire question concerning the ontological status ("reality") of aesthetic qualities is spurious—they no more "exist" in the world than mathematical equivalence does. Qualities are attributed by human beings to their perceptual experiences in terms of their already confirmed constructs. Thus aesthetic qualities themselves do not have existential status independently of human interpretive cognitions. (This, however, does not preclude that aesthetic constructs may have objective counterparts in the world, much as other species of constructs have. But the objective counterparts of aesthetic constructs are not aesthetic qualities, but patterns and sequences of "physical events.") Once the doctrine of "immaculate perception" has been overcome, naïve realism can no more reassert itself in aesthetics than it can in any other sphere of empirical cognition.

The epistemology of aesthetic cognition concerns the relationship of aesthetic constructs and the relevant objects and events in the subject's existential surroundings. A meaningful aesthetic experience—aesthetic *cognition*—comes about when the perceived surroundings match the already confirmed aesthetic constructs. This match can be established through one, the other, or a combination of, the two cybernetical processes we distinguish in systems philosophy, namely, through negative-feedback self-stabilization or positive-feedback self-organization. Both involve a purposive, goal-directed interaction of subject and environment.

The above assessment seems to conflict with traditional aesthetic theories in ascribing an active response to the subject. Many aestheticians appear to deny this. Fry, for example, tells us that art is connected with the "second imaginative life" which differs from the ordinary life experience in that herein no practical action is called for. It is in consequence of this that our whole consciousness may be focussed upon the perceptive and emotional aspects of the experience. "Art, then, is an expression and a stimulus of this imaginative life, which is separated from actual life by the absence of responsive action."[38] Bullough proposes his much discussed "psychical distance" to account for the fact that in art we seem to put the phenomenon "out of gear" with our practical self. The working of psychical distance, he says, "has a negative, inhibitory aspect—the cutting-out of the practical sides of things and of our practical attitude to them—and a positive side—the elaboration of the experience on the new basis created by the inhibitory action of Distance."[39]

Use is in action and beauty is in contemplative meaning, assert or imply a host of aestheticians, including DeWitt Parker, Collingwood, Croce, Cassirer, and others. Art produces no activity which transcends the experience itself. Do these opinions, presumably founded on intimate experiences of art and on their critical study, mean that the concept of an active response is false and misplaced in regard to art? A careful examination of the issues shows that they do not. The response these aestheticians refer to is of the common-sense gestalt-cognizing variety, where some aspect of the perceptible environment is brought into correlation with the goal-pursuing subject. If such response did obtain in regard to art, aesthetic cognition would reduce to ordinary perceptual cognition: the art-object would be no more than a sense-object having a specific relation to the subject in the light of the latter's biological need patterns, projects, and social or other purposes. But when we cut out such responses in art, we do not by the same token eliminate *every* response. The response which applies to art (and to all aesthetic objects) is one which is designed to bring about, or simply to maintain, prolong, and intensify an aesthetic experience. The role of such activity as a vital component of aesthetic theory is clearly recognized by Stephen C. Pepper who, unlike traditional aestheticians, affirms that "consummatory satisfaction" in the quality of an experience is a typical instance of aesthetic value. It is an *activity* of a goal-directed "selective system" (a self-regulative feedback system arising from the transactions of organisms with their environments): "an active process maneuvering for the maximum delight in terms of both duration and intensity for this quality of that experience." Pepper emphasizes that "the big thing to notice here is that the activity is self-regulative, internally normative within the sphere of its own activity. A consummatory field in

[38]Roger Fry, *Vision and Design,* London, 1920.
[39]Edward Bullough, "Psychical Distance as a Factor in Art and an Aesthetic Principle," *British Journal of Psychology,* V (1912).

action is a selective system correcting the movements of the organism by its own drive dynamics for the maximum satisfaction available for the quality of the transaction."[40]

Positive- and negative-feedback response in the context of art and aesthetic enjoyment means a purposive striving to bring about, or to optimize, a meaningful aesthetic experience by maintaining, prolonging and sharpening the match between the perceptions of the subject and his pre-established aesthetic constructs. It ranges from such minute acts as keeping the eyes focussed on the art object, holding the head in optimum position for stereophonic hearing, touching the exhibited surface, sampling the taste, the perfumes, etc., through acts involving the elimination of sensory and imaginative experiences that conflict with the relevant perceptions (e.g., suppression of light reflections on a canvas by centering attention on the colors, suppression of traffic noise and coughs in the concert audience by concentration on the music, and suppression of irrelevant thoughts, desires and fantasies by attention to the seen or otherwise perceived art object), to the more interesting varieties of such responses which involve genuine artistic creativity. The latter is called for when the subject's perceptual experience is inadequately meaningful in the light of his existing aesthetic constructs. In that event he is impelled to produce new constructs, fresh ways of looking at experience. These new constructs may code purposive responses, leading to the embodiment of the constructs in some environmental "control object." We then encounter an instance of genuine artistic creativity.

"Art"

Similar to science, the collective endeavor of a population of aesthetically constructing natural-cognitive systems ("artists") can be examined as the multi-individual system in which such constructions are typical ("art"). In viewing aesthetic cognitions in the context of populations of artists jointly constituting the art of their epoch, new insights can be won of the here discussed processes.

Now, the identification of "art" with typical modes of cognition and creativity in a community of persons is more problematic than the similar identification of "science." There are great individual variations among any group of artists and art-lovers. Unlike science, the "rules of the game" are implicit and need not be strictly adhered to: there is no immediate counterpart to a stated scientific method in the arts. Behavior is less consensual, more individualistic. Thus there is no such thing, for example, as

---

[40] Pepper, "Autobiography of an Aesthetic," *op. cit.*, p. 281.

an entire community practicing purely self-stabilizing "normal art," although the concept of a "normal *science* community" is a good approximation. In any actual art community there are likely to be persons who accept a given style as their "paradigm," and others who reject it. Consequently the multi-person extension of our epistemic framework cannot be to art *communities* as such, but to *types of artists*. We can talk of "normal-art type artists"—more simply *conservative artists*—and of "crisis-art type artists"—or the *avant-garde*. A restatement of the nature of the art experience in these terms is illuminating, because it exposes remarkable, and usually undiscerned, parallels between stylistic change in the arts and paradigm shifts in the sciences.

"Conservative artists" (and this term is used here to include not only professional artists but all conservatively creative persons) constitute that segment of the art community which is concerned with maintaining an already established style. This style is the basis of their artistic activity; it functions analogously to a paradigm in science. Conservative artists do not seek stylistic innovations—their creativity consists of adopting the style for their own artistic purposes. Theirs is an essentially "puzzle-solving" activity. They take a style, and use their skill and ingenuity in devising new techniques and new topics for it. Thereby they extend the range of application of the aesthetic constructs proper to that style, and refine them. For example, impressionistic art takes the cluster of colors evolved by Van Gogh, Cézanne, Renoir, and its other founders, and uses it in pictures with different subject matters and new techniques of putting it on canvas. New works of art result from such activity, but not new styles. The new works extend the impressionism of 19th Century French painters to a wider range of subjects, and refine its application by exploring further varieties of perspectives, color combinations, textures, and so on. The same process takes place in impressionistic music. The cluster of sounds evolved by Debussy, Ravel, and Fauré are extended by their followers and applied in new compositions. And what goes for impressionistic art goes, *mutatis mutandis,* for all other styles as well, in all the many branches of the arts.

The invariance of the conceptual framework used in the foregoing assessment of stylistic change in art and paradigm shift in science is evident. A "style" in art is the functional analog of a "paradigm" in science. Both are construct-sets, lending meaning to experience and coding conative responses to it. (In science, however, the construct-sets are abstract: they refer *beyond* the plane of perceptual experience; while in art they are concrete, being mainly an evaluation and interpretation of events *on* that plane.) The world of perceptual experience is constructed as the "natural universe" in science, and as a meaningful, "felt reality" in art. In both cases the construction proceeds by purposively acting in regard to one's perceptions and observing the difference in the resulting percept-patterns. Insofar as the sets of constructs pre-exist and control the actual experience, the latter is negative-feedback controlled and self-stabilizing. Purposive responses are produced thus, that the resulting changes in perception converge upon a match with the constructs. A theory within the paradigm is confirmed by the match in science, and a work within the given style is created and appreciated in art. The reality-constructions of art and science are hereby reinforced: the natural universe mapped by the theory acquires an additional degree of reality for the scientist, and the kind of world suggested by the style to which the art-work belongs gains additional degrees of cogency and conviction for the artist. But patterns of experience in a changing societal and cultural setting can become modified, and styles bounded by fixed techniques and principles of expression can find themselves left behind. Much like scientific hypotheses, artistic styles can lose their validity when new patterns of experience supervene over the old ones. Thus a style which incorporated adequate aesthetic constructs at one time can find itself working with forms and techniques which, at a later time, appear inadequate to many artists. At such times, stylistic change is called for and is normally initiated.

Drawing our parallel with science in the framework of art as a cognitive discipline, we can say that a style which incorporates adequate aesthetic constructs in an art-oriented sub-group in a culture represents the *paradigm* for that group. The members of the group extend and refine the style, but do not basically revise it. The picture changes when the style is no longer felt by the members of the group to express their personal felt experiences. The conservative "normal-art" members become revolutionary "crisis-art" innovators. They are out searching for a new paradigm—a style which could map with more adequacy and greater faithfulness the patterns of felt experience in their culture. The kind of activity undertaken by the avant-garde has much in common with the activity of scientists during a period of crisis: there is a scramble for new ideas, new ways of expression and new techniques, and in this rather haphazard activity many experiments with new styles are undertaken. In the art of this period, novelty is itself a value, and it can come to be so highly prized that it becomes a fad: it is sought for its own sake. But such "purely experimental" works normally create but a temporary stir; they seldom stand the test of time. Novelty coupled with a basic idea, which grasps in some hitherto unexplored manner what most members of the community feel and are trying to express, is what is required for an art experiment to become a lasting success. If it does, it may lead to the establishment of a new style—a new paradigm which thereafter will be imitated and explored in thousands of versions. Great artists of the avant-garde, such as Picasso,

Schönberg, Wright, and others, evolve from lonely innovators to masters of an accepted paradigm in a flourishing and fashionable movement.

These are highly general schematizations concerning the structure of artistic revolutions, designed to exhibit factors which the processes share with sudden shifts in other areas of systematic human cognition. Nothing is said here about the specific *content* of a style, and why one style is adequate in one cultural setting and not in another. These are issues for art critics and specialized aestheticians to discuss and clarify. The above remarks furnish the framework, however, in which decisions concerning excellence can be made (e.g., as that which grasps and communicates the artist's discovered meaning of the patterns of experience in his culture), and in which subsequent styles in the history of art can be linked in a consistent manner (stylistic dynamics being referred to the adequacy of a given style to match contemporaneous experience in its cultural setting). Stylistic dynamics exhibits *negative* feedback within an adequately functioning paradigmatic style, and *positive* feedback in the phase of exploration which follows wide-spread dissatisfaction with it. In the latter case, deviations from the style are progressively amplified in the search for something new to replace it. When a suitable novel style is found, it becomes accepted, first by the avant-garde, and thereafter by an increasingly numerous public; it then becomes stabilized in the many attempts to perfect and extend it.[41] It is likely to remain dominant as long as it expresses the existential reality of the contemporaneous art public. But with further shifts in patterns of experience (and sometimes merely in the dominant tastes and fashions), the new style itself may become *passé,* and the positive-feedback phase of the cycle becomes operative again. By means of these processes, the art-oriented members of given cultures map into their cognitive structures many experiences which escape commonsense and rational systematization, and can control these experiences by creating works of art which clearly and coherently exhibit the meaningful patterns.

A style which is accepted and admired poses no problems: it is not questioned. Instead of penetrating analyses, its adherents engage in dogmatic assertions concerning it. They tend to look upon those contesting the style as heretics or (what is worse) uneducated boors. That the style and its masters represent the highest embodiment of art is not doubted; debate only centers on whether this or that work truly exemplifies the greatness inherent in the style. Artists and art lovers use the paradigmatic style to

---

[41]As Paluch says, "Each movement in 20th century art has begun as a scandal and ended as part of the establishment—so dada, surrealism, futurism, cubism, abstract expressionism and pop art." Stanley Paluch, "'Meta-Art,'" *The Journal of Value Inquiry,* **5** (1971), 276.

satisfy themselves, and when they discuss it, they discuss its merits, rather than analyze its basic rules and implicit assumptions. This, too, is strikingly similar to the self-assured non-reflective mode of procedure in normal science. But the breakdown of a style, like that of a scientific paradigm, brings with it discussion of a quite different sort. Basic ideals are questioned, dogmatic attitudes are discarded, and questions concerning the very purposes and nature of art are raised. Art works and art styles are confronted with experience as lived and felt by people. The avant-garde, dissatisfied with traditional modes of expression, is thrown back on the rock-bottom question of their *raison d'être* as artists. As scientists re-examine the presuppositions of their discipline, so artists scrutinize the foundational problems of their art. As scientists meet in frequent and intense symposia, so artists congregate in studios and cafés to discuss where they are and where they are going.

There is philosophical analysis in crisis-art as well as in crisis-science. But, unlike in science, the verbal exchanges and commentaries in art show but a poor picture of the basic processes. For, whereas the theories of science can be expressed in ordinary verbal symbols and, failing that, in the symbols of an artificial language, the aesthetic constructs of artists are expressible and communicable only in their works of art. Whatever is said about them will be but a poor approximation of the meanings. Moreover, scientists are required, by the very nature of their rational theoretical enterprise, to be consciously aware of what they are doing, but artists are not: explicit awareness of the creative and appreciative processes tends to interfere with their functioning. Thus the true issues in crisis-art may not be consciously recognized, although what emerges into consciousness in an avant-garde group is symptomatic of them. Correspondingly, what is explicitly said at such times may not *describe* what is actually taking place; it will, however, be an *effect* of it. The real issue is the crisis, due to the inadequacy of a traditionally accepted style to reflect felt reality for the artist, and the consequent demand for new modes of expressing and communicating experience. And the manifestations of this issue include the seemingly haphazard activity of trying out innovations in quick succession, and the simultaneous concern with what art is all about. Thus philosophical analysis in crisis-art is not so much a re-thinking of basic presuppositions, as it is in science, but a *refeeling* of them (if the reader will forgive the neologism). The avant-garde artist not only feels his experience intensely; he is also intensely *aware* of feeling it. He engages in a particular species of self-reflexion: in the "refeeling of his feelings." He is testing them, for he is no longer sure of any mode of creative self-expression. He is prompted to constantly check back on his felt experiences in executing his creative work.

Read remarked in a perceptive passage: "Research, experiment–these words describe the efforts of all the great artists that fall within these seventy years—Millet, Courbet, Manet, Degas, Monet, Pissarro, Renoir, Rodin, Whistler, Seurat, Van Gogh—it is all a persistent effort to correlate art and reality . . . Constable, Cézanne, Picasso—Hegel, Husserl, Heidegger; these names represent parallel movements in the evolution of human experience."[42] The attempted correlation of art with reality through research and experiment is typical of art in periods of rapid stylistic upheaval; it is then that artists seem to resemble phenomenological and existentialist philosophers in their speculations and introspections. Whether artists congregate as flower-children in California hippie live-ins, or whether they meet as bohemians in the cafés and night spots of Paris, they are engaged in trying to understand their experience through a stylistic innovation which they do not quite know, for it does not quite exist. Their search guarantees, however, that, if not next year, then a few years hence, new styles will replace the old, and our culture will continue to be enriched by adequate and satisfying modes of beholding and communicating the ineffable and shifting, but nonetheless lived and meaningful aspects of human experience.

In summary, then, we should note that the epistemology offered on these pages is the epistemology of interacting, highly evolved dynamic open systems. It is based on the principle that cognition is an ongoing process of adaptive interaction in the course of which the human being extends his cognitive loops into the far reaches of the universe. Knowledge, on this view, is neither imposed by the objective world upon the passive receptors of the perceiving subject, nor is it the arbitrary product of human fancy and convention. It is the outcome of man's persistent attempt to lend meaning to his experience by evolving, in slow but accelerating succession, commonsensical gestalts and the manifold constructs of the arts and the sciences. Notwithstanding specific differences between the thus emerging cognitive modes, their common functional context and origin assures sufficient isomorphies to overcome the vexing but specious dualisms of reason and feeling, science and art, and practical common sense and abstract reasoning.

---

[42] Sir Herbert Read, *The Philosophy of Art*, Greenwich, Conn., 1967, p. 27.

Chapter 12

FREEDOM: FRAMEWORK FOR A PHILOSOPHY OF MAN

Contemporary philosophical arguments tend to proceed on the tacit assumption that one can meaningfully discuss the perennial problems which confront all thinking men, equipped with nothing but good common sense and faultless reasoning. Debates on freedom have often taken this tack. Whether man is, or is not, free has been debated as though any reasonable person using proper language and logical arguments could resolve the issue. Unfortunately, the problem involves factors concerning the ties of determination between man and his surroundings which do not lend themselves to overt scrutiny, but require research and experiment to uncover. Realizing this, others (mostly scientists) have made sweeping generalizations from the freedom and determination found in domains of nature less complex than the human, to human freedom itself. Postulations of the freedom of men on the basis of the indeterminacy of subatomic phenomena is one (no doubt an extreme) form of this alternate type of approach.

In this study, the above fallacies will be avoided by setting forth the method hitherto followed. This means that we shall not assume that the problem can be adequately debated solely in the light of the knowledge of a common-sense observer, however faultless his logic may be; and we shall not generalize from types of relationships holding in regard to other domains of nature in assessing the degrees of freedom of man. Rather, we shall make the necessary presuppositions: the "primary presuppositions" that the world exists and that it is open to rational inquiry, and the "secondary presupposition" according to which general types of order characterize this world.[1] Consequently we take man to be a part of the real empirical world, and use a general theory (of systems) as the explanatory framework for coming to reasoned conclusions about him, including his

[1]Cf. Chapter 1, above.

freedom. In this chapter, therefore, we shall explicate: (i) the meaning of human freedom in the light of our general theory of systems, (ii) the adequacy of the thus gained concept of freedom to furnish an explanation of the experience of exercising free choice, and (iii) the compatibility of our concept of individual freedom with the manifest determinateness of social systems.

I. THE SYSTEMS PHILOSOPHICAL CONCEPT OF FREEDOM

Man, on our theory, can be biperspectivally observed and described. When introspected upon, man is a cognitive system, constituted by mind-events. In an external analysis, he is a natural system of physical events. When we consider human freedom, we must explore both these perspectives and then synthesize them in a general concept of man as a psychophysical, natural-cognitive system.*

(i) *Freedom of cognitive systems*

The constitutive mind-events of cognitive systems have been identified as patterns of *percepts,* sets and organizations of *constructs,* and correlated patterns of purposive acts or *conations.*[2] In what, we may ask, consists the freedom of such a system?

Freedom, we shall argue, presupposes a decision among alternative modes of thought and action; freedom, then, is the same as freedom of choice. It resides in the ability of a subject to choose x, when given the alternatives x, y, and z. If the choice of x is not determined by any factor other than the subject himself, he is free in that respect. We shall now consider to what extent this concept of freedom applies to cognitive systems.

Cognitive systems are unable to choose their own percepts and percept-patterns; these are "given" and are assumed to be determined by physical events in their environment. But cognitive systems do possess freedom in *constructing* their given percept-streams. Systems as complex as the human possess multiple sets of constructs in regard to most percept-patterns. They have a *choice* if, for percept-pattern N, constructs x, y, z apply equally. Then the decision, which of these constructs to choose in regard to N, rests entirely with the systems. It is not determined by their environment, which is only responsible for the pattern N.

---

*Since the discussion will henceforth focus on human beings, interpreted as one species of natural-cognitive systems, I shall bow to the male anthropocentrism of current linguistic usage and refer to such systems in the singular as "he."

[2]Chapter 5, "Theory of Cognitive Systems," above.

Consider the percept-pattern which the external analyst would identify as *water*. (Introduction of this physicalist evidence is merely to identify the percept-pattern in unambiguous terms—something that is not possible from the immanent viewpoint.) The percept-pattern could be non-interpretively described as undulating, shimmering light-patterns, possibly with changing colors and outlines. Suppose further that the cognitive system possesses at least three major varieties of constructs which match with this pattern, and that these constructs belong, respectively, to the gestalt, scientific, and aesthetic variety. A genuine choice is possible, then, for the cognitive system: the pattern is undetermined as to which of these matching constructs is used in regard to it. For example, this pattern can be *seen as* the familiar everyday liquid which has the capacity to assuage thirst, to wash off dirt, dissolve soap and salt, and so on. The perception of the pattern as the gestalt *water* implicitly involves seeing the pattern *as* all these things. Thus the immanent observer would describe the cognition of the pattern (shimmering, undulating colors with changing hues and outlines) *as* water: a liquid having the capacity to assuage thirst, wash off dirt, dissolve soap, and so forth. This is a fair account of perceptual cognition in the cognitive system of the writer and, he presumes, may be a fair account in that of anybody else.

Now, the cognitive system can be assumed to have other, different constructs, also matching with the above percept-pattern, and one of these constructs may be the *scientific* concept of water. The construct in question consists of a series of interrelated sub-constructs, which include molecules in the specific combination H_2O, with the possible addition of others (if the perceptual pattern is being had of natural, as opposed to pure water), with a freezing point at 0° C., boiling point at 100° C., and a maximum density of 1,000 grams per c.c. at 4° C. Further sub-constructs entering into the scientific construct *water* could be those of the theories of liquid and solid state physics, thermodynamics, inorganic as well as organic chemistry, and others. Presented with the percept-pattern described above, the cognitive system is free to construct it in such terms, provided that he possesses the constructs in question (i.e., has scientific training).

Finally, we may assume that the system possesses constructs of a still different kind, and that these constructs also match with the percept-pattern. If the system in question is a poet (or a person capable of poetical cognition), he may analyze the pattern in terms of the intense and meaningful feelings evoked in him by water as it figures in a pond on a hazy summer afternoon, on a wind-swept sea coast, in chattering brooks, and so on. Water, in other words, may be an *aesthetic* construct in the system. In that event the pattern will connote an emotive and spontaneous cognition of the existential relations of the system to that part of the environment which he also knows in terms of gestalt (and perhaps also scientific) categories. Now, the percept-pattern which the system identifies as water is a "given": it is determined by the physical environment. However, the choice of the applicable type of construct with which to decode the perceived pattern is not determined by the system's environment, but by the system himself. Since we do not appeal to supernatural determination, there is nothing *outside* the system which could determine this choice. Hence we conclude that the system has the freedom to choose from among its applicable alternative constructs. This freedom is by no means trivial. Cognition is not merely the extraction of meaning from raw data, but also a responding to the data. Each of the constructs mentioned in the above example correlates with different sets of purposive acts. The gestalt-construct is unlikely to correlate with the composition of poems; the science-construct with washing one's hands, and the aesthetic-construct with either one of these. Of course, which particular variety of response is produced depends on the entire set of constructs at the command of the system, including not only constructs of the same species, but all others as well. Except for short periods of time under highly controlled circumstances, people do not operate on the basis of a single species of constructs. A tremendously wide range of responses is available to cognitive systems, and the choice between them is determined by the confluence of all constructs in the systems, i.e., by the total cognitive equipment evolved in them. Moreover, most constructs and correlated conations can also be activated in the *absence* of the matching percept-patterns: cognitive systems are capable of acting on the basis of imaginative representation and abstract thought.

Thus, the immanent observer could conclude that the cognitive system is capable of free choice: when given percept-pattern N, he can choose constructs x or y or z . . . , and activate the correlated and jointly determined response patterns x′ or y′ or z′ . . . ; furthermore, he can represent these constructs and reactivate the conations in the *absence* of N. Consequently, the immanent observer is entitled to state that the system has freedom. The latter is free to choose and activate his constructs and response patterns, both in the presence and, through imagination and abstraction also in the actual absence, of given percept-patterns, being limited only in regard to the actually "given" percepts. And even "given" patterns can be manipulated by producing the relevant purposive acts. Only how the patterns modify under such conditions is beyond the system's control—that, it seems, is determined by the environment. However, the immanent observer could hardly argue that this fact robs his concept of freedom of meaning. Freedom of the kind where all events are self-determined by the system's choice is appropriate only to God—and to dreams. (Although there is still some question whether *all* events could be determined either by a God or a dreamer.) Agents of the human variety can be free in an environment of constant and predictable modifiability, as long as they are free to interpret and act in relation to it.

(ii) *Freedom of natural systems*

Natural systems are systems of physical events constituted by energy transfers and processing. Their input is channeled through a complex structure, defined by state-variables, and issues in a response that reacts on the environment. The human natural system consists of a large and flexible dynamic structure which permits alternate processings of most varieties of sensory inputs. The number of cells, and the associations between them, is so large that considerable redundancy is created: the same stimulus, reaching the system via the exteroceptor organs, can be channeled in many different ways, and can activate a large number of different behavior patterns. An almost infinite variety of responses can be produced to any given stimulus by a system of the order of complexity of the human brain. Walter has calculated (and has demonstrated with the help of artificial models) that there are seven possible ways of behavior for a creature with only two "brain cells" or interconnected elements in its decision center.[3] With six such units there are enough modes of behavior to provide a new experience every tenth of a second throughout a long lifetime. The formula for behavioral variation is approximately $M=2^{(n^2-n)}$, where M is the possible modes of behavior and n the number of elements in the decision center (brain). The number of behavior patterns available to man is truly astronomical. If we assume that the number of active elements is of the order of 1,000 ensembles of homologous neurons capable of dynamic combinations and permutations, the number of possible behavior patterns is of the order of $10^{3,000,000}$. This truly fantastic number refers to the "degrees of freedom" of the human brain in regard to the signals fed into it. It defines the freedom of the human being with respect to his empirical experience. No wonder that scientists conclude that there is no physiological limit to the power of association: "for an animal such as man, anything that affects the nervous system can come to 'mean' anything else."[4] Of course, the things which it actually (as opposed to theoretically) *can* mean depend on the kind and number of circuits evolved by an individual in his higher nervous centers, through which he decodes his perceptual inputs. The greater the redundancy in the neural organization regarding the habitual patterns of per-

---

[3]W. Grey Walter, *The Living Brain, op. cit.*, Chapter 5.
[4]*Ibid.*, Ch. 6.

ception, the greater the system's freedom to choose his own pathway in processing the signals and coordinating responses with it.

Neural organization includes circuits which can be autonomously activated, in the absence of signals conduced from the exteroceptors. These circuits are complex phase sequences (Hebb) with relative independence built into them. Due to his ability to activate these circuits of neural nets autonomously, the human natural system gains an additional degree of freedom—the freedom to process information generated by himself.

Natural systems, the external observer can conclude, are adaptive systems, which evolve complex organizations as a function of rendering themselves compatible with the kind of world in which they find themselves. Some of these organizations can, in the course of the systems' existence, run free, exploring the systems' own considerable degree of freedom. However, the degrees of freedom inherent in human systems are those acquired in phylo- and ontogenetic adaptations to the environment. Hence, their actual freedom is not total, but relative to the history of their environmental adaptations. This, the external analyst might conclude, may or may not deprive the concept of human freedom of meaning. But it is the best he can do in regard to human beings: open natural systems which require controlled transactions with their environment, and command elaborate organizations which, even when they run free, represent functional solutions to the problems of open-state self-maintenance. (This holds true, according to Piaget, even for abstract logico-mathematical operations.)

(iii) *Freedom of natural-cognitive systems*

Two different reports have been obtained of the freedom of biperspectival systems. Viewed "from the inside" *qua* cognitive systems, they are said to be free in the choice and activation of their constructs and responses. Inspected "from the outside" *qua* natural systems, they are free to process information and to respond to it within the degrees of freedom acquired in the course of their history as adaptive open-systems. Now, if the here exposed general theory of natural and cognitive systems is empirically sound, then the two reports refer to one and the same system, seen from different vantage points.

What a given observer finds, however, is determined not only by the objective state of the system, but also by the amount of information available at the given vantage point. The immanent observer has access to *percepts,* cognitive *constructs,* and *conations.* The external observer has access to *input,* neurophysiological *organization, output,* and *environment.* These correlate as follows:

Cognitive system (mind-events)		Natural system (physical events)
Percepts	←→	Segment of *Input* (neural signals which, following high-level integrations, emerge into the cortex and are there available for analysis)
Constructs	←→	Segment of neural *Organization* (cell-assemblies in the higher cognitive centers of the cortex)
Conations	←→	Segment of *Output* (efferent signals fed back to the cortical centers through proprioceptors)
———	←→	*Environment*

Thus the two observers may attribute some of the differences in their reports to the different accessibility of information at their points of observation.

They might agree on the following. It is true in both perspectives that the system is free to choose the way he processes the information received of his environment. Here the system is limited only by the redundancy of his own neural organization (=construct-sets). It is also true in both perspectives that the system is free to act on the basis of his freely chosen routes of information processing. The choice of response is determined only by the organization of the nervous system (the set of all constructs). It is likewise true in both perspectives that the inputs (=percepts) are modified by the responses and thus that, from the system's viewpoint, the actual input (=percept-patterns) are co-determined by the system and the environment. However, the "physical environment" is an assumption for the immanent observer while it is a reality for the external analyst. Finally, it is true that the system can run autonomously, in the absence of actual inputs (percepts). The system is free in all these respects, and to the extents indicated.

Then, however, the external analyst can point out that the neural organization of the system (a segment of which correlates with construct-sets) is a product of the interactions of system and environment in the system's history. Thus the degrees of freedom of the system, in all his choices, is limited by the kind of organization warranted by his past existence in the environment. In this sense, the system can only choose that which his past interactions with the environment has made available to him. Reinterpreted in this way, constructs appear as historical products of natural selection and empirical learning, coded into the circuits of the developing nervous system.

The internal analyst is unable to offer an opinion on these arguments, since he has no access to the relevant information. He can only point out that, from his vantage point, the system acts freely. It is not determined by what he takes to be the physical environment—namely the unperceived *source* of the percepts. Thus the immanent observer concludes that there is free choice within the system. *As far as he can know,* there are no external determinants of the system's choices of meanings and purposive responses.

The external analyst can cogently reply that the freedom of choice observed by the introspective analyst is fully explicable within his physicalist framework. Due to the immanent observer's lack of information concerning the historical genesis of the system's constructs and of the correlated conative patterns, the degrees of freedom possessed by the system appear self-determined. The external analyst, however, has information which shows these degrees of freedom to be products of prior interaction of system and environment. Thus, while he maintains that the system is not self-determined but co-determined with his environment, he can also explain, and even predict, that the immanent observer will report that the system has what appears to be an entirely self-determined "free" choice. And to this the immanent observer, with his more limited fund of information, has no reply.

There is no contradiction in asserting that an immanent observer reports free choice, and an external analyst reports interdetermined choice, *in regard to the same system.* There would be no contradiction in this even if the external analyst reported strict causal determination. MacKay made this clear in his "Principle of Logical Indeterminacy." He supposes that neurophysiology is so far advanced that all workings of the brain could be made available to observation, and that the brain is as mechanical as a clockwork. He then argues that even in this hypothetical case, "the denial of human freedom in general would be not only unfounded but demonstrably mistaken."[5] The mistake would consist in disregarding the fact that a change must take place in the physical brain (according to mechanistic brain theory itself) when knowledge is acquired by the human being. The position hypothesized by MacKay holds that every cognitive state is one-one correlated with every brain-state (i.e., that there is a rigorous isomorphy between the sequences of brain-events and mind-events). Therefore, what a person believes (cognizes or holds to be true), will make a difference to

[5]Donald M. MacKay, *Freedom of Action in a Mechanistic Universe,* Cambridge 1967.

his brain-state. If the person believes, at time t_1, that he is facing true alternatives and has not made a decision between them, then this state of his mind is accurately paralleled in the state of his brain. The fact that an external analyst, observing this brain-state, perceives deterministic factors which enable him to *predict* what choice the person will make at time t_2, does not render the subject's report of indeterminacy at t_1 invalid. In fact, the opposite is the case: if the subject at t_1 would believe that he is predetermined in his choice at t_2, his brain-state would be different from what it actually is. Thus any attempt on the part of the external analyst to convince the immanent observer that what the latter believes is a free choice is in fact predetermined by causal factors makes the analyst's explanation self-falsifying. It brings about brain-states in the subject which no longer correspond to what he observes.

The Principle of Logical Indeterminacy advocated by MacKay shows that it is entirely possible for the immanent observer to conclude that the cognitive system observed by him is free, and at the same time for the external analyst to report on even the strictest form of causal determination in the *natural* system. The amount and type of information available to the former justifies an inference to self-determination and freedom, without contradicting the assertion that the processes in question are entirely mechanistic when viewed by an outside observer. Now, the here presented theory does not warrant the assumption of clockwork determinism. Rather, it postulates a complex interdeterminacy, which renders every choice made by the system analyzable into factors determined by past interactions with the environment. Since in such a system brain-events and mind-events are biperspectival correlates, reports of free mind-events do not contradict reports of interdetermined brain-events. In fact, if the mind-events were found to be deterministic, we would be dealing with a different system, and the brain-event reports would have to be re-examined. The system is such that his mind-events appear free when introspectively viewed. That they correlate with interdetermined brain-events does not contradict the manifestation of this freedom.

II. THE ADEQUACY OF THE SYSTEMS PHILOSOPHICAL CONCEPT OF FREEDOM

We conclude that *interdetermined natural systems correlate with self-determined cognitive systems in biperspectival natural-cognitive systems.* The latter are interdetermined, although to themselves they appear self-determined. Does this concept do justice to the experience of human freedom?

To answer this question, I shall review four principal types of theories advanced to account for human freedom of choice. The first holds choices to be externally determined; the second to be entirely undetermined; the third to be determined by some factor incorporated in the agent (his body or mind); and the fourth, agreeing with the here advocated theory, views choices as self-determined within a range of alternatives evolved in the past history of the environmentally interacting agent.

First, we have the strict ("hard" or "mechanistic") determinist. He assumes that every choice by the system is determined by a confluence of factors in the environment. The system himself is but a part in a clockwork universe, and has no choices available to him. He must obey the laws and principles of the mechanistic universe. (Introspective reports of having made free choices do not contradict this position, as we have seen.) Would such a system possess freedom in any meaningful sense? The answer is, *no.* Such a system cannot be regarded a free agent, and be held responsible for his actions since, whether known to him or not, he was determined to act as he did. External compulsion is the case, and it precludes attributing freedom to the system in any but an illusory sense.

Second, we consider the directly contrary viewpoint, which holds the system entirely free to choose among alternative courses of action. This position denies that there are any causal and deterministic factors responsible for the choice: it is an exercise of "absolute free will." We inquire, then, whether or not the free will may incorporate in itself a principle on the basis of which the choice is effected. If this is denied, we get, at time t_1, an equal probability that x, y, or z, . . . will be chosen at t_2. If so, however, the choice is random. Assigning numbers to the possible choices, we could substitute a random-number generator in place of the "free will." Statistically, the outcome should be the same whether we did or not. Because if there is any noticeable difference between the distribution of probabilities with which the free will made the decision, and that produced by the random-number generator, the free will would incorporate a factor which makes it determinate in regard to certain (more probable) choices. Thus the adherents of the "absolute free will" view must choose between randomness in its operation, or some degree of determination. If randomness is opted for, we may leave the argument where it is: such a concept of freedom is appropriate to carefully randomized games of dice, and to devices such as random-number generators; not however to human beings who, at the very least, show *some* purposiveness in *some* of their actions.

Third, we are left with the position that the choice effected by the system is *self*-determined. We can subdivide this hypothesis according to whether the system is self-determined in virtue of (i) the "body," or (ii) the "mind." Some evidence needs to be marshalled to make each of these assumptions hold water.

(i) With respect to the assertion that the body incorporates all factors determining the action of the system, it needs to be shown that the body is free from dominant influence–or, at least, that such influence can always be eliminated. It is not easy to see how this could be accomplished on the level of the human organism, since such things as breathing pollen and sneezing would have to be attributed to the free choice of the organism; i.e., one would have to assume that the organism *chose* to sneeze, and was not *caused* to do so. The myriad ties which bind the organism to its living environment, including bacteria, sexual mate, social companions and the substances necessary for the sustenance of life, would have to be taken as presenting undetermined alternatives among which the organism is free to choose. This assertion could hardly be upheld in the face of modern biology, medicine, and individual and social psychology. Compulsions do seem to act on the organism from the outside, and to determine at least some of the resulting behavior.

(ii) The argument is more sophisticated when the factors of indeterminacy are ascribed to man's mind or spirit. Potent arguments could be marshalled in support of the "sovereign mind" hypothesis. It could be said that the mind is not identical (or even correlative) with the brain, but that there is a sovereign will or other principle which directs and controls mental processes. In this sense, mind is not reducible to the organism but remains somehow encased in it, a separate entity. One-way causality would have to be postulated : from sovereign mind to body, but not from body to sovereign mind. There is nothing which conflicts with such a concept in experience. Scientific experience is presumably of the body, and psychological evidence, particularly that gained through introspection, is of the (presumably sovereign) mind. When I experience my own volition, I experience my sovereign mind as it exercises its free (externally undetermined) will. The effects of the will are transmitted to the body and, through the body, to the external environment. I can thus be entirely responsible for my actions—they were not in the least determined by an external agency.

The argument furnishes an explanation of freedom. The question is, however, on what basis? What, precisely, are the grounds for such an argument? A large segment of contemporary philosophers and humanists might reply, "the experience of free choice." Having the power to decide for ourselves what to do is such a universal experience that many people would never question its truth. When options are presented to us, it seems perfectly clear and evident that the outcome is undetermined until we make up our minds how to determine it. The "sovereign mind" principle gives a plausible account of this self-determination. The choice is not random, but based on principles operative within each of us. These principles are undetermined by any outside factor, thus the choice is an entirely responsible one. That such choices are in fact made by all normal people is the assertion of the argument. And that they are evident in introspection appears to be the empirical ground for it.

In the history of philosophy, many qualified answers have been given to the question concerning the nature of this irreducible mental principle. Descartes thought of it as a mental substance, contrasted with the mechanistic universe of material substance. Kant identified it as the *a priori* mental structure making up transcendental apperception and imposing logical constraints upon our knowledge of the external world. I shall give in more detail one example which, coming from a philosopher who in almost all other respects pioneered the kind of systemic thinking upon which this book is based, shows particularly vividly the power of the free-will experience to affect theory formulation. The reference is to Whitehead, in particular to his notion of God as the originative source of human liberty.

God and the quasi-Platonic "eternal objects" of Whitehead's metaphysics fulfill several functions, and I am not suggesting that their warrant lies entirely in the explanation of the experience of human freedom. But these categories were *also* offered to fulfill that function. Consider the following passage from *Process and Reality:*

> in the case of those actualities whose immediate experience is most completely open to us, namely, human beings, the final decision of the immediate subject-superject, constituting the ultimate modification of subjective aim, is the foundation of our experience of responsibility, of approbation or of disapprobation, of self-approval or of self-reproach, of freedom, of emphasis. This element in experience is too large to be put aside merely as misconstruction.[6]

The experience of oneself as determining his own destiny, and the correlative judgments of approval of oneself and of the chosen or rejected alternatives are, Whitehead says, factors in experience which are too evident "to be put aside merely as misconstruction." Whitehead does not do so; he postulates the doctrine of *creativity* as their explanation. Freedom lies in eliciting the relevance of eternal objects in the synthesis of the data of actual experience. The basic entities of his metaphysics, "actual entities" (or "occasions") import these Platonic forms from the realm of non-actual reality—the domain of real, immutable, eternal objects. This is also the domain constituting the "primordial nature of God." The capacity of actual

---

[6]*Process and Reality, op. cit.*, p. 74.

entities to elicit the relevance of these eternal objects is proportional to the coordination of their components in an integrated totality, which defines the degree of the entity's "subjective intensity." "Each occasion exhibits its measure of creative emphasis in proportion to its measure of subjective intensity." Creative emphasis is negligible in the case of entities of slight subjective intensity—these relatively unstructured entities receive and transmit the data of the external world, and do not put the stamp of their own creativity on them. Nevertheless, each entity performs a synthesis of all data received from its contemporaneous external world by means of objectivifying them with the "forms of definiteness" represented by eternal objects. Thus "each concrescence is to be referred to a definite free initiation and a definite free conclusion." The human being, having a very large degree of subjective intensity, has the capacity to occasion the "ingression" of highly complex and varied eternal objects. Man is free to objectify his data, transmitted in his actual world, through manifold concepts, feelings and ideas. Herein lies his freedom.

Now, as the discussion in Section I of this Chapter has made perfectly clear, I am in full agreement with Whitehead on the last named point. However, Whitehead postulates a Platonic realm of God and eternal objects with the one-way causal connection to actuality mentioned earlier : eternal objects can ingress in actuality and thus qualify its course, but actuality does not affect them. (An exception is Whitehead's conception of the "consequent nature of God," which exhibits the traces of the difference the ingression of eternal objects has made in the creative evolution of the actual world.) In regard to eternal objects, it is made abundantly clear that they are "externally related" to actuality, i.e., are themselves immutable and unqualified by their exemplifications in the actual world. In the freedom of actual entities to "conceptually prehend" eternal objects we encounter the perfect example of an introspective sensation of freedom hypostatizing an irreducible principle of transcendent reality.

The question is, whether to account for this element of human experience demands such a frankly speculative principle. The decision to adopt an explanation is not merely a decision to explain a selected item of observation or experience. If that alone would guide our selection of explanations, *ad hoc* theories and principles would multiply beyond count. Hypotheses which adopt cosmological principles as means of accounting for introspective evidence are particularly prone to error: the danger is very real that they end up with "mind-forces" and "love-principles" as furnishings of the objective world. An immutable realm of externally related Platonic "eternal objects" is another principle which one should hesitate to adopt purely as an explanation of the introspective sensation of freedom (I do not suggest, of course, that this was Whitehead's *only* ground for eternal objects. But it was, I submit, one of them.) If explanatory principles are available which could account for the sensation of freedom without at the same time taking us beyond the bounds of the actual world, we should explore them and, if they prove to be consistent with the facts, adopt them in preference. This leads us to the next and last type of theory accounting for freedom:

Fourth, we take the position that factors of self-determination traced to the past interactions of system and environment adequately account for the system's actual freedom of choice. The theoretical framework capable of accounting for freedom of choice in an environmentally interacting entity was outlined by the writer in an earlier work. The relevant passage follows:[7]

> Fully deterministic and fully libertarian viewpoints must measure up against absolute standards of freedom. Freedom implies the independence of a subject from all forms of interdetermination, and determinism implies the full absence of the potential of autodetermination. Both views taken in this absolute sense are without justification. Freedom, like determination, depends on degrees and not on absolutes . . .
>
> We can rationalize the issue by means of the law of causality. This law, we maintain, has universal validity and every particular is determined by its application. Now interdetermination implies a dual relation between cause and effect: a cause producing an effect and the effect, acting as cause, resulting in a further effect. A determines B and B determines A. A in this example is the general (the *totality* of the cosmic substance) and B is the particular (the particle, or unit of substance, upon which the general acts).[8] Therefore all aspects of the cosmos add up to a specific effect which is manifest in B as some modification of its previous state of being. Since B is in constant effective relation with A, a specific cause originating from A produces a specific corresponding alteration upon the relevance of B to A. Therefore the reciprocity of the causality connecting A and B consists in this: as a result of a cause emanating from A, B manifests a modification in its relations to A, which modification *itself* can be regarded as the cause produced by B, acting on A, and resulting from the effect of the primal cause (A acting on B). Hence every cause gives rise to an effect and every effect in turn acts as cause. Consider now that the causal principle of a specific effect may be manifested in different forms: acting as cause we can have

[7]Ervin Laszlo, *Essential Society: An Ontological Reconstruction*, The Hague, 1963, pp. 100–104 ("Freedom").

[8]The arbitrariness of the scheme consists in the fact that we must abstract B (the subject) from the cosmic sphere of interdetermination and scrutinize it in abstraction. [Footnote in the quoted text.]

mechanical impulses, chemical agents, etc., up to, and including events which are apprehended by the subject (i.e. perceived either directly through the senses or conceptually through ideational processes). Each of these different means of producing an effect qualifies for the predicate *causal,* since it originates outside the subject and produces a definite effect upon the latter. However, the subject also acts as causal agent with respect to the rest of the world. As a result it is wrong to say that the sum total of external causal factors determines the subject fully and absolutely; in fact, the sum total of all external factors determines the subject *in conjunction* with its own being.

How can a subject effectively determine itself in an interdeterminational relationship? The answer must be, through the modification of the prime cause in the reciprocal cause, i.e. by qualifying the original impetus into a specific reciprocal cause corresponding to the exigencies of its own inner structure. The reciprocal cause will then be more than a direct, unconditioned transmission of the prime cause; it will signify a measure of the interrelation's active determination on the part of the subject. Consequently we must hold that the inner, structural organization of the subject determines the degree of its autonomy.

A concept of the universe as an interdetermined network of mutually qualifying causes and effects assigns freedom to particular entities in processing their inputs ("prime causes") and producing outputs ("reciprocal causes"). The more factors of interdetermination the entity has internalized, i.e., the more it is in control of the sphere of the universe wherein it finds itself, the freer it is. Man, endowed with a complex nervous system with practically unlimited possibilities of associating inputs and deriving meaning- and response-patterns from them, is the freest entity known to him. The instrument of his freedom is his brain-mind, which lets him choose, from among a vast number of alternative modes of information-processing, the one he wishes. And he can reflect on his own processes of handling the information and producing the responses, and render himself partially independent of his environment. Man can both know his world, and know that he knows his world. He can act or not act on the basis of the levels and varieties of his cognitions. If "truth" is interpreted as man's manifold cognitions of the world around him, then, indeed, truth sets man free. The more he knows, the more freedom of choice he has, and the more he knows that he knows, the more he can make purposive use of his freedom.

Freedom for man is thus freedom of his brain-mind. More exactly, it is the freedom of a decision center which has true alternatives at its disposal and can choose between them in the light of its own principles. Thus the present (fourth) position rejoins the foregoing (third) one in viewing the mind as the agency of freedom. But this freedom is not attributed to some transcendent and naturalistically undetermined principle, but to effective knowledge eliciting true alternatives of choice. And such knowledge is held to be historically acquired in real, interdetermined transactions between the agent and the world around him.

If the here outlined fourth position concerning freedom is a meaningful one, there is no good reason for defending the "sovereign mind" principle. Libertarians resort to such principles because they are dissatisfied with accounts based on immanent causation, connecting the will as agency to the objects of its decision. Consequently they end up with postulating "transempirical" entities, such as sovereign minds acting as uncaused causes. (In a penetrating analysis, Tennessen has shown this conception to prevail not only among metaphysicians, but also in the views of analytic philosophers, e.g., in L. J. Russell.[9]) The notion of a "transempirical ego" endowed with a "transempirical power center" underlies arguments defending the indeterministic position, holding freedom of choice to be the occurrence of an uncaused decision. But why resort to transempirical realities, whether of logical-analytic, or of metaphysical, varieties, when concepts of the empirical world of real interdeterminacy are sufficient to account for the facts in question? Consciousness need not contain transempirical power sources, or be constituted as a "sovereign power center" in order to provide the ground for choices which can be felt as self-determined and uncaused by external agency. We can view consciousness as constituted by items of empirical cognition (representations of environmental states), and items of reflective cognition (an ascending series of representations of representations of environmental states), and assess free choice as given by decisions based upon the variety and combinativity of these cognitions. The more empirical cognitions there are, the more choices are present on the level of reflective cognitions. And the more items of reflective cognitions are given, the greater the possibilities of choice on meta-level reflections.

Natural-cognitive systems define their own degrees of freedom. The more you know, the freer you are to decide to act or not to act, and the more you know that you know, the freer you are to think about your own knowledge and to act or not to act on that basis. All that is required for freedom and responsibility is that one should have choice, in the sense that one could have done otherwise than he did. (The usual analytic claim concerning the logic of responsibility is: "If X is responsible for A, then it must be possible that X could have done other than A.") And when a

---

[9]H. Tennessen, "On Free Agents and Causality," *The Journal of Value Inquiry*, V (1970). The paper referred to is L. J. Russell's "Ought Implies Can," *Proc. Arist. Soc.*, **36.**

man acts on the basis of his empirical and meta-level reflective cognitions, he could always have acted otherwise than he did, for his constructions and cognitions of his environment are not dictated by his environment, but by his present cognitive (=cortical) organization, which evolved over a long series of past interactions in phylo- as well as ontogenesis. The choices of the human agent are built into his present psychophysical organization, and that organization, reflecting the outcome of past interactions, has sufficient autonomy built into it to give rise to, and (when known) to provide an explanation for, the most indubitable sensations of having made self-determined, free choices.

III. INDIVIDUAL FREEDOM VS. SOCIAL DETERMINATION

The conclusion, that a man can act or not act depending on his own autonomous decision, that he himself determines his behavior, rather than being determined by his environment, derives from a consideration of freedom from the viewpoint of the individual. When we shift the level of analysis to the social system, the single individual becomes a coacting subsystem, and his autonomy appears to conflict with the functional determinacy of the social system. Freedom for the individual may mean randomness for his society, and we broach the familiar dilemma of *freedom vs. organization.*

I believe that it can be shown that there is no necessary incompatibility between individual freedom and determinate social organization. The argument here must rest, for the time being, on conceptual premises: actual experience in these matters is too indefinite, unrepeatable, and open to conflicting interpretation to be of much use. But conceptual premises should not be underestimated : if we hold with the possibility of a general theory of systems, we must allow that observations of dynamic patterns in one realm may have significant applications in other areas, where similar observations are *de facto* unfeasible. Now, the relevant theoretical principle in regard to the present problem is the "holon-property" of systems. In Koestler's words, "the single individual constitutes the apex of the organismic hierarchy, and at the same time the lowest unit of the social hierarchy. . . . No man is an island, he is a holon."[10]

Systems on the most diverse levels of the microhierarchy turn out to possess holon-properties: atoms, cells, biological organisms, and even social systems. There is no *a priori* reason to believe that man would constitute an exception. And if he *is* a holon, then the conditions which satisfy his

---

[10]A. Koestler, "The Tree and the Candle," *Unity Through Diversity, op. cit.*

systemic requirements are compatible with those which render his social behavior functional and determinate in regard to his superordinate social system. This, at any rate, is the hypothesis. Not being able to appeal to decisive empirical evidence either for or against it, I shall attempt to show that, at least in principle, the hypothesis can be maintained without contradictions.

Our task is to propose a model whereby the constraints introduced by the systemic connectedness of the social system can be examined in regard to the autonomy and self-determined freedom of the individual. I believe that the conceptual model most suited to meet this requirement is that which considers the degree of organization of the social system against the degree of communication of its component subsystems. If subsystem communication is reduced to zero, the whole system has zero level of organization; i.e., we are dealing with a limiting case in which the whole system ceases to be a system and becomes a heap of independent components. At that point the whole has no properties over and above the sum of its parts; the distribution of the latter is random and the state of the whole is thermodynamically the most probable one. Now, if communication between the subsystems is intensified to some conceptually determined maximum, the organization of the whole system reaches optimum levels: the system is negentropic and disposes over non-summative system-properties of its own. Thus the entropy of the system and the communication of the parts (or the negentropy of the set of alternatives open to the parts) is inversely related: the more entropic the system, the less communication there is among the parts, and *vice versa.*

What does this model tell us about actual conditions pertinent to the intrinsic freedom of human beings? It says, first of all, that the more highly organized the social system in which the individual finds himself, the more he is in communication with his fellows in that system. "Communication" is used here in the specific sense in which engineers use it to define the level of coaction in a system: it means the determination of the behavior of the entity by the messages it receives. Communication transmits messages (or "information") and being "in-formed" means being made determinate in behavior. (According to Quastler, for example, "information" means definiteness, form, order, lawfulness, specificity, as contrasted with disorder, randomness, uncertainty and generality.[11]) As a consequence, the more two or more components communicate, the more information they pass to one another, and thus the more they determine each other. This is precisely the sense in which Thayer uses the concepts 'com-

[11]H. Quastler, *Information Theory in Biology,* Urbana, 1953.

munication' and 'information.' "To the extent that any living system—such as a human—is accurately and completely communicated-to by its environment, that living system is perfectly controlled by that environment. . . . For example, if there were but two persons on earth, and their intercommunication was both fully accurate and complete, those two persons would be perfectly in control of each other, or perfectly controlling each other (neglecting other physical aspects of their environment for the moment)." Thayer concludes, " 'Effective' communication is thus indistinguishable from 'effective' (satisfactory degrees of) manipulation and control."[12]

Thus if in highly organized societies persons are highly in-formed by one another through large channels of communication, then they are made determinate to the corresponding extent. Human beings in such societies are "controlled" or "manipulated" by the social system and its institutional structure. They would appear to become increasingly like Skinner's rats and Weber's "bureaucrats"—mere cogs in the machine.

The here sketched model applies to systems of oscillators, capacitors, and similar clearly "in-formable" subsystems. Does it apply, however, also to human beings? Here I believe that some modifications become necessary.

Human beings are not constituted as S-R arcs—there is no simple linear causality between their inputs and outputs. Rather, their inputs impact upon a highly structured internal organization, and the output is determined as a processing of the input by that organization, i.e., as some modification of the autonomous behavior-pattern of the system.[13] Consequently, information introduced into such a system does not issue in behavior in a linear causal chain, but acts upon the system's organization, and the summation of the two results in the observable behavioral effect. This is directly contrary to linear-causality input-output systems where inputs determine the output proportionately to the information carried by them.

Let us take a concrete example. A relatively simple machine, such as a typewriter, has a direct input-output correlation: its print-out is determined by which keys are depressed on the keyboard. The legendary monkey who needs 10^n number of years to write out the complete works of

---

[12]Lee Thayer, "Communication—*Sine Qua Non* of the Behavioral Sciences," *Vistas in Science*, D. L. Arm, ed., Albuquerque, 1968.

[13]Cf. von Bertalanffy: "Even without external stimuli, the organism is not a passive but an intrinsically active system. Modern research has shown that autonomous activity of the nervous system, resting in the system itself, is to be considered primary. . . . A stimulus (i.e., a change in external conditions) does not *cause* a process in an otherwise inert system, it only modifies processes in an autonomously active system." "General Theory of Systems: Application to Psychology," *Social Science Information*, VI, 6 (1967).

Shakespeare is introducing relatively little "message" into the machine compared with the amount of "noise." His typing activity is random. However, an expert typist reduces error (noise) to a minimum and produces, through purposive communication with the machine, a highly determinate print-out. Thus simple typewriters are rendered determinate by the typist's communication. This is no longer true in an unqualified sense for teletypes, or machines hooked up with computers. Here the print-out is co-determined by the relevance of the initial communication to the system's internal organization, including its memory stores. The print-out will be largely an outcome of the internal organization of the machine, as the information introduced through the input relates to it. And even such machines do not function as good analogs of complex organic systems. Systems of the latter kind are more like "learning-machines"—they can actually use the input to reorganize their proper structure, reaching increasing levels of differentiation of subsystems and integration of the differentiated components. Their autonomy increases, rather than decreases, as new nets and loops form, alternate pathways develop, and the effect of the information conveyed in the input upon their manifest behavior progressively loses any semblance of linear causality. What the systems do, is determined more by their own organization than by the messages introduced into them.

We can now integrate these various threads of argument and theory into a systems philosophical conception of man in society. We are dealing with systems on different hierarchic levels. Man, as well as the series of social systems of which he is a part, are concrete systems, with the typical systemic properties of ordered wholeness, self-maintenance in stationary states, self-organization to states of higher organization-level, and two-faced holon characteristics. The immediately pertinent fact here is the tendency in both human and social systems to reach higher levels of cybernetic stability. In man, this involves an evolving nervous system with the higher cognitive centers becoming superimposed upon, and hierarchically integrated with, the pre-existing lower ones. In society, it involves the overall differentiation of the institutional structure, despite the multiple and diverse forms it may take, and notwithstanding cases of regression, and partial and uneven development. Thus we get increasingly organized human subsystems in increasingly organized social suprasystems. Our concern is with the *de facto* limitations this situation places upon human freedom.

A highly differentiated social structure imposes correspondingly specific constraints upon the individual role-carrier. These constraints are forms of messages through which the individual is informed of his social setting and is placed in communication with the rest of the social system. If the individual system has a limited learning capacity (due, for example, to physiological limitations on brain size, as in insects), the increasing level of communication becomes a linear behavior-determination: the individual is rendered determinate in his actions. Such determination by environmental constraints evolves through phylogenetic evolution, as inputs take on the form of trigger-stimuli, imprinting patterns, etc., and the internal organization provides various release mechanisms of predetermined form. Particular inputs become associated with particular behavioral outputs: the individual is molded by his environment and is determined by its conditions. Not so, however, for systems endowed with high-level learning capacities. The messages become stimuli for their adaptive self-organization, evolving a largely autonomous internal structure and making the individual less, rather than more, determined by his environment.

Now, this does not mean that the foregoing model is inapplicable and the laws of communication and information theory do not hold here. There is a sense in which the individual is made more determinate by the information communicated to him in a highly differentiated social system; however this determination is not *output determination,* but *decision-making determination.* What is being determined by the information in the individual is the structure of the dynamic processes through which he *handles* the information, and not his *responses* to the information. Hence whereas relatively simple systems become cogs in the superordinate system, complex systems with learning-capacities retain and even evolve their autonomy. They do become more determinate in a highly structured situation, in the sense that their randomness is reduced and their purposiveness increased. But this form of determination involves the decision-making capacity of the systems, and thus it *contributes* to their freedom rather than restricting it.

Contemporary examples are not difficult to find. Western technological societies provide multiple channels of communication between their members. Information is passed between the latter concerning biologically, as well as intellectually, emotionally, aesthetically and theologically pertinent matters. Such societies are typically highly differentiated in their institutional structure, with roles becoming well-defined in scope and pattern. The role-carrying individual is expected to master the requirements of his position and to use his judgment in carrying out his tasks. The more complex the task, the less can it be entrusted to simple S-R sequences of instructions and activities. The individual must discriminate the relevance of the messages communicated to the goals and objectives entrusted to him, and optimize the chances of fulfilling his tasks. In complex situations he is expected to "grow into the job," i.e., learn to handle tasks and read signals which are partially new to him. In fact, every social role carries with it some amount of "training," which is nothing if not learning to discriminate and to make judgments and decisions. (Here rote-learning of purely mechanical skills is disregarded for the moment as untypical of complex social role requirements.) Thus communication in a differentiated technological society functions as a means of determining the decision-making capacity of individuals, rather than their immediate behavior. Behavior-determination is typical of insect societies and other multi-organic systems, where high degrees of plasticity and self-determination are ruled out by evolutionary and physiological factors; in *human* societies, behavior-determination is encountered only on the lowest levels of a hierarchy, where the learning-capacity of individuals is either at a minimum or has not had a chance to develop. (A case in point is the private in the army who is told not to think, but to obey.) However, decision-making determination is the rule on all medium and high strata of human societies, where the learning capacity of the individual role-carrier is a precondition of his effective functioning. Ideally, such determination is effected in the process of higher education, where not patterns of behavior, but patterns of information-handling and decision-making are to be taught. Having been "in-formed," the educated person is more determinate in many respects than the uneducated one: he perceives, and is able to respond to, a wider range of situations; he can assert his individual judgment in regard to many more consciously held goals. Randomness and noise are reduced, purposiveness and signal-activity are increased. Given our definition of individual freedom (as the self-determination of the given person by means of his historically evolved adaptive organization), social differentiation is not anathemic but propaedeutic to individual freedom: it stimulates learning, and therewith widens individual autonomy.[14] We thus get a parallel, mutually reinforcing evolution of differentiated complex systems on the individual and on the social levels. Highly differentiated modern societies tend to stimulate the learning capacities of their members, and highly learned individuals tend to create increasingly complex social roles.

An asymmetry enters our model, however, when we compare the degree of intrinsic freedom of a high-level cognitive individual with the constraints of his social role, which requires him to make only one out of a large number of possible decisions—namely, the "right" one. Whereas the educated person *could* act as royalty, and also as a bank-robber, if he so chose, he is required to act in accordance with his own social role. Social sanctions of the most diverse kind ensure that he does not transgress its limits with impunity. Hence it may be (and in fact often is) argued that

---

[14]This is a point Skinner completely misses in *Beyond Freedom and Dignity* (New York, 1971).

highly organized societies shackle the individual, drastically limiting his intrinsic freedom.

In assessing this problem, we must revert to our conceptual paradigm. If man is a holon, then constraining him to act as a functional subsystem of his society is not a limitation of his *de facto* freedom, for, in possession of evolved decision-making capacities, he could, but *would not*, choose to act otherwise. That which appears to be a social constraint becomes identical with the exercise of an autonomous judgment: the individual wills, by his own decision, to be a functional part of society. What is eliminated in this account is merely what Isaiah Berlin termed the "negative freedoms": the freedom to commit unnecessitated acts.[15] Such acts, however, while theoretically possible for the real human being, signify a system malfunction in practice. Behavioral freedom for man is the freedom to act on the basis of his own decisions, reached through the purposive processing of information in relation to definite fundamental objectives. The latter include the many goals of self-maintenance of conscious open system in complex hierarchical environments. It corresponds to Berlin's "positive freedom" and the "freedom to do the good" of the Hellenic philosophers. It involves the capacity to discriminate, to value, to decide, and to act accordingly. Freedom is self-determination, not random indeterminacy. Hence to be free means to be aware of constraints as well as of possibilities, and to act by weighing all factors to the best of one's ability. It means committing the necessitated act, while having a real choice in rejecting the unnecessitated one.

There need not be any incompatibility between determinate social systems and individual freedom. If man is a holon, i.e., has the capacity to adapt to his environment both as an organic whole and as a societal part, then living in a society, however highly organized, provides him with the structures to which to adapt his motivations and behavior. This is not a passive adaptation, for man is not determined by his environment but by his own decision-making capacities. Thus the well adapted individual *chooses* the mode of action which is functional for the social system, as an expression of his self-determination—the self-determination of a relatively autonomous part within a larger whole. This does not mean that *every* social system is supported by its members as a function of their self-determination. Some systems constrain individual development and randomize social roles, others freeze them into inflexible bureaucratic structures. It is well to keep in mind, however, that a social system which stunts individual freedom is just as much an aberration from systemic norms as any other type of system which reduces the function of its parts below

---

[15]Isaiah Berlin, *Two Concepts of Liberty*, Oxford, 1959.

their maximum potential. But a system of this kind is relatively unstable—whereas the well evolved social system, like every other species of evolved natural system, is cybernetically highly stable. This means that the system is composed of relatively autonomous and differentiated, yet functionally integrated sub-systems. Social systems of this kind stimulate the decision-making capacities of their human members and enable them to evolve into individuals with a full complement of properties, and the freedom to exercise them.

Nature, in building hierarchies, balances subsystems within suprasystems. Man, as part of nature, is an integrated module in this hierarchy. The presence of superordinate systems is not incompatible with his intrinsic degrees of freedom, for he is neither a passive part, nor the arbitrary ruler of his environment, but an integrated and yet highly autonomous component in the intricate web of existence.

Chapter 13

VALUE: FRAMEWORK FOR A NORMATIVE ETHICS

Concepts of "the Good"

Plato posed the question concerning that which always is and has no becoming; and also of that which is always becoming and never is. He replied that the things apprehended by intelligence and reason are always in the same state: they represent Being. That, however, which is conceived by opinion with the help of sensation and without reason, is always in process of generation and perishing: such things are Becoming.[1] The changeable conceived with the help of the senses must have been created by some agency and on some plan, since it comes into being and passes away. The unchanging, on the other hand, is eternal and not subject to creation. The question is, then, on what pattern or model was the changeable created? If the world is indeed fair and the artificer good, Plato held, then it is manifest that he must have looked to that which is eternal. And, having been created in this way, the world has been framed in the likeness of that which is apprehended by reason and mind and is unchangeable. As Being is to Becoming, so truth is to belief.[2]

The unchangeable pattern, apprehended by pure reason rather than by sense, is the Form; and the highest Form (or perhaps the sum of all Forms) is the Good. It is both a cosmological principle and a norm of human behavior. And this principle must be grasped with the intellect, for sensations only convey copies of it, never the original.

The contemporary philosophical temper mistrusts cosmological-ethical principles of this kind; it prefers to analyze the meaning of moral words (such as 'ought,' 'right,' 'good,' 'duty') and the concepts or states of affairs to which these words refer (e.g., wisdom, pleasure, stealing, polygamy, etc.). The case for confining the field of ethics to analytical and logical inquiries, it is often held, is that ethics has usually been considered to be a part of philosophy, and such inquiries are much more akin to inquiries

---

[1]Plato, *Timaeus*, Jowett translation.
[2]*Ibid.*

generally included in philosophy than the normative, cosmological kind exemplified by Plato's argument. Thus logical-analytic ethics stands in much the same relation to morals as does the philosophy of science to science.[3] It is evident, however, that such arguments beg the question: they presuppose the kind of arguments which are "generally included in philosophy" as the premiss which proves the conclusion (that analytic meta-ethics, and not normative moral theory, is true philosophy). The difficulty with logical analytic meta-ethics is that its no doubt rigorous analyses are often deployed to illuminate uninformed statements and attitudes. Meta-ethical philosophers analyze the meaning of words and concepts as they are used in everyday contexts, without seriously facing the possibility that such contexts do not furnish an informed source—one worthy of a painstaking analysis. The classical moralists, nowadays disparaged by meta-ethical philosophers, never derived their prescriptions from what their contemporaries professed as goals and values, even if (as Aristotle) they paid much attention to their explicit statements. Moral theories were the result of theoretical inquiries based on some considered notion of the nature of man and the natural and social world in which he lives. When we consider what odd and misguided notions people often entertain about themselves and their world it becomes highly questionable whether their moral statements constitute the indicated foundation for philosophical ethics. Every empirical science has learned not to analyze ordinary common-sense pronouncements on its subject matter, but rather statements based on careful methodological inquiry. Is it not more reasonable, then, to follow this tack also in ethical theory and derive the meaning of the moral terms 'good,' 'ought,' (etc.) from informed scientific theories concerning man, nature and society, and not from the highly unreliable sources of everyday usage?

The standard objection to a recommendation such as the above is that science does not offer moral statements, only factual ones. And thus science is irrelevant to the area of values: one can never legitimately derive an "ought"-statement from an "is"-statement. This argument, still strong in analytic philosophy, has been discarded (if often tacitly) in most scientific quarters. Since the sciences of cybernetics, and systems and information theory have been born, it has been possible to *describe* mechanisms ("systems") which operate on the basis of programmed goals, which their overt behavior strives to realize. Such mechanisms range from target-seeking anti-aircraft guns and sonar-guided torpedoes, to automatic pilots in aircraft and the sophisticated servomechanisms which sprout forth in

---

[3]Cf., for example, J. O. Urmson, ed., *The Concise Encyclopaedia of Western Philosophy and Philosophers*, London, 1960, p. 136.

ever-increasing numbers in modern technology. These systems are not "value-free"; they incorporate the standards or norms toward which they strive, sometimes in remarkably human-like manner (e.g., *M. speculatrix*, which is endowed with both positive and negative tropisms, discernment, self-recognition, the recognition of others of its species, and internal stability, all as a function of pursuing its goal).[4] But, the objection goes, these "values" are in fact programs built into the systems by human beings, so that it is illegitimate to speak of them as values possessed by the systems themselves. This objection leaves out of account that once a goal has been programmed into a system it is effectively functioning in it and defines its intrinsic value, regardless of the *origin* of the goal in the history of the system. Hence it would seem that even a man-made machine can be objectively described in terms of goals and purposes. But the real reason why the "fact-value gap" is obsolete is that human beings—as indeed all living things (in the present view, *all natural systems*)—are *themselves* goal-directed systems. Their "programs" are known as "genetic codes," "instincts," or "conscious purposes" and the mechanisms whereby they are sought as "homeostasis," "the autonomic nervous system," and the analytic centers of conscious thought in the cortical hemispheres. The above terminology is to exhibit the fact that human beings and servomechanisms are alike *insofar as* they both strive after goals the norms of which are incorporated in their control centers.

This fact has prompted a new literature on values, coming from the science-philosophy interface, and including such men as (originally) R. B. Perry, John Dewey, and (lately) Rollo Handy, Abraham Maslow, Stephen C. Pepper, and the present writer. Pepper points out that these all agree in finding the locus of value in certain activities of the organism in relation to its environment. After pointing out that Handy calls these activities "transactions" (in *Value Theory and the Behavioral Sciences*) and that I call them "feedback circuits" (in *System, Structure and Experience*), Pepper says,

> Purposive behavior furnishes an excellent example of a selective system. Here a dynamic agency (call it a need, a drive, a desire, or an interest, arising either from changes within the organism, like hunger or thirst, or from external stimulation, like a sudden downpour of rain or a nail in your shoe) presents a pattern of tensions with accompanying conditions of satisfaction. In appetitions these acts lead to instrumental and terminal goals and often a consummatory act yielding pleasure. In aversions there are acts of avoidance of objects of apprehension

---

[4] *The Living Brain, op. cit.*, Ch. 5.

> and actual pain terminating when there is relief from these tensions.
>
> Here is a dynamic structure of activities. The structure institutes a norm on the basis of the conditions of satisfaction intrinsic to the specific need or drive or desire. Acts and objects are selected as correct or incorrect in proportion as they serve toward the attainment of the conditions of satisfaction for the dynamics of the purposive structure. It is a selective system. And values of various kinds spread out along the route of these transactions. There are positive and negative conative values—and as goal objects are anticipated these are potential objects of value—and there are also objects of potential value instituted in the environment. There are frustrations and achievements and pains and pleasures, all closely bound up with the intensification or relaxation of tensions in the patterns of the purposive transactions. And in the process of achievement the purposive dynamics selects toward the shortest path.
>
> There is no question that such a selective system institutes values and sets up norms for the good and the bad, the better and the worse, within its range of application.[5]

Maslow goes even further in fusing facts and values. In a paper appropriately entitled "The Fusion of Facts and Values," he proposes the "inclusive generalization, namely that increase in the factiness of facts, of their facty quality, leads simultaneously to increase in the oughty quality of these facts. Factiness generates oughtiness, we might say."[6] This striking statement is backed by analyses from the realm of human personality, and organismic as well as cognitive functioning. For example, Maslow refers to Goldstein's assertion that a damaged organism is not satisfied just to be what it is, namely damaged, but strives, presses and pushes—it fights and struggles with itself in order to make itself into a unity again. It governs itself, makes itself, re-creates itself. Likewise with the cognitive aspects of perception. Perceived "facts" are not static; they are not scalar but vectorial (having not only magnitude but also direction). "Facts don't just lie there, like oatmeal in a bowl; they do all sorts of things. They group themselves, and they complete themselves; an incomplete series 'calls for' a good completion. . . . Poor gestalten make themselves into better gestalten and unnecessarily complex percepts or memories simplify themselves." From these considerations Maslow concludes that "Gestalt and Organismic psychologies are not only is-perceptive but they

---

[5]Stephen C. Pepper, "Survival Value," *Human Values and Natural Science,* Ervin Laszlo and J. B. Wilbur, eds., New York, 1970. (A similar argument is advanced already in Pepper's *Sources of Value,* Berkeley, 1958.)

[6]Abraham H. Maslow, "The Fusion of Facts and Values," *American Journal of Psychoanalysis,* **23** (1963).

are also vector-perceptive . . . instead of being ought-blind as the behaviorisms are, in which organisms only get passively 'done-to,' instead of also 'doing,' 'calling for.' "[7] These dynamic characteristics of facts, "vectorial qualities," fall within the semantic jurisdiction of "value" and bridge the dichotomy between fact and value, conventionally held to be a defining characteristic of science itself.[8]

The above propositions, coming from a psychologist, include reference to "facts" as they are perceived and interpreted by cognizant human beings. But it is the human beings themselves who make the facts dynamic, endowing them with their vectorial character. Thus the fusion of facts and values, described by Maslow, resides in the last analysis in the dynamic vectorial quality of organismic, gestalt-perceiving and cognizing *human beings*. Which is precisely what Pepper, this writer, and those referred to in this chapter, maintain.

When the telic nature of physical and biological (and even certain artificial) systems is acknowledged, and not confused with final-cause teleology and entelechy, descriptions of states in the systems include the renowned moral term *norm*, and descriptions of their functions include such concepts as *goal, purpose*, and *preference*. The way is opened for a return to normative and naturalistic ethics, offering definitions of moral terms in the light of invariant theoretical constructs. We shall review two examples of such ethics, both of which show marked similarities with Platonic moral theory and equally marked dissimilarities with contemporary schools of analytical meta-ethics.

(i) Kenneth Boulding suggests that we normally believe that the evolutionary process carries us not only to more organized systems, but also to "better" systems.[9] Hereby a generally monotonic relationship is implied —which does not, of course, have to be linear—between degree of organization and "goodness." Both organization, and the value coordinate, are clearly vectors. They go together: when organization increases, so does goodness (generally). Boulding, like Spencer, reaffirms the basic long-run optimism of most religious and even secular faiths, that the course of evolution toward higher organization is also a movement toward the "good."[10]

Now, Boulding defines "organization" as an ordered structure capable of behavior and perhaps of growth as well. All organizations are essentially

---

[7] *Ibid.*

[8] *Ibid.*

[9] Kenneth Boulding, "Some Questions on the Measurement and Evaluation of Organization," *op. cit.*

[10] *Ibid.*

"role-structures"—open systems with a throughput of components consisting of lower-level organizations, in which the components are forced by the related roles around them to play a certain role in the organizational structure. Atoms, molecules, cells, organisms, and social structures are examples of "organization." In fact, Boulding holds that conceptualizing organization is the great quest of the sciences: organization is ordered structure, in space and time, and this is what scientists are looking for, whatever their field of investigation may be.

It is clear, therefore, that organization is not a phenomenal characteristic of experienced nature—not something perceived as such through the senses. Open systems with throughput, in which the components play fixed roles under constraint of the total structure, are not things we can *see*. These are scientific constructs, products of a rational mind which formulates the principles and then locates them "in nature" in the context of informed observation. Furthermore, the concept of organization is an invariant, calculable to degrees or levels with the aid of other invariants, such as entropy and information content. But the phenomenal world is variable, and persists in a state of 'blooming, buzzing confusion.' Thus in Boulding's view "goodness" correlates with, and is defined in reference to, an invariant theoretical construct much like Plato's *Forms* (although Boulding does not hold goodness itself to be a Form).

(ii) Consider now a second view of this kind, proposed this time by a biologist. James Miller describes organisms as open systems in steady states and defines the range of stability of the systems as "that range within which the rate of corrections of deviations is minimal or zero, and beyond which correction does occur. Inputs of either energy or information which, by lack or excess of some characteristic, force the variables beyond the range of stability constitute *stresses* and produce *strains* within the system."[11] Strains and stresses, Miller tells us, may or may not be capable of being reduced; this depends on the adjustment resources of the system. The existing strains are called the "values" or "utilities" of the system, and the relative urgency of reducing specific strains represents its "hierarchy of values."[12]

Miller's definition of values also involves an invariant scientific construct: the steady state of the system within its range of stability. What the system (in this case, the living organism) experiences is but a variable set of strains and stresses, which appear with diverse degrees of urgency, depending on their relation to the norm which defines its range of stability.

---

[11]James G. Miller, "The Organization of Life," *Perspectives in Biology and Medicine*, **9** (1965); the relevant passage occurs also in "Living Systems: Basic Concepts," *op. cit.*

[12]*Ibid.*

Values are experienced as variable strains and are defined in reference to the invariant norms of the system.

We get a significant parallel between these contemporary science-based ethics and the classical views of Plato. Goodness is traced to an intellectually discovered principle which is nevertheless thought to be rooted in nature. It is unchanging and hence Plato would characterize it as Being. Experience gives us the many changing things to which goodness applies in differing degrees. These things could be characterized as Becoming. The points of agreement thus extend to naturalism in ethics and rationalism in theory of knowledge. But in some respects contemporary views tend to be more cautious than the classical Platonic approach: Boulding speaks of a *correlation* between goodness and level of organization, and, for Miller, values are defined by strains *correcting toward* a goal or norm, rather than constituting entities or states in themselves.

The view I shall advocate here has points both of agreement and of disagreement with the above-cited systems theoretical value theories. It agrees with them (and also with Plato) in locating the sources of value in nature and in analyzing it through intellectually constructed and empirically applicable concepts. It disagrees with Boulding on the specific point of a generally monotonic correlation of goodness with level of organization, and with Miller in refusing to hold that *normative* values can be defined in reference to strains experienced by a system.[13] In positive terms, I shall consider normative values to be correlates of certain states of a system within the system-environment continuum, defined by the degrees of adaptation of the system to its environment, rather than by its level of organization. Thus I agree with Boulding that such values are correlates of states in organized systems, and with Miller that the capacity of the system to adjust to strains and stresses is a determinant factor in the attribution of values.

The reasons for the above agreements and disagreements will become apparent in showing how transactional states in natural-cognitive systems correlate with "goodness" and possess "value" for the systems. I shall show this by deducing the pertinent theorems from the general theory of systems here under consideration, rather than constructing a value theory on the basis of ordinary language and experience. Thus I shall not present an analytical value theory, but show how one can give empirical meaning to the classical notion "*the Good*" and the modern concept "*value*" in reference to a stated theory of man, nature, and knowledge, based on information provided by the contemporary sciences.

[13]Miller cautions us that he uses *value* in the descriptive, rather than the normative sense, but then suggests that normative values are defined by the consistency of the system's goals with the norms established in the suprasystem. Hence he does offer a normative definition of the strains which define the system's "hierarchy of values."

Values and systemic states: (i) *Cognitive systems*

Cognitive systems are open systems of mind-events, adapting to their environment by bringing about a match between their "cognitive organization" (sets of constructs) and their perceptual input. Using conative (behavioral) patterns as the means of exercising negative-feedback control, the systems restructure their environment to bring about the match of their perceptions and conceptions. Faced with persistent anomalies in the perceptual field, cognitive systems switch to positive-feedback in hypothesizing new sets or combinations of constructs, i.e., evolving their own cognitive organization as a function of obtaining the required match. Through these alternating or parallel processes, cognitive systems adapt themselves to their environment, and adapt their environment to their own form and level of organization. These concepts are familiar from previous chapters and need no further elucidation here. Our present concern is with the question: are there states of cognitive systems which can be said to correlate with "goodness" and "value"?

Being cybernetically self-stabilizing, cognitive systems have an intrinsic norm: adaptation to their environment. Since the state of the physical environment is signalled through percepts, *a state of the system in which its percepts match its constructs is a value-state.* By this I mean that if the cognitive system knows his own norms and reflects on the degree to which they are approximated in his actual situation, he correctly identifies those of his states as the most valued ("good") in which his actual perceptual experience is optimally matched by his construct sets. The match in question is not a momentary or fortuitous coincidence of some constructs in the system and the actual patterns of experience. Rather, the match is a state brought about by the system through self-stabilizing and self-organizing activities representing an adaptation to the *factors* of change governing conditions in the environment. Thus such a match does not ignore the future: it is the outcome of predictive–extrapolative adaptation to the limits of the system's capacities. It is never an absolute match—one that could never be broken—but always an optimum one—one that is as permanent and perfect as the system can make it. But the relativity of the correlated value does not connote its subjectivity: different systems may have different capacities for adapting to their environment, yet the degree of their individual adaptation is reflected in an objective state.

The states with which value correlates for cognitive systems range over a spectrum extending from minimal to optimal value. *Optimal value* correlates with the cognitive organization through which the system can successfully predict and compensate for most of the relevant environmental variables. Such an organization maps conditions in the environment with the highest degree of precision attainable by the level of development of the system. Systems that evolve such optimal organizations have maximum freedom in the high level of redundancy of their construct-sets, and have maximum leeway in tolerating, and purposively correcting for, environmental changes. Such systems are optimally adapted to their environment, and their states of adaptation signify their states of value. States of *less value* are those where experience is not fully matched by cognitive constructs but is within the threshold of adaptability of the system—for example, when the perception of an unfamiliar object or event calls for learning by the system. *Negative values* attach to those states of the cognitive system wherein he is unable to meet the challenge of the environment: the experience remains anomalous—chaotic and unintelligible. Such a state, if prolonged, leads to the disorganization of the cognitive system, as experiments on sensory overload, deprivation and randomization demonstrate.

(ii) *Natural systems*

Natural systems may be open or closed systems. A system equipped with adiabatic walls, which transmit neither thermal heat nor other forms of energy (matter), is a fully closed system. Closed systems are idealizations, however: they do not exist in nature, although they can be relatively closely approximated in experiments. Concrete natural systems of the "closed" variety are special cases of open natural systems; ones in which the inputs and outputs are temporarily zero. There are no concrete natural systems equipped with adiabatic boundaries; their walls are at the most diathermal, allowing the exchange of heat and energy, although preventing that of matter (energies with rest-mass). Such systems have inputs and outputs when energetic conditions in the surroundings exceed the penetrability of their walls. They are adapted to their environments if the thermal radiations in their surroundings do not disorganize their structure but can be assimilated to the fixed forces governing their steady-state parameters. Lastly, there are systems the boundaries of which allow the exchange of both thermal and other forms of energy, including matter. These systems are *open*. They are adapted to their environments only if the input and output is so correlated that their state-variables remain relatively constant.

The question of value-correlation arises for all types of systems, but it is non-trivial only for the relatively open varieties. Fully closed systems, if they would exist in nature, would be either "adapted" to their actual environment, or they would be destroyed by incompatible mechanical forces. But open systems (or relatively closed systems, temporarily opened) can exist in states of various degrees of adaptation. Such systems interact with their environment by means of a throughput of energies in diverse forms and, since they perform work, they necessarily increase their state of entropy. The increase can be balanced, however, by the importation of energies which are usable by the systems to perform work. Open natural systems can impose a gradient on the flow of the energies: their input is more organized (less entropic) than their output. (E.g., food is more organized than waste products; consumer goods than garbage.) Now, open systems are optimally adapted to their environment when they regularly obtain all the free energies from their surroundings which they require for performing work (i.e., to maintain themselves). Adapted open systems can transform energy radiations in the surroundings for their own purposes, namely, for counteracting thermal disorganization coming about with the wear and tear on their components. Systems which can compensate for changing conditions in their environment have a higher level of cybernetic stability than those at the mercy of fortuitous circumstances. Different degrees of value correlate with the various states of adaptation of the systems. *Optimum value* for natural systems correlates with that state of organization through which they are capable of predicting, and compensating for, the greatest variety of relevant environmental changes. Since natural systems are found on diverse rungs of the hierarchy of evolution, their optimally adapted state—and hence their state of value —is relative to their level of development. But, as with cognitive systems, relativity does not connote subjectivity: a system's state of adaptation, while dependent on its level of attained cybernetic stability, is reflected in an objective state. Values are scaled down from the state of optimum value to those which represent less extensive and intensive forms of adaptation of the system to the environment. *Negative value* correlates with such states of the systems wherein they cannot utilize energies available in the environment. In such states their entropy increases and organization decreases. If the states persist, the systems are condemned to disappear.

(iii) *Natural-cognitive systems*

The foregoing analyses of cognitive systems and natural systems do not imply dualism, notwithstanding the distinctness of terminology and the separate handling. As before, the two types of systems can be brought together as different aspects of biperspectival natural-cognitive systems. Thus dualism can be avoided, without telescoping physical events into mental events, or folding mental events into physical events.

The biperspectival reading of the above sections yields the conclusion that the states of natural-cognitive systems correlate with value corresponding to their level of adaptation. Values thus range from an individually determined optimum, correlating with the most thorough adaptation of which the system is capable at his actual stage of development, to the absolute minimum (a negative value), where the disorganization of the system is imminent. Specifically, the state of the *cognitive-system* (i.e., of the natural-cognitive system viewed from the immanent viewpoint of mental events) which correlates with optimum value is a *state of sufficient information input*. The patterns of percepts had by a system have sufficient information-content if they correspond to the system's cognitive organization; in that event the constructs endow the patterns with meaning and they function as representations of conditions in the system's environment. A cognitive state where the percepts match the constructs is one of lucidity and satisfaction. Not merely rational meanings are involved in such a match—aesthetic and religious meanings may also come about. Whatever the nature of the comprehension, it is one that is meaningful for the system, and represents a state of psychological adaptation.

The *physical* aspect of natural-cognitive systems which correlates with optimum value is a *state of sufficient energy import*. Here "energy" refers to free energy utilizable by the natural system to offset his own entropy production. Under such conditions the system effectively counteracts thermal disorganization and, not only does not lose organisation, but may actually gain it (if the dissipation function, ψ, is negative, i.e., less than zero).

Restated in intuitive ordinary-language terms, the "state of sufficient information input" (*cognitive systems*) means that the mind is not confused but operating clearly and to the fullest extent of its manifold capacities. The individual in such a state has access to experiences which correspond to his requirements for rationality, beauty, and transcendental significance; and is neither burdened with paradoxes and incomprehensibilities, nor deprived of patterned experiences of sufficient complexity and regularity. The "state of sufficient energy import" (*natural systems*) means, in turn, that the biological organism of the individual, including his nervous system, is well supplied with the necessary foods and stimulants. The right patterns of energy flow through the appropriate channels; the individual thrives.

The two perspectives can be brought together as follows. *A cognitive system in a state of sufficient information input is identical with a natural system in a state of sufficient energy import*. The differences between them are due to the viewpoint of observation: the cognitive system is described by the introspective observer, and the natural system by an external analyst. When the systems are identified as one biperspectival natural-cognitive system, we find that, in the combined state of "*sufficient information input/sufficient energy import*," the system is healthy and prospering: mentally he is characterized by cognitive satisfactions and physically by the fulfillment of energetic requirements. He attains a state of optimum value in fulfilling all relevant biological, psychological, and sociocultural needs.

Value criteria: adaptation vs. organization

The above theory of value finds a natural norm in adaptation, viewed simultaneously as a mental and a physical state. It differs from the Spencerian type of evolutionary ethics, championed by a number of biologists and social scientists (e.g., Boulding), in that it correlates value with *appropriate,* and not with *complex,* organization. The system in the best adapted state is maximally fitted to the changes and chances of life in his particular environment; he is not necessarily the most highly organized. The fallacy of attributing value to the complexity, rather than to the fitness, of organization is a natural one in man: it makes him the most valuable being on earth. Evolution seems to conduce toward complex organization, and man is the most complexly organized living system. Hence man is the goal of evolution. The syllogism is a simple one, but it is based on specious premises. Their fallacy has been pointedly brought out by Russell. "What might be called the biological theory is derived from a contemplation of evolution. The struggle for existence is supposed to have gradually led to more and more complex organisms culminating (so far) in Man. In this view survival is the supreme end, or rather survival of one's own species. Whatever increases the human population of the globe, if this theory is right, is to count as good and whatever diminishes the population is to count as bad."[14]

The conclusion Russell draws from "complex organization is the good" complements that drawn by Boulding. Integrated, they are to the effect that, since complex organization is the good, and man and the societies of men are the most complex organizations on earth, the good is the greatest number of societies organized to the greatest complexity. The *falsity* of this concept is evident when we consider that unlimited population increase is never the natural resultant of evolution,[15] and its *inadequacy,* when we contemplate the state of affairs that would come about when the largest possible number of people live in the largest possible number of societies of the greatest possible complexity. We could rightly object, with Russell, "what humane person would prefer a large population living in poverty and squalor to a smaller population living happily with a sufficiency of

---

[14]Bertrand Russell, *Authority and the Individual,* Boston, 1949.

[15]See Chapter 14, below, concerning literature on natural population regulation.

comfort?"[16] Now, even if we hold, with Boulding, that the number of surviving individuals is to be limited by the social organization capable of maximum complexity (and thus able to provide all individuals with a sufficiency of comfort), we would still make a speculative (and false) interpretation of the goals of evolution, which are never complexity or numbers *per se.* Processes in nature, I suggest, warrant the assignment of value to the *appropriateness,* and *not* to the *greatest number* of the *most complex* organization. It is possible to find a perspective from which the decrease in the level of complexity (measured in entropy, or bits of information) of populations of natural entities appears to be of value, and it is possible to argue that that viewpoint is not intrinsically better or worse than the opposite. Some supernatural spirit endowed with rationality could argue with as much conviction that the even distribution of energies throughout the universe is of ultimate value and is the absolute good, as it could argue that the convergence of energies into maximally numerous and organized systems is the highest value and the absolute good. In either case, the assignment of goodness and value is unverifiable by the observed events. The events show a progression in some sectors of the universe toward maximum states of entropy and in other, delimited regions, toward increasingly high local negentropy, and neither the one nor the other compels us to regard its end-products as the ultimate good. On the other hand, from the viewpoint of existing negentropic entities, values are definitely instantiated in the universe. Such entities substitute cybernetic for structural stability, and cybernetic stability means the capacity for effective dynamic adaptation to the structures and forces of the environment. Hence for cybernetically stable entities the valued states are the adapted states. The level of organization achieved by existing systems is a *product* of their adaptations in their evolutionary history, and not their *goal.* Although the two generally go together–a high level of organization usually means a high level of intensive adaptation to the environment–organization-level is wrongly viewed as a final cause of the process and hence wrongly attributed teleological value. The basic goal of natural-cognitive systems is their continued existence, and that is assured through adaptation in the *appropriate* state of organization—not necessarily the highest one. A hydrogen atom is not intrinsically "less good" than an amoeba for being less organized, and an amoeba is not necessarily "less good" than a human being for the same reason. If a hydrogen atom is adapted to the medium in which it finds itself—say, in an interstellar dust-cloud—it is appropriately organized for

---

[16]Russell, *op. cit.*

its environment: a higher level of organization, such as that of an amoeba, would make it maladapted there. And the same goes for an amoeba in regard to its medium, and for man to his. The point is that adaptation is the goal of cybernetic stability, and thus it is the level of adaptation, and not the level of organization, which correlates with value for the systems.

Values of social adaptation

Cybernetically stable natural-cognitive systems adapt to their environment, and their environment may include sets of systemic relations which jointly constitute one or more higher-level systems. The given system relates to the suprasystems as part to whole. If the system is adapted to its environment, it is, by that token, adapted (also) to the suprasystems in which it is a component.

Now, human beings exist in various stages of social adaptation. A minimal level of adaptation by the representative majority of individuals in any society is the minimum condition of that society's existence; if individuals tended to be socially unadapted, i.e., a-social, society as such would not be possible. However, social systems are a fact and can be recognized to exist in the same sense in which individual organisms exist: as concrete natural systems.[17] Thus we would expect that social adaptation in individuals be well-nigh universal, and that it achieve expression as some universally shared basic value-system. This assessment is now corroborated by cultural anthropology.

In the last Century, hypotheses of universally applicable norms were advanced by such writers as E. B. Tylor and Lewis H. Morgan. Each society was held to express values appropriate to its state of development on a universal evolutionary scale. Corresponding to savagery, for example, Morgan posits the values of promiscuity and incest; correlates of the next stage, barbarism, are values of polygamy, and of civilization, monogamy. Similar universal value-scales can be read off Marx' and Engels' conception of history, where morals are relative to the material conditions of life, but these in turn are objective, and follow objective laws of history. Hence different societies pass through their particular configuration of class-conditioned moral valuations according to a universally applicable dialectical scheme of evolution.

Now, in the early part of the 20th Century, anthropologists came to reject universal values as anthropomorphic schematizations of investigators who argue out of their own culturally acquired norms. The diversity of

---

[17]See Chapter 5, "Social Systems," above.

values in different cultures was recognized and held to be insuperable; cross-cultural comparisons were dismissed as subjective and without objective foundation. Anthropology concentrated on differences among cultures and espoused total cultural relativism. This school was spearheaded by Franz Boas and his disciple, Ruth Benedict. The latter concluded her *Patterns of Culture* by stating that "all cultures have equally valid patterns of life." No comparison between them is possible since different cultures "are travelling along different roads in pursuit of different ends, and these ends and these means in one society cannot be judged in terms of those of another society."[18]

Cultural relativism was in turn superseded by a new direction in cultural anthropology. The contemporary trend is to gather empirical data to locate cross-cultural universals: norms and values shared by all people, everywhere, notwithstanding differences between their particular cultures. These cultural universals are fundamental invariants to be located by theory construction, and inferences from and to the phenomenal findings. Linton, for example, states that "behind the seemingly endless diversity of culture patterns there is a fundamental uniformity."[19] Note that the diversity of culture patterns is said to be "seeming," i.e., phenomenal, and that *behind* it there is to be found a *fundamental* uniformity.

The cross-cultural universals turn out to be basic societal values which represent the *sine qua non* of social existence. Linton points out that "the roles of individuals within the family are everywhere culturally prescribed. Parents . . . are expected to assume responsibility for the child's physical care and training. Conversely, the child is expected to render them respect and obedience . . . and to care for them in old age. The failure on either side to live up to these obligations is severely reprehended." [20] Kluckhohn reaches analogous conclusions and emphasizes different aspects of what he calls the "universal categories of culture." "No society has ever approved suffering as a good thing in itself . . . No culture fails to put a negative valuation upon killing, indiscriminate lying, and stealing within the in-group. . . . Nor need we dispute the universality of the conception that rape or any achievement of sexuality by violent means is disapproved. This is a fact of observation as much as the fact that different materials have different specific gravities . . . Reciprocity is another value essential in all societies."[21]

---

[18]Ruth Benedict, *Patterns of Culture,* Boston, 1934.

[19]Ralph Linton, "Universal Ethical Principles: An Anthropological View," *Moral Principles of Action,* R. Anshen, ed., New York, 1952.

[20]Ralph Linton, "The Problem of Universal Values," *Method and Perspective in Anthropology,* R. Spencer, ed., St. Paul, 1954.

[21]Clyde Kluckhohn, "Values and Value-Orientations in the Theory of Action," *Toward a General Theory of Action, op. cit.*

The currently emerging position is summed up by Firth. "Some common factors are discernible in the basic requirements of all societies, so certain moral absolutes exist . . . Morality, then, is not merely subjective. It is objective in the sense of being founded on a social existence which is external to the individual, and to any specific social system."[22]

It is now found that such values as courage and self-control, group recognition, equitable judgment from those in authority, reciprocity in exchange, deference of children to parents, incest taboos, regulation of kinship relations, and so on, are universal. Moreover, they are typically *societal* values: they are the bonding forces of the primary social systems directly based on the socially relating individuals. They reappear in the laws, regulations and mores of more complex societies, where the primary subsystems are groups, organizations and institutions, each with its own set of bonding rules. Without universal social values, social systems would either dissolve into self-concerned individuals, or survive as purely arbitrary institutions superimposed as constraints on spontaneous individual behavior. But social systems do not vanish, nor do they seem to be purely conventional and arbitrary structures constraining individuals.

The discovery of universal social values supports the thesis that men, as all natural systems, are holons, adapted both as physiological-psychological wholes to their manifold subsystems, and as social (economical, ecological, political, cultural) parts to the societal systems in their environment. Proper adaptation as a whole represents health for the human being, both physical and mental; and proper adaptation as a part signifies social adjustment. (These states of adaptation are not disjunctive, however, and a socially unadapted individual is as unlikely to be physically and mentally healthy as a physically and mentally sick person is likely to be socially adapted.) The integrated systemic state in which the individual is adapted both as *whole* and as *part* is the state with which *optimal value* correlates: in that state the individual is a fulfilled biocultural being, as well as a functional member of his society.

Adaptation and reform

Adaptation as whole and as part to man's organic subsystems and social suprasystems is not synonymous with passivity and the maintenance of the *status quo*. In every instance, adaptation is a dynamic, continuous activity in which the system fits himself to the environment in a manner designed to secure his own optimum existential interests. Such adaptation may require the active restructuring of some *part* of the environment, as a precondition of achieving a state of adaptation to the *whole* environment. For example, in regard to organic subsystems, it may become necessary to restructure

[22]Raymond Firth, *Elements of Social Organization,* London, 1956.

some organs or tissues in order to assure the coordinated functioning of the whole. The more drastic of these interventions are the surgical operations, where parts can be cut out, transplanted, or otherwise radically modified in the interest of organic health. A direct counterpart of medical manipulations is the sphere of reformist and revolutionary activity in society. Just as medical interventions restructure part of the organic subsystems in order to attain a better coordinate functioning of the whole organism (and as psychology and psychiatry may attempt to restructure part of man's cognitive and emotive systems in function of assuring a better balance in the mind and personality system as a whole), in the same way goal-directed social and political activity can be designed to restructure a part of man's social suprasystems with a view toward assuring a more thorough adaptation of the individual to his general socio-cultural environment.

Adaptation on the level of the human being means having the degrees of freedom, types of experiences, and social relations appropriate to his own genetic and cultural specifics and endowments. Any social system which suppresses the realization of man's potentials of organic and cultural growth is a system that "alienates" its members, i.e., produces maladaptation for them. To "adapt" to such a social system is just as desirable as to "adapt" to a tumor in the brain. But, since neither is a necessary element in subsystem and suprasystem organization, both can be restructured, given the technical know-how and a knowledge of the relevant functional norms. Purposive adaptation by a human being presupposes, therefore, a knowledge of the norms, and the relevant technical skills to bring them about, both in regard to his own biological and psychological subsystems, and with respect to his socio-cultural suprasystems. In both areas, adaptation may involve gradual reform ("therapy") or more drastic revolutionary ("surgical") intervention.

The norms of proper systemic functioning can be discerned in both cases by an analysis of the largest relevant whole. In medical science this whole is the physical organism and its relevant environment. In psychology and psychiatry, the whole in question is the mind and personality structure, and the pertinent social and cultural setting. And in the sphere of social and political activity, the norm is supplied by the analysis of the hierarchy of social systems and their relevant supra-societal environment. Each of these systems is an open system, coacting with other systems in the environment and jointly participating in suprasystems. Thus their adaptation is the adaptation of holons: facing toward their parts, as wholes, and toward larger wholes, as parts. If a given system in the hierarchy is dysfunctional, it imposes arbitrary constraints both on its subsystems and its suprasystems. Thus adaptation to any given dysfunctional system is synonymous with maladaptation to the hierarchy of systems as a whole. Such a system must be reformed or radically restructured in order that the levels of adaptation of all systems within it, and all systems in which it participates, can be readjusted to the norm.

If we take the individual person as our holon, and his immediate social or socio-cultural suprasystem as the dysfunctional suprasystem, we get reform or revolution (depending on the degree of dysfunction) as the course of action indicated in view of adaptation. For example, if an educational organization, such as a university, is dysfunctional, it imposes states of maladaptation on its students and faculty. To adapt to such a university is analogous to adapting to a disease in the organism : it is not adaptation, but defeat. If the university is alienated from the rest of the society, it imposes alienation on all those who accept and adopt its standards. Consequently it makes its members maladapted to the larger social systems in which they, too, participate within the multiple hierarchical chain. The ultimate social norms to which adaptation signifies physical and mental health (and thus correlates with optimum value), are furnished by the largest discernible systems in which the individual participates. For example, whether one's adaptation to any given university signifies a truly adapted state, or one of alienation, can be seen by considering the level of adaptation of the university to its immediate socio-cultural environment, and then considering the level of adaptation of the systems of this environment (e.g., the state, province or nation) to their own (inter-state, inter-province, or international) environment. Optimum values will correlate with the adaptation of the individual to that university which in its turn is well-adapted to the entire socio-cultural environment, i.e., to the emerging global social, cultural, political, economic and ecological system. Much as the norms of the functioning of every single cell in the body are furnished by the coordinate functioning of all cells constituting the body, so the norms of the functioning of any given social system are furnished by the coordinate functioning of all such systems within the global socio-cultural community.

Adaptation is a dynamic and goal-directed process. Total adaptation, unattainable in practice but delineable in theory, consists of the coordinate functioning, via a multiple hierarchical series of systems, of every particle of every atom in the biosphere. If such were the case, we could "cut" the hierarchy at the level and specific location of any system, and find that its subsystems are fully adapted to it, and that it is fully adapted to its suprasystems. But in reality, of course, systems exist in various, less-than-optimal states of adaptation. Hence the level of adaptation of any one system is conditioned by the level of adaptation of all in its hierarchical sequence. It is in the "interest" of each system to assure the optimum adaptation of all others with which it interacts. Thus it comes about that the adaptation of one system may involve, as its precondition, the restructuring of some other system. "Adaptation" is not the concept of a passive, quietist state, but that of a dynamic, goal-directed one.

Value misidentifications

If optimum adaptation, as part and as whole, correlates with value, and if each man, *qua* cybernetically stable natural-cognitive system, is genetically and culturally programmed for optimum adaptation, each man strives for value without fail. How could value-mistakes be made, then?

The problem of value-mistakes, as that of error in general, is a complex one, requiring consideration of the characteristics of the judging and evaluating entity. Simple systems do not make errors (at least, not detectable ones); the behavioral options open to them are restricted. They consistently behave in a way that is conformant to their intrinsic goals, built into their very structure. Complex systems have considerable leeway in choosing their patterns of behavior, and the choices become prone to error. Their freedom consists in applying complex sets of constructs to each of their recurrent types of perceptual experiences, and sets of constructs can be applied which do not lead to the factual environmental relationships which are the intrinsic goals of the system. Error is the correlate of freedom; it is the price paid for complexity in interpretation and choice in behavior. Errors in valuation arise due to the misapplication of construct-sets by human beings, when they pursue intrinsically value-optimizing, but consciously misapprehended goals. The basic value orientations are life, and the satisfactions corresponding to the norms of sociocultural, conscious beings. Optimum adaptation for a high-level biocultural system means a life of biocultural and psychosocial adaptation; one wherein the environmental conditions satisfying the intrinsic strivings of the person are maximally attained. This is the intrinsic ideal, striven for in virtue of the genetic and social "programs" of the high-level natural-cognitive system. That system, however, is also an intensely *conscious* system, mapping in himself a record not only of his active patterns, but also of his interpretations and decisions. The system, on the human level, not only knows and acts; he also knows that he knows and acts. And it is on the level of reflexive consciousness that error can enter. For it is always possible that the conscious representations of basically sound goal-strivings might become distorted through misidentifications of the relevant environmental conditions, and even of the goals themselves. Instincts can hardly be said to err, except in permitting a misidentification of the conditions pertinent to the unfolding of the motivated activity. But conscious representations of instinctual goals can also err in misidentifying the goals themselves. The textbooks of psychology are filled with examples of misapprehended goals, and thought and action patterns. To enlarge upon them would require a treatise in general psychopathology; one which applies not only to the psychotic members of the species but, to some extent, to *all* its members. By way of illustration, I shall choose here one typical instance of value-mistake, due to a misidentification of intrinsic norms. Human beings, like other natural systems, are Janus-faced entities: wholes when facing toward their own constitution, and parts when facing toward the social systems in which they live. One frequent misidentification of goals results from a disregard of the latter. Individuals tend to set and pursue their conscious value objectives *qua* individuals, rather than *qua* individuals *in society*. A common-sense representation of one's environment discloses a substantive individual, as against the rest of the world, including other men and other entities in shifting relationships.[23] Thus the individual, acting on the basis of these misleading conceptualizations, pursues his goals in disregard of the fact that he is part of higher-level systems. He acts *qua* isolated, fully autonomous individual. Insistence on rights, as *human* rights, in abstraction from any given social, moral, legal and political system, is one indication of such a mistaken goal-setting on the theoretical level. More basic expressions of the same misidentification include acts of egoism, in which the individual places his particular interests, or even those of his immediate family or associates, above that of the wider social groups of which he is a part. Anti-social character and behavior patterns are the result, and these patterns are pathological, given the hierarchically integrated nature and situation of human beings. On the opposite side of the coin, value-misidentifications take the form of martyrdom and unreasonable heroism in the service of a social cause. To die for one's tribe or country when one's death is not a precondition of the survival of that collective entity is just as much a misidentification of intrinsically healthy goal orientations as to refuse to serve it. As a hierarchically embedded system, man is committed, by his very nature, to optimize the conditions of his own existence (as a hierarchical whole), together with optimizing the existence of the social systems in which he lives (as a hierarchically integrated part). It is obvious, however, that recognizing these commitments of one's basic nature and choosing the indicated course of action in particular situations is a highly complex matter, calling for error-free conceptualization and practical decision. Little wonder, then, that human beings frequently make mistakes, both in identifying their own intrinsic goals, and in choosing the routes of behavior leading to their realization.

---

[23]See the discussion of the anthropocentrism intrinsic to the "existential bubble," Chapter 9, above.

The task of a *descriptive* ethics is the empirical task of recording people's conscious value identifications and correlative patterns of behavior. The task of a *normative* ethics, then, is to explicate the values which correlate with the functional states of active open systems in given situations. The objective of the present chapter is to provide a framework for a normative ethics but, in so doing, it must account for the differences between its postulated norms and the actual values and behaviors forming the subject of empirical description. Since our theory is not generated by inductively generalizing the values consciously held by people, but rather represents an integration of data from a wide variety of scientific fields, the differential between the advocated natural norm ("optimum adaptation of natural-cognitive systems to the environment"), and the observed diversity of values is not detrimental to its cogency. The differential follows from a closer examination of the structure of high-level natural-cognitive systems. Such systems are intrinsically directed toward biological, cultural and social goals issuing in their own, as well as their society's, self-maintenance and self-organization (endurance and evolution); but, in pursuing these goals, they are prone to commit errors which may result in the destruction of their own existence, or even that of their society. Although natural norms are programmed into each system, and are evident both in the self-regulated integration of the psychophysical organism, and in the universal values which bond all social systems, human beings are complex self-reflexive systems, and hence can misidentify the norms, and mistake the factual conditions leading to their satisfaction.

Systems value theory and meta-ethics

The property "goodness" has been attributed to the state of optimum adaptation of natural-cognitive systems, and the question now arises whether adaptation itself is good. The answer is, of course, that it is good inasmuch as existence for the systems is good; adaptation is a precondition of existence. The "open-question" can then be pressed by asking, is *existence* good? If this question is answered by a disembodied Reason—a non-natural, alienated self, intuiting goodness or looking in on the processes of being and becoming from some position *sub specie aeternitatis*, the answer will entail an infinite regress—whatever is said can further be questioned with regard to goodness. But if the question is answered by a natural system, maintaining himself in the natural world against great odds, the answer is an unqualified (though not necessarily an explained) *yes*. Natural systems are programmed to maintain themselves in a potentially hostile medium, in states of fantastic improbability. Such an entity, when faced with the question, "is existence good?" correctly answers "it is," since his entire constitution is such that it is directed toward maintaining and improving it. For example, when the present writer says "*my* existence is good (and of value)" he may be expressing a deep-seated attitude, but is not, for all that, making a meaningless non-cognitive pseudo-statement, as philosophers such as Ayer charge. He is attributing "good" ("value") to "existence," but he can show that doing so has an objective basis—not in the sense-perception of "goodness" to be sure, but in the total complex of mental, physical and societal states which add up to the cybernetic stability of body and mind. The value of existence is programmed into the writer's DNA coded cells, which instruct the production of millions of cells in his organism to operate feedback circuits maintaining body temperature, ion concentration, energy supplies, and the structural configurations, the parameters of which define his organism. Whenever states of this gigantic array of organized systems exceed the threshold of the genetically coded norms, the system activates intricate error-correcting processes which bring it back to the programmed range. It is within this range that biological existence in his *milieu* is assured. The value of human existence is also programmed into the writer's sets of cognitive constructs, which lend meaning and relevance to the sensory information he receives of his environment. When the level of information exceeds or falls below the norm given by the range of applicability of the constructs, strong motivations appear toward reestablishing the normal range through negative feedback error-reduction. The transgression of the genetic norms appears in the cognitive system as the feeling of stresses and strains with accompanying pains and discomfort. The overstepping of the cognitive norms appears as a disturbing anomaly in experience: a puzzle, a psychedelic overcharge, or sensory deprivation and monotony. These sensations signify that the system is striving toward re-establishing his states of adaptation to the environment. These are the states in which existence is maximally assured and the deviation from which causes strains and stresses in the "mind" as well as in the "body." Thus when this writer professes a disposition toward maintaining his own existence he is verbally reporting on feelings which arise as signs of the operation of multiple feedback processes in his organism. Likewise, when he says in regard to an item in his experience—whether perceptual, cognitive or conative—that it *ought* to be thus and thus, he is referring to a norm from which that item represents a deviation. (The case is analogous to that of an automatic heating system when the temperature falls below the "norm" to which its thermostat is set: if the system could reflect on its own operation and could verbalize it, it would report that the temperature *ought* to be raised by the furnace the *n* degrees by which it has fallen below the setting.) If environmental conditions are less than optimal, the disposition toward existence manifests itself in feelings of anxiety, strain and possibly pain, which add up to verbalizable convictions that things ought to be different. I may or may not be right in identifying the way things ought to be—my cognitive constructs are prone to error. But with experience I learn to make my construct-systems more reliable. Nevertheless, *independently of my knowledge of it,* there is a set of states at which I exist in optimum adaptation to my surroundings, and this state correlates with maximum value for me. "Correlating with value" means here that if I knew the state in question, I would correctly identify it as that in which I have optimum chance of existence. Thus, being an existence-directed, self-regulating natural-cognitive system, I would rightly attribute highest value to this state.

The present framework for a normative ethics overcomes the over-confident bias of classical objectivism, which held that values are objective, empirically verifiable "facts," without sinking into the morass of unconditional skepticism, holding that the mental states of evaluators are the sole criteria of values and are, moreover, entirely private. In the here advanced views, valuations by real persons are relative to their cognitive capacities; but the relativity of manifest values does not entail the relativity of *normative* values. Value-norms can be discerned, not by an analysis of the mental states underlying manifest valuations, but by theoretical considerations pinpointing the parameters of the psychophysical organization of the evaluator. Mental states may or may not be mistaken : they represent correct value-identifications if they lead to activities maintaining or improving the system's level of adaptation, and they misinterpret normative values when the activity motivated by them decreases the adaptation-level of the system. In any case, in order to achieve rational certainty of identification, detailed information must be had of the complex regulative processes which define the norms of the human condition. Value judgments, if they are to be reliable, must be based not only on a mental state and feeling, however healthily goal-directed these may be, but also on an explicit theory of man as a psychophysical entity, interrelating in determinate normative modes with his natural, cultural, and societal surroundings.

This is not a meta-ethical argument for the meaning of moral terms, and thus it is not an argument for naturalistic ethics as it is usually conceived. Much more is this an attempt to update the classical type of moral theory by basing its normative prescriptions on theoretically postulated and empirically confirmed system-states, rather than on commonsensically observed and argued "facts."

Chapter 14

SURVIVAL:

FRAMEWORK FOR A NEW AGE ETHOS

Toward a general orientation for action

We are at a cross-roads. We can set forth our traditional activities and pursue our accustomed goals; or we can set out on new paths, using our creative imagination, restrained by reason and experiment. We can either come to a new assessment of mankind's role within the natural order of things, or very likely face extinction. This is the choice before us. I shall not debate which of these courses is the right one; I hold this to be self-evident. What I would like to do is to outline those ideas which can guide our steps in seeking a new order, in which man does not exploit nature, but fits himself into it for the benefit of all concerned.

Ideas become a material force, Marx said, when they take hold of the minds of men. He was right—without ideas that people can believe in, no purposive goals can be pursued on the larger level of social action. Thus what we need today is a new morality—an ethos which does not center on individual good and individual value alone, but on the adaptation of mankind, as a global system, to its new environment. We are dealing with emergent realities; no longer with isolated groups of men, but with a systemically interdependent world community. It is this level of analysis which we must keep before our eyes if we are to inspire large-scale action designed to assure our collective—and hence also our individual—survival.

Sociologists have clearly recognized the importance of the theories which dominate the minds of the decision making groups in society. In the last resort, it is some model of the world—a world-view or "ideology"—which, explicitly or implicitly, guides the processes of social change, especially when traditional modes of thought and patterns of behavior break down. Hence it is of paramount importance today, to offer a cogent set of ideals capable of taking hold of people's minds and, through their minds, guiding their decisions in matters of fundamental policy.

The importance of social ideals can hardly be overestimated. They can make the difference between victory and defeat in war; they can prompt heroism in the face of overwhelming odds; they can cancel out the accumulated weight of centuries of tradition. Witness their effect on the Arab-Israeli and the Vietnam conflicts; their power in the achievement of the Bolsheviks in the early years of the Soviet republic, and the accomplishment and victory of the Chinese Reds under Mao. To fight without a social ideal is to surrender to defeat before the test; the question "why do I fight?" must have a clear-cut and acceptable answer.

And fight we must. Not one another, but *for* one another. Not with violence, but with perseverance and courage. The fight is not for national victory, of one people over another, but for the persistence of the kind of life which we call human, and which incorporates our finest traditions and achievements. This is not a fight *against*, but a fight *for*. And it can be lost just as readily as any fight against an enemy. Defeat here will be marked by the extinction of our species or, at best, by the downfall of human civilization and its replacement by a form of existence at which we can only shudder. It may be the ant-society of maximum regimentation described in Orwell's *1984*, or an uncoordinated, aimless physical survival in a state of near chaos. Victory will signify the continuation of civilization as we know it, with material and spiritual resources available to all. The alternatives before us are having self-centered, non-symbiotic social systems, unable to cope with the challenge of the new environment, de-differentiating or else maintaining themselves by freezing their misfunctioning institutions into bureaucratic rigidity, or achieving a *world* system of mutually symbiotic societies, mapping the new conditions into a flexible institutional structure and dealing with change through constructive reorganization.

If this fight is to be won, we need an orientation for action. It must inspire as well as guide. It must say what is at stake, and what we can do about it; and also, that it is worth doing. I suggest that such a social ethos can be keynoted by the ideal of a *reverence for natural systems*. In the following I shall outline what I mean by it.

"Reverence for natural systems"

The earliest systematic speculations of mankind regarded the universe as a sphere of order and harmony, and considered it man's duty to fit himself into it with his conscious purposes. Pythagoras, Plato, and their followers dominated Western natural philosophy until the rise of Christian theology with its other-worldly emphasis. While in the Orient, philosophy and religion fused in the appreciation of nature as "self-thusness" (*tzu-jan* in Chinese and *ji-sen* in Japanese); as something *for-itself*, rather than *for-us*, in the West there arose out of Christianity a Puritan ethic of hard work for human well-being and the glory of God, which was premised on *using* nature for human ends.[1] With the contemporaneous rise of science and the scientific attitude typical of the mechanistic Newtonian world-view, utility was conceived as the finest goal of knowledge. Bacon wished to "subdue and overcome the miseries of humanity" by discovering in the womb of nature "secrets of excellent use."[2] Hereby man's attitude to nature changed from one of *reverence* to that of *exploitation*. Whereas Eastern man continued, at least until our day, to revere nature for what it is in itself, Western science and technology appreciated it mainly for its use. Our theorists have even produced justification for it, in holding nature to be brutish and dumb, and man intelligent, just and good. "No one," said Mill, "either religious or irreligious, believes that the hurtful agencies of nature, considered as a whole, promote good purposes, in any other way than by inciting human rational creatures to rise up and struggle against them." And he added, "All human action whatever, consists in altering, and all useful action in improving the spontaneous course of nature." Man must be perpetually striving to bring nature into conformity with his own "high standard of justice and goodness."[3]

What supreme arrogance! What *folie de grandeur*! Few men realized the extent of the mistake until today, when our attempt to bring nature up to our own "high standards of justice and goodness" threatens to pervert its order to an extent which precludes our own existence. The use of nature for our egoistic ends had far-reaching consequences in altering the relationships which form the preconditions of our very being. These relationships took millions of years to evolve, and they could only evolve in balance, for any imbalance would have had to either be rectified, or result in disorganization. Thus if there is organization in nature, it is because it could not have been otherwise, given the thermodynamically improbable, fantastically adaptive, and delicate nature of the entities that make up our environment, including ourselves.

Now, the "norm" of nature, if we can correctly discern it as it applies to the microhierarchy, is not the existence or demise, the presence or

---

[1]See Max Weber's somewhat controversial, *The Protestant Ethic and the Spirit of Capitalism* (New York, 1958); L. White's "The Historical Roots of Our Ecologic Crisis," *Science*, **155**, No. 3767 (March, 1967), and Paul Ehrlich's widely read *The Population Bomb* (New York, 1968).

[2]Cf. *Selected Writings of Francis Bacon*, H. G. Dick, ed., New York, 1955.

[3]J. S. Mill, "Nature," *Readings in Philosophy*, J. H. Randall, Jr., J. Buchler, E. Urban Shirk, eds., New York, 1950.

absence, of any particular system *per se*. Natural norms always refer to the set of relationships in which given systems are relata. Nature builds role-structures (to use Boulding's term), not particulated individuals. The particular exhausts its significance in its relation to the structure. The norm, ultimately, is given by the highest suprasystem: the hierarchy. It is that hierarchy which is the significant resultant of millions of years of piecemeal evolution, and not any particular individual or species. This, I believe, is a lesson to be learned from the slowly emerging new knowledge of our age. It is a lesson which must be learned fast, if the pupil is to survive his graduation.

Evidence for this hypothesis can be restated here under two general headings: (i) evidence concerning the subordination of the individual to the intraspecific *social collectivity,* and (ii) evidence concerning the subordination of the social collectivity to the interspecific *ecological hierarchy.*

Under (i) we may mention the gradual submergence of isolated individuals in increasingly socialized communities, as a concomitant of evolutionary development, and the consequent dependence of individuals upon the collectivity. Although here, too, there are exceptions, they are rare among species with nervous systems sufficiently evolved to communicate with one another; integrated societal organization, as Mather affirms, is commonly displayed among vertebrate animals other than the more primitive members of this phylum.[4] Organization among them may be either for procuring food, common defense against predators, safeguarding common territory, or for constructing shelters. Cooperative organization in its many forms (symbiosis, parental care, etc.) places definite controls on individuals at the same time as it confers benefits on them. One member of a collectivity becomes dependent on the others, and cannot pursue his own goals to the unconditional detriment of the rest. Individuals may even be sacrificed for the good of the collectivity—as in the case of lion cubs which do not get to feed until their elders are satisfied, or of insects which die when their social or reproductive function has been performed.

Evidence concerning (ii) includes the various cyclic processes, such as the carbon and the nitrogen cycle which interrelates atomic and chemical compounds in the soil and air with biochemical substances processed in plants and animals and, using all products without surplus and waste, recycles energy in forms usable by all systems within the biosphere. The whole of nature, Sears suggests, can be conceived as a "self-repairing,

[4]Kirtley F. Mather, "The Emergence of Values in Geologic Life Development," *Zygon,* **4,** 1 (1969).

constructive process" representing a type of equilibrium that "approximates an open steady-state."[5] (Figure 14.)

The checks and balances which maintain the equilibrium of the ecological hierarchy are particularly striking when they impose constraints on animal behavior. Traditionally, basic drives and motivations were thought to be egoistical, or having at the most the good of a social collectivity or a species as their goal. But recent evidence indicates that there are constraints built into the very constitution of individual organisms in the form of stress, and subtle changes in body chemistry, which prevents them from upsetting the balance of the suprasystem in which they live. This is the phenomenon of "population regulation" in accordance with the distribution of energies and information in an ecosystem. Some species have developed drastic forms: lemmings, snowshoe hares and some species of shrews are the prime examples. Other species (such as mice) merely abort their offspring beyond the normal population density, or inhibit sexual acts, or produce states of overexcitation and stress.[6] It is now becoming evident that individuals and species do not pursue even their basic reproductive goals in full independence, but manifest restraints whereby they keep within the bounds of balance in the integral ecology of nature.

Adaptation in a natural system presupposes the presence of constraints, limiting his individual degrees of freedom, and making his behavior consistent with the superordinate system in which he finds himself. Adaptation in a system on one level is coordinate functioning in the system on the next. Nature builds systems by adapting parts in wholes, and the wholes as parts in superordinate wholes. Thus nature's norm is the optimum functioning of the hierarchical suprasystem containing all subsidiary systems: it is this system which imposes the final constraints on the degrees of freedom of all subsystems.

Most systems within the biosphere effectively manifest and obey the constraints whereby their existence and behavior remain consistent with the chain of energy and information processing in their total environment. Only man has upset the balance of his environment by favoring his own species over others, transgressing upon natural, and largely instinctual

---

[5]Paul B. Sears, "Utopia and the Living Landscape," *Daedelus*, Vol. 94, No. 2 (1965).

[6]For descriptions and theories of population regulation, see Charles Elton, *Voles, Mice and Lemmings* (Oxford, 1942); David L. Lack, *The Natural Regulation of Animal Numbers* (Oxford, 1954); Paul L. Errington, *Of Predation and Life* (Ames, Iowa, 1967); and Lawrence B. Slobodkin, *Growth and Regulation of Animal Populations* (New York, 1961).

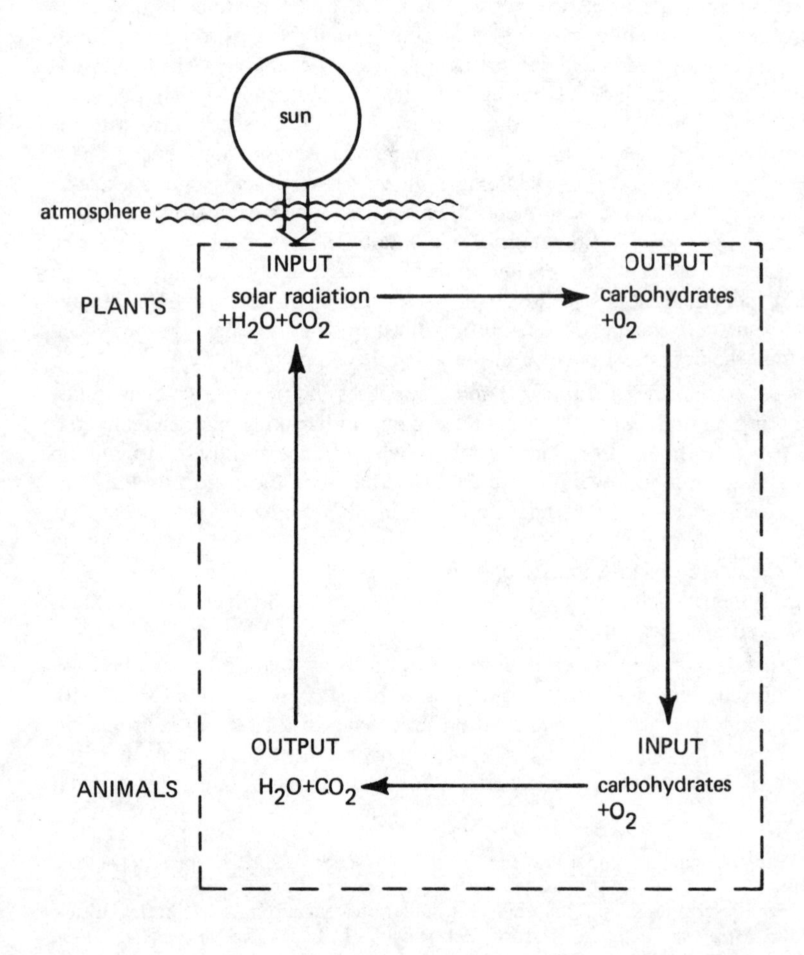

Figure 14

Terrestrial nature as a self-maintaining system, using solar energies and maintaining a balance between its various components.

constraints on aggression, population increase, and environment pollution.[7] For some fifteen thousand years now, the harnessing of fire, water, and later of steam, coal and nuclear power, and the exploitation of other species of organisms, plant as well as animal, for man's own purposes, interfered with the organization of the earth's supremely but delicately balanced ecosystems, but did not disadvantage human existence. These processes are cumulative, however, and they are accelerated by modern technology. It is our misfortune to live at a time when the accumulated effects of man's arbitrary recycling of nature to his own ends begin to tell, and to cut into the survival potentials of the species itself. Our "natural sins" are not new—new only is their acceleration through our egoistical use of technology and the feedback of effects in the form of a survival threat. Ever since man recycled a bit of the natural hierarchy by planting a crop in one place and depositing his wastes in another, he has been working for his own immediate advantage which now, in the long run, turns into his own general disadvantage. Viewed from this awesome perspective, technological progress, the pride of mankind, appears as a tremendous blunder—an egoistical boomerang. But it is too late now to undo the work of the past hundred years, not to mention that of the past fifteen or twenty thousand. Rather, we must try to recognize the effects of having overstepped the limits, and transgressed on the checks and balances of nature, and correct our future activities to reduce the margin between human survival and technological-social development. We can do this, if at all, only through the rediscovery of classical Hellenic and Oriental modes of thought as the inspiration for a new-age ethos which *reveres* the natural order, rather than *exploits* it. The healthy instincts of ancient philosophers are now supported and supplemented by the concrete findings of the contemporary sciences, permitting us not only to put our faith in this ethos, but also to approve it with our intellect.

Albert Schweitzer called for reverence for life. On this basis we must call for reverence for the level-structure of the microhierarchy, including all systems on all its levels, from atoms to an emerging planetary culture, economy, and ecology. We must envision the biosphere as a whole—the

---

[7]It may be, however, that some of these natural constraints have not been lost, even in modern technological civilization. Coale calculated that if all unwanted births would be eliminated in the United States (by providing birth-control information, contraceptive devices, and safe and free abortion), the number of children born would bring the fertility to just about replacement level, stabilizing the population. Although Coale tends to think that this might be fortuitous, it may also be that natural controls are still operative, although masked by social and moral constraints. Cf. A. J. Coale, "Man and His Environment," *Science* **170,** No. 3954 (October, 1970).

earth as a "spaceship," with mutually reinforcing or mutually destructive interdependencies.[8] The emergence of a planet-wide interdependent ecology is an imminent reality, and it needs to be dealt with. We must regain our implicit natural values; our instinctual and long-buried adaptation to the order of nature in the microhierarchy.

I believe that we can express the recovery of our intrinsic natural values in requesting a *reverence for natural systems*. It is a reverence for our own kind, when our vision is wide enough to see ourselves not only in our children, family and compatriots, and not even in all human beings and all living things, but in all self-maintaining and self-evolving organizations brought forth on this good earth and, if not perturbed by man, existing here in complex but supremely balanced hierarchical interdependencies.

"Reverence for natural systems" and the new mentality

In our rapid success in manipulating nature, we have come to esteem bits of knowledge more than encompassing insight. We have come to believe that to know more than is called for in bringing about an immediate goal does not pay; that it can be disregarded. We have come to look at man's relation to nature as that of a child to his toy : you press here if you want it to go, and there if you want it to stop. But nature is not merely an object to be manipulated by limited willful experiments. It is an interconnected, ordered network of events in which every "button" is connected to every "function." And while this in itself is an assumption,[9] it is one which we must make if we are to stabilize our roles not merely *vis-à-vis*, but *within*, nature.

There is beauty in such an assumption. Order is the highest ideal of the human mind; order in thought motivates science, order in feeling inspires art, and order in the course of human existence is a mainspring of religion. The ordered is the meaningful, the knowable; the chaotic, the nonrecurrent is the puzzle and the source of anxiety. The latter is the element to be reduced to order by action, or by the discovery of overlooked ordered aspects. Yet order, stolidly repeating and unrelieved by novelty, would be boredom and stultifying to the mind. Such are simple orders, given, unchanging, and stubbornly present. Such, however, is not the order of nature. The order of nature is a dynamic order, of ever-renewing patterns, of novelty, of creativity. Yet it is order, for the events are not haphazard but follow their own inner logic. To discover this logic is the great adventure of the cognitive mind.

---

[8]Cf. Buckminster Fuller's *Operating Manual for Spaceship Earth*, Carbondale, Ill, 1970.

[9]See the "secondary presuppositions" in Chapter 1, above.

The order of nature, as we are now coming to appreciate and recognize it, is the best source of inspiration of social morality we can have. Nothing man-made can compare with it. The starry heavens above and the morality within were what amazed Kant, and we can now see the connection between them. The starry heavens, as indeed all nature, is not the mechanical, soulless machine of Newton's natural philosophy. The moral agent confronting it is not its spectator, passively marveling at its workings. Man here, and nature all around, are part of an embracing network of dynamic, self-regulating and self-creative processes. To marvel at nature is to marvel at ourselves, for although there may be others like us in the vast reaches of the cosmos, we surely are one of the most remarkable examples of dynamic order brought forth in this universe. To know this is to admire the matrix out of which we arose, and to want to preserve what we are. We do not need elaborate fantasies or tales of legendary creation. The panorama is before our eyes: wherever we look we see creative novelty arising in ordered sequence out of a prior template or generative matrix, as system obeys system and coacts with vaster system, in autonomous yet symbiotic patterns of functional behavior.

To be part of nature is to have a reason for existence. To be one of the most evolved systems in nature is reason enough for self-confidence and the wish to live and propagate. We are not alone: we are in nature. We are a part of the tremendous, balanced hierarchy which distributes roles and values to all things, from atoms to international federations. Thus, to live and to propagate ourselves, we must safeguard both the atoms and the international federations, and all things in between, to the best of our ability. For we do not live and propagate *on* them or *in virtue of* them, but *with* and *within* them, in multiple ordered relationships.

"Reverence for natural systems" expresses this insight. It gives explicit form to an attitude which already pervades the mind of today's younger generation—this is evident in its insistence on universal and more meaningful human interrelations, closer ties with nature, the abolition of egoistic, materialistic values, and the bringing about of a communal society based on love and mutual understanding. This mentality is no longer Baconian; it is not mechanistic, manipulative, and exploitative of the natural order.

There is no cause for despair. The problems are many, but the groundworks for meeting them are being laid. "Mankind," as a decisive control organ of a global socio-ecosystem, exhibits remarkable adaptive potentials. Almost as a form of preconditioning, for meeting the challenge of the future, the vital attributes of a differentiated and diversified, and yet functional and integrated world-society are emerging in the attitudes and preferences of progressive young people. Together with a growing need for increased levels of human communication—the *sine qua non* of social integration—they have a new awareness of the environment and of the population problem, of the need for love and justice, civil rights and consumer protection. There is a new perception of the earth as a finite and readily misused spaceship, with limited resources and systemic interdependencies. Young people still differ greatly, and sometimes violently, in their ideas of the future society, but their emotions, instincts, and attitudes converge progressively. No longer is there the great unbridgeable gulf between collectivist Communism and individualist Democracy. The theories clash, but the practical aims are but so many variations on a common theme. The mystic oneness of the Orient, the classless humanism of the Marxists, the group-sensitivity of American youth, are different expressions of a common concern, now emerging into world consciousness: the concern for living together, in systemic wholeness, on a crowded, instantly communicating, and economically and politically interdependent planet.

The groundworks are being laid, but the concerted conscious expressions are still lacking. Special theories lack breadth of application, and hence guidance value, and general theories are either defended on faith, or are denounced for being founded on faith. Yet, as Poincaré said, every man carries a picture of the world in his head. Today that picture must be a basically sound one, lest misguided human actions, embedded in global interdependencies, propel mankind on a path to disaster. The contemporary *Weltanschauung* must have reliable foundations. Contemporary science, when its operational compartments and positivist interpretation are overcome, can provide that foundation. It gives the concept of a coherent systems universe, built up of relatively autonomous wholes, functionally integrated in hierarchical interdependencies. Its insights can be integrated in a general theory of systems, with its normative ethics correlating value with the mutual adaptation of dynamic, cybernetically stabilizing systems (cf. Chapter 13). The ethics and natural philosophy of this new world view can help explicate and justify an emerging supranational social ethos: *"reverence for natural systems."*

The new generation is ready to understand this concept intellectually, and to embrace this ethos emotionally. Its adaptive capacities can be made conscious, and hence vastly enhanced, by a commonly accepted theoretical and ethical framework, for it is around such general and yet concretely meaningful principles that concerted patterns of purposive behavior can crystallize.

Chapter 15

ULTIMATE PRINCIPLES: FRAMEWORK FOR A METAPHYSICS

Criteria for metaphysics

Metaphysics is perhaps the oldest of human cognitive endeavors: it shades imperceptibly into myth in antiquity, and into philosophy and science in modern times. Metaphysics, as religion, gives answers to questions concerning the ultimate nature of reality. It differs from religion in that its propositions rest not on faith, but on the intrinsic coherence, simplicity and adequacy—in one word, the *elegance*—of its answers. Hence great schemes of metaphysical thought are relatively impervious to empirical confirmation or disconfirmation; their validity is largely determined by the appeal of their principles. It is perhaps fair to say that the great metaphysical theorists of history have, between them, exhausted the number of intrinsically appealing modes of thought. They have produced metaphysics which conceived of the universe as the perfection of being, and also as the flux of change. They offered process-theories which envisaged reality as progressing from chaos to order, or recycling itself in eternally recurrent patterns. Monism, dualism, and pluralism; materialism and idealism; spiritism and mechanism, have been explored in depth and in breadth, making much, some, or no appeal to contemporaneously available empirical data.

The contemporary temper tends to seek connections between ultimate metaphysical principles and empirical experience. Hence it tends to restrict the number of intrinsically appealing metaphysics to the number of empirically interpretable ones. But, whether a metaphysical scheme is empirically interpretable or not, depends in large measure on what one means by "empirical." When we go directly to the flux and flow of sensory experience, and its interwoven elements of memory, feelings, and volitions, we get such diversity that almost any form of order can be abstracted from it. Direct sensory and "lived" experience justifies Heraclitus as well as Plato, although the constitutive role of the rational intellect is more pronounced in the latter than in the former. But experience can be coerced to testify for flux as well as for being, for mind as well as for matter. It is when we measure our metaphysics against some already established categories of reality that we find criteria for selecting between alternative schemes. Hence a Newtonian conception obviously ill accords with spiritism; the foundation for a spiritist theory is better sought in the *telos* of a Drieschian biology, or the *id* and *superego* of Freudian psychology. Now, if we assume that empirical conceptions themselves can be ranged on a scale of better to worse, and if we also assume, on the basis of arguments presented in this book, that the general world-view here called *systems paradigm* is the best general conception available today, then we get criteria selecting for one particular kind of metaphysics. If this metaphysics is also intrinsically appealing in virtue of the elegance of its postulates, it will be our indicated choice. Let me, then, offer some pointers and speculations as to the basic notions which would constitute the foundations of such a metaphysics.

Pointers toward a systems metaphysics

We choose a monism, tempered by an evolutionary, categorial pluralism as our basic framework. We reject dualism and pluralism as foundational principles; we have no systems-theoretical warrant for conceiving of the universe as composed of inert matter, in-formed by some externally related Spirit, or set of ideal Forms or possibilities, nor can we cogently consider it as a layer-cake made up of co-existing but externally related types of entities, some of which can be defined as "physical," and others as "biological," "psychological," or "sociological." Rather, we are led to conceive of the universe as a giant matrix out of which arise the many phenomenally distinctive entities. We can conceive of no radical separation between forming and formed, and between substance and space and time. We do not have existing substantial things, located at discrete points in space and time. Rather, the universe is conceived as a continuum, defining both space and time, and the spatio-temporal events which disclose themselves to empirical observation. The latter can be thought of as "stresses" or "tensions" within the constitutive space-time matrix, emerging within the phenomenal field as tips of an iceberg above water. Hence there will be no action-at-a-distance, despite causal or functional correlations of spatially and temporally distant events. The connections are propagated within the cosmic matrix, limited by a constant, such as the speed of light in vacuo. Interaction can thus be defined by the light-cones of Minkowski and Weyl.

Within the range of interaction defined by the cones, the cosmic matrix evolves in patterned flows, one actualized flow conditioning the emergence and development of the rest. Some flows hit upon configurations of intrinsic stability and thus survive, despite changes in their evolving environment. The flows represent recurrent sets of events which jointly constitute the invariance of the flow: these we call *systems*. The actualized configurations of stability–acting as constraints in the systems–are realizations of possibilities intrinsic to the cosmic matrix. Likewise, the constraints acting *on* the systems, and both limiting their degrees of freedom and providing their vital medium, represent possibilities actualized in the continuum, externally to the systems. Hence the possibilities in the cosmic continuum become actualized in the selective evolution of systems, adapted to their environment. The evolution of systems is the evolution of the continuum into increasingly adapted, and (in part) increasingly complex and individuated, modules. This process is but the accommodation of the singularities of flows in the continuum, within its general flow-patterns. In the final analysis, it is the matrix which orders itself, bringing about a build-up of organization in some sectors, at the expense of smoothing out complex flows in others. But in those areas where flows complexify, the matrix forms multidimensional fields, with elaborate hierarchical patterning.

The events which engage our attention are the observable points of the continuum, and especially those where flows are either constant or complex. Constant flows of relative simplicity furnish the events investigated by physics, such as the many radiations, including light and quantized energy transfers. Complex flows with high levels of cybernetic stability are the objects of biological, psychological and sociological inquiry. In each instance flows are mapped by a theoretical model insofar as they exhibit an invariance. The self-maintaining invariances give us the concept of "entities"; in addition, those which have perceptually observable properties furnish the referent of "things." We, of course, as cognizant human beings, are likewise flows, of a highly complex, self-maintaining and organizing kind. This consideration can explain why some flows are perceptually observable for us and some are not: those which are, are relevant to the maintenance of the flow-invariances we ourselves incorporate, while the unperceivable ones are not.

The phenomenon of mind is neither an intrusion into the cosmos from some outside agency, nor the emergence of something out of nothing. Mind is but the internal aspect of the connectivity of systems within the matrix. It is there as a possibility within the undifferentiated continuum, and evolves into more explicit forms as the matrix differentiates into relatively discrete, self-maintaining systems. The mind as *knower* is continuous with the rest of the universe as *known*. Hence in this metaphysics there is no gap between subject and object—these terms refer to arbitrarily abstracted entities.

The present doctrine incorporates possibility in actuality, rather than placing it on a separate, transcendent plane. What *is*, is a partial realization of what *can be*. The latter is the sum of the potential strata of stability within the cosmic matrix. Within that matrix we find the properties which come to be realized, one after another and in sets, as the actualization of one possibility conditions the actualization of others. The last vestiges of the passive "neutral stuff" or "inert matter" conceptions of classical thought are hereby discarded. The universe is *causa sui*. The manifest orders represent the perceivable phases of its self-evolution, not determined by any outside force. There are only internal relations. Platonic *ideas*, or Whiteheadian *eternal objects*, are rejected as uncalled for; likewise the notions of a transcendent God or other Deity. Ordering is from within, but proceeds from the non-perceivable continuum toward increasingly discrete particulars. The latter are not *fully* discrete (since systems are but individuated sets of points along the continuum), but are increasingly perceivable as such since, in the course of their evolution, they acquire integrated properties which can interact with human sense organs.

The observed panorama is a process wherein entities form the relatively stable flows. Such flows recur in many different forms and under diverse circumstances, and are subsumed under general categories. The latter code the invariances which the inquiring mind (likewise an element arising in the process) abstracts by selectively screening the observed data. Hence we are dealing with two kinds of "universals": the concrete flow-patterns in the natural universe, and the abstract-general categories whereby they are coded in the mind. Yet it is fallacious to conclude from this that what is given is only the former, and relegate the abstract universals to the shadow realm of a "transcendent ego" or "constitutive consciousness." Abstract universals are as much an element in process as the concrete flow-pattern universals are. An abstract universal is real as an element in the complex natural-cognitive system in the brain-mind of which it arose; a *true* universal is also *applicable* as an abstractive coding of a concrete flow-pattern outside that system (or *in* that system, if it is an element in one's self-knowledge).

There are abstract universals which have no concrete counterparts in nature. Such are the sensory data, and the thereby constituted visual, acoustic, (etc.) gestalts. "Whiteness" is a sensory universal, an element in mind. Yet there are no concrete flow-patterns which could be said to be white, apart from their perception as such by some other flow-pattern entity with an evolved sensory apparatus. Whiteness, hardness, silkiness, etc., are mental correlates of physical flow-patterns in the brain of the complex natural-cognitive system. They do not have extra-system counterparts "in nature" other than in the brain-minds of other systems. The latter, however, are not the *objects* of these sensory universals, but their *analogs*.

Sensory universals have no "objects." Yet they signal states of "objects" (i.e., of systemic flow-patterns) to the subject-system, and enable the latter to respond to objective conditions in the environment. Hence such universals are *functionally* and not *representationally* true, analogously to the "truth" embodied in a punched tape carrying information to an automated machine. The punch-holes do not represent the objects to which they refer, like a photograph, but code it in a form relevant to (i.e., decodable by) the machine.

Concepts of universals, of knowledge, possibility and actuality, can be cogently developed as general metaphysical principles, using systems theory as a foundation. But such notions are not uniquely entailed by any systems model, and hence remain more "open" than the latter. Nevertheless, if ultimate questions are asked, ultimate (but hypothetical) answers can be given by extrapolations from general systems theory. Such hypothetically ultimate theories constitute *systems metaphysics*. Some investigators will doubt its importance for the acquisition of knowledge, and others will point to its vital heuristic role.[1] This may be left for each investigator to decide for himself. I have argued here to the *feasibility* of systems metaphysics, and not to its *necessity* within the cognitive endeavor.

---

[1]Bohm suggests that "metaphysics is fundamental to every branch of science." Although it is not a "well-defined study, on top of which we erect a towering structure of physics, chemistry, biology, sociology, and so on," it is nevertheless "something that pervades every field," and "conditions each person's thinking in varied and subtle ways." Hence "what is needed is the conscious criticism of one's own metaphysics, leading to changes where appropriate and, ultimately, to the continual creation of new and different kinds." (Bohm, *op. cit.*) Waddington concurs; he argues that "a scientist's metaphysical beliefs are not mere epiphenomena, but have a definite and ascertainable influence on the work he produces." (C. H. Waddington, "The Practical Consequences of Metaphysical Beliefs on a Biologist's Work: an Auto-Biographical Note," *Towards a Theoretical Biology, op. cit.*)

SYSTEMS PHILOSOPHY: CONCLUSIONS

Much of the *malaise* on the contemporary scene comes from the rift between the sciences and the humanities. Whatever the many factors may be which enter into the determination of this rift, the majority of English and American contemporary philosophers contribute to it, even if only inadvertently. They examine the perennial questions of philosophy as if no empirical sciences existed—or at least as if their inquiry was prior to the theories of such sciences. Questions concerning knowledge, freedom, mind, obligation, value, etc., are pursued by these philosophers on the assumption that whatever evidence is available to a consistent and critical thinker in his daily life, is sufficient to decide—at least to discuss cogently—the issues. There is an almost exclusive reliance on everyday information and on the language in which this information is stated.

Both of these assumptions have come to be rejected in the sciences as insufficient, and reliance on them as obsolete. Commonsense knowledge, however critically examined and logically expounded, is regarded as but the layer of prejudice which, Einstein said, impedes scientific progress. And ordinary language is found entirely inadequate to symbolize the meaning of the concepts which constitute the scientific knowledge of our time. Thus it comes about that scientists look at philosophers with incomprehension mixed with awe and resulting in bewilderment; philosophers appear to be doing much of the time what scientists have rejected as naïve and transcended, yet they are the inheritors of the awesome wisdom of the ages and stand, presumably, on the shoulders of their gigantic predecessors. Philosophers in turn tend to consider scientific theories as based on assumptions which must first be examined, or to dismiss them as grasping but a limited sphere of reality, and giving partial or inadequate answers to the perennial questions to which they alone seem to have access. (Unfortunately, the access to these undoubtedly essential questions turns out to be via a logical scrutiny of "consciousness," or of everyday information.) As a result, contact between scientists and philosophers tends to be brief and mutually irritating.

I confess that I do not see a scintilla of evidence in anything produced by philosophers analyzing the contents of consciousness, commonsense, or ordinary language, which would justify their efforts to come up with theories that could be read as serious recommendations to believe what the objects of knowledge, the nature of mind, beauty, religious significance, or moral values, may be. To try to piece together a concept of the world, of man and of mind, and of the adventures of mindful man in the world, on the basis of information obtained by observing one's own mental processes and what oneself and others are doing, is like trying to reconstruct the principles of automotive engineering from the observation of what one thinks and does when he drives a car. Such an attempt is foredoomed to failure, especially when we consider that an automobile is simplicity itself compared to the dynamic, non-mechanistic processes of nature, man, and mind.

The contribution of philosophical analysis, whether linguistic, conceptual, or phenomenological, to our knowledge of the nature of man and the world, becomes real only when it is placed within the context of our scientific (i.e., imaginatively postulated and rigorously tested) knowledge of natural events. *Ignoring* such knowledge, linguistic and conceptual analysis becomes the analysis of insignificant concepts and propositions, and phenomenology the solipsistic burrowing in "worldly" and "nonworldly" phenomena and intentional-constitutive processes.[1] But *within* the context of a general knowledge of the fundamental principles of our sciences of man and nature, and sharing their "primary presuppositions,"[2] rigorous piecemeal analyses have as their valid and valuable task to lay bare the fine structure of cognitive and behavioral processes, and thereby to fill in the existing scientific framework.

When the above suggestion is put to the average analytical philosopher, his standard reply is that, whether or not the advocated general knowledge of principles is important and necessary, it cannot be acquired by a philosopher due to the complex technical nature of the scientific fields wherein it is gathered. Even a single science requires the better part of a lifetime to master, and philosophy itself is a complex technical field necessitating full-time study. Hence the expert professional philosopher cannot be expected to master fields other than his own.

The average analyst is wrong, however. He forgets that, as Happold said, synthesis "does not mean that a man should strive to know everything about everything. Such would be an impossible task. Nor is it necessary; for out of the coordination of the separate items of knowledge in each particular field of knowledge there emerge key ideas, general principles and

---

[1]Cf. the author's *Beyond Scepticism and Realism,* The Hague, 1966.

[2]I.e., that the world exists, and that it is open to rational inquiry (cf. Chapter 1, above).

unified notions, which an intelligent man, who has trained his mind to do so, can without difficulty understand and assimilate."[3]

The present work sought to demonstrate that an understanding of the general characteristics of the dynamic sets of events which constitute us and our world can be achieved in reference to simple universal postulates, exemplified throughout the range of terrestrial organizations—and possibly beyond. Nature, Whitehead said, may not come as clean as we can think it. But, we should add, we can only understand any of it if we can think nature itself cleanly. The human mind grasps repeating patterns against vaguer backgrounds of order wherever it turns and whatever its training. It is an instrument of finite knowledge, and the primary patterns upon which it focusses must be adequate. This is the key characteristic of paradigms. They must be interpretable for the full range of experience, for they can give consistent meaning to the "blooming, buzzing confusion" which surrounds us throughout our conscious existence.

The most consistent as well as most general paradigm available today to the inquiring, ordering mind is the systems paradigm. Explicated as a *general theory of systems,* both "physical" and "mental," and applied to the analysis of human experience and its problems, it constitutes *systems philosophy*. This, then, is the conceptual framework within which classical philosophical analyses of experience, knowledge, art, religion, valuation, freedom, morality, dignity, and many other facets of philosophical interest, can take place. For within this framework the anthropomorphic bias of commonsense experience is neutralized: the systems philosophical paradigm takes man as one species of concrete and actual system, embedded in encompassing natural hierarchies of likewise concrete and actual physical, biological, and social systems.

If systems philosophy is to become a working paradigm of contemporary thought, it must be pursued by a growing number of informed and perceptive investigators. Systems philosophy does not invalidate any scientific or philosophical question by branding it meaningless: it seeks to solve, not dissolve, problems. There is ample opportunity within its conceptual framework for collaborative work by empirical scientists and philosophers of the most diverse interests. By sharing a common general framework, theories advanced by investigators on diverse fields can become mutually relevant and fertile. Thus the present communications-gap in the disciplines may be overcome, without forcing individual insight into a preconceived mold. For not any one formulation of systems philosophy, but its general methodological and conceptual presuppositions, are its decisive constants. Specifi-

[3]F. C. Happold, "The Challenge of a Leap-Epoch in Human History," *Unity Through Diversity, op. cit.*

cations of the paradigm, and of any of its applications, can come and go; they can benefit from the interplay of many minds, with mutually relevant information. But the systems philosophical framework can remain: it can be held that (i) the understanding of the human being and the world around him is possible in reference to a hierarchy of dynamic systems, defined in terms of their organizational invariances of state and function, and that (ii) the concept of such a systems hierarchy is the optimum framework for the interpretation and integration of analyses of empirical matters of fact.

Incomplete, imperfect, and tentative as it is, the work presented on these pages is offered as evidence that this is not merely a program, but a feasible theoretical endeavor. When not only a single investigator, with his limited fund of information and insight, but many workers with different backgrounds and interests engage in doing systems philosophy, the errors of commission and omission of the present work will be overcome. Let this stand, therefore, not as a statement of systems philosophy, but as an introduction to it, and an encouragement to others, with superior or complementary knowledge, to offer their insights to specify the systems philosophy paradigm, and to extend it to the many areas of philosophic and scientific interest.

APPENDIX

SYSTEMS PHILOSOPHY AND THE CRISIS OF FRAGMENTATION IN EDUCATION

*Jere W. Clark**

An intellectual specter is haunting the world today. It is the specter of contradictions. Evidences of these contradictions are to be found in an interrelated network of educational, social, biological, and physical trends which are posing evermore serious threats to the lifeblood of human civilization.

Gyorgy Kepes has beautifully documented the view that these contradictions are frustrating intelligent, well-intentioned, hard-working—but ineffective—educators in their efforts to update our educational system. In his editorial introduction to *Structure in Art and Science,*[1] he describes the situation as a crisis in communications.

> Those working in cultural and scientific fields today recognize that there is a crisis in communication which is due to fragmentation of experience and the dispersion of knowledge into many self-contained disciplines, each with its own ever-growing, increasingly private language. Our sense of a cohesive world has been endangered by this communication crisis. Despite our richness of knowledge we are suffering cultural impoverishment; immense new vistas have been opened to us by science but we have failed to utilize our new technology fully or to share it wisely.

It is against the background of this specter of contradictions—this crisis in communications—that the significance of the broad conceptual framework outlined by Professor Laszlo in the present volume can best be interpreted and appreciated. A quick glance at a few of these frustrating

---

*Jere W. Clark is Professor of Economics and Director, Center for Interdisciplinary Creativity at Southern Connecticut State College. He also serves as Chairman of the Task Force on General Systems Education, Society for General Systems Research. This study is adapted, by permission, from two previous papers, "The Crisis of Fragmentation," *International Associations* (February, 1970), and "Quantum Leaps in Education," *Connecticut Industry,* Vol. 47, No. 5 (May, 1969).

[1]Vision & Values Series, New York, 1965 (jacket cover).

contradictions will set the stage for identifying and assessing the prospective impact of the present volume on the education system of the future.

* * *

As the volume of accumulated information grows at a geometric rate, our understanding of the key dimensions of wisdom seems to be fading further and further into the horizon. The confusion resulting from a superabundance of unassimilated information leads us to wonder if this explosion of information might possibly prove to be a curse rather than a blessing. As society becomes more complex and interdependent and requires ever greater degrees of cooperation, we see more groups threatening to throw "monkey wrenches" into the system. Finally, as our global society moves ever nearer to the point at which survival dictates that we become more open-minded and flexible, we are freezing up, becoming more rigid and inflexible—thereby creating even more anxiety about the future.

Fortunately, aroused citizens around the world are not content to stand idly and wait. Restless men and women of vision have been mobilizing for action for more than a decade. These groups seem to share the belief that to try an experimental change—even if it should fail—is preferable to standing, waiting idly for someone else to do something about a slowly deteriorating situation.

Unfortunately, most of these groups seem to be using an *obsolete compass* and an all but *irrelevant map* in their efforts to plot a new course for themselves and society. In a world of complexity and rapid change the *compass of absolute values*—even if new absolute values—is not likely to help a great deal. Such a compass tends to lead groups to strike at symptoms rather than causes.

Nor is a *static map* or blueprint likely to lead them to a better position in a dynamic, relative world of change. The map which is needed has more in common with the Copernican map of the universe than with that of Ptolemy. Furthermore, the use of the Ptolemy map in an intellectual world of Copernicus concentrates efforts on the relatively superficial, static mechanics of control as apart from the dynamic flows of information which make meaningful control possible.

Frustrated as are many of these groups, another large segment of our society may be even more frustrated. This group is made up of well-intentioned people—especially professional people—who are aware of the need for fairly drastic change and who are not aligning themselves with any of the action-oriented groups. They are simply saying, "What can I do?—it is the system that needs changing. I obviously can't change the system."

Still another group, or actually a group of groups, has been working conscientiously, patiently, and persistently to update the education system by developing and refining such practices as programmed instruction, computer-aided instruction, simulation, team teaching, sensitivity training, and rapid-reading instruction. Although these groups also feel "trapped" by their education system, most of them continue to confine their efforts to provincial maneuverings within the present system.

Valuable as are these innovations in education, they are being wasted to a great degree because of their failure to provide the needed kind of perspective. What the space-age student needs is a simple interdisciplinary model that will help determine what information he requires in virtually any situation, and serve as a framework for organizing and using that information. He must be viewed as an *intellectual navigator* rather than an *intellectual "squatter."* Time was when "squatter's rights" to a static field of knowledge in a relatively static, simple, highly specialized and pre-charted world were acceptable.

In other words, the emerging trends toward rapid change in a complex, interdependent, and relatively unknown world are rapidly changing the requirements of education. The present-day student must not only learn to navigate, but learn to do so on his own in the relatively strange, dynamic, interdependent, and complex intellectual, social, and physical worlds of his lifetime travels.

The failure of the approaches criticized earlier to provide the panoramic, dynamic perspective required stems largely from the fact that these methodologies are being used within an outmoded organizational framework of knowledge. The *real need* of the day is to *restructure knowledge* into simple, integrated, and flexible patterns which are broadly applicable to wide ranges of phenomena.[2]

This need is dramatized by Peter Drucker in *The Age of Discontinuity.*[3]

> The most probable assumption is that every single one of the old demarcations, disciplines, and faculties is going to become obsolete and a barrier to learning as well as to understanding. The fact that we are shifting from a Cartesian view of the universe, in which the accent has been on parts and elements, to a configuration view, with the emphasis on wholes and patterns, challenges every single dividing line between areas of study and knowledge.

---

[2]For an elaboration of this thesis, see Jere W. Clark and Anthony J. N. Judge, "Trans-disciplinary Aids to Education and Research," a Study Paper published in both English and French by the Union of International Associations, 1 Rue aux Laines, 1000 Brussels, Belgium.

[3]New York, 1968.

The following two statements from Jerome Bruner's book, *Toward a Theory of Instruction,* reinforce the same view:

> If there is any way of adjusting to change, it must include . . . the development of a metalanguage and "metaskills" for dealing with continuity in change.[4]

> The first response of educational systems under such acceleration is to produce technicians and engineers and scientists, but it is doubtful whether such a priority produces what is required to manage the enterprise. For no specific science or technology provides a metalanguage in terms of which to think about a society, its technology, its science and the constant changes that these undergo with innovation.[5]

In their timely introduction to the two-issue special, "Designing Curriculum in a Changing Society," R. W. Burns and G. D. Brooks also emphasized the need for updating our 19th Century patterns of knowledge.

> Basic curricular reform must deal with the structure of *content* first... The central problem is the curricular content, which is not much different today from what existed in the schools of the 1800's.[6]

Given the necessary kind of knowledge reorientation, educational curriculums can be developed which would enable the ineffective educational innovations of recent decades to take on new life. And *this is where Laszlo's proposal for systems philosophy comes in.* Systems philosophy can provide a conceptual basis for adapting the fundamentally outmoded parameter—the knowledge structure—of the educational system to the needs of the hour.

Thanks to this emerging systems philosophy, it now seems possible to develop curricula which will enable the student to learn simultaneously several skills, processes, and concepts, quicker, cheaper, and more meaningfully than he can learn any one alone.

The implications of this expectation—if it should be realized—are enormous. This means that the education process might be used throughout the entire world to help the undereducated peoples in the developing countries and in the ghettos of technically advanced countries to by-pass the traditional educational programs and move right into the educational opportunities of the space-age. These related developments in education

---

[4]Cambridge, Mass., 1966, p. 35.
[5]*Ibid.,* p. 33.
[6]*Educational Technology,* April, 1970. (Italics in original.)

and research should do much to help people generally to become masters —rather than victims—of the emerging technological systems which are shaping our society.

Having an explicit proposal as well as a general over-view of the field of systems philosophy in this volume—and a very readable one for intelligent laymen as well as for professionals in the field—is indeed fortunate. The further fact that this volume was written by one of the more prominent philosophers of our times is also valuable in helping to give to the general systems field added credibility while there may still be time for the masses of educators to gain an orientation to the field. In this context it should be pointed out that the scholarly and authoritative manner in which Laszlo points the way toward bridging the gap between the philosopher, the scientist, and the humanist, adds even more to the value of his pioneering book.

One of the most obvious payoffs coming from the integrated education patterns made possible by systems philosophy is the better-educated worker in industry, education, government, and the professions generally. The time seems to be approaching when the added flexibility, versatility, vision, and social concern will be indispensable to success in terms of job security as well as in terms of continuous professional growth and development.

A less obvious but more important payoff is likely to come in the form of renewed emphasis on the dignity and freedom of the individual. As our society becomes more dynamic, complex, and interdependent, the need for coordination becomes more crucial. Events of each month that passes seem to indicate more and more that the basic types of choices are narrowing down to two general categories.

The first and most obvious choice is to rely primarily on direct, explicit controls imposed by a central government with the idea that it can use modern technological means to coordinate the activities of people by prescribing more precisely what they can and cannot do in the interests of survival. In that case, the responsibility of the education system would tend to become that of conditioning people's minds to welcome being relieved of responsibility for making important decisions for themselves. Over time individuals would tend to become impersonal but safe cogs in a great machine.

The other choice—one for which businessmen have been pressing for decades—is to use the new education system with its supporting technology to help people generally learn to appreciate the higher values in life, especially individual freedom. The chief problem posed by this approach to coordinating society, however, is that it would continue to leave individual members free to throw their own monkey wrenches into the system. If we are to use this approach, we must begin a major movement to educate people generally to appreciate the nature and importance, not only of freedom, but also of the new technology and new modes of systems creativity making them possible.

This trend toward increasing destructive power of individual groups of persons began centuries ago. In the days of the horse and buggy, when most communities were basically self-sufficient, the refusal of men to work on the pony express or on a steamship line would not likely cripple the economy of the entire nation, for example. Today, however, if a few engineers working in electric power companies were to go off the job for a few days, especially in the hot weather of summer or the cold weather of winter, our entire national economy would soon be in jeopardy. Experience is showing, in other words, that as the number of specialties increases within a society, the greater is the degree of dependence of any given specialty on the proper functioning of other specialties. This tends to "lock people" into a belief that the simplest way for them to get a special advantage, such as a pay raise, is to threaten to sit down and make the whole society "sorry" it did not give in.

The growing frequency with which buildings are bombed and planes are hijacked is but an extension of the same trend toward total "negative power" for small groups. Improved communications networks also are extending the trend toward concerted action at the national and global levels for more and more interest groups.

Because this combination of trends is leading more and more toward a crisis of crises, it becomes all the more important that philosophers, such as Laszlo, not only work to integrate and simplify patterns of knowledge, but to make the fruits available to the average person.

If we are to continue to use civilized means of societal coordination as we move toward the "world system" (at the apex of Laszlo's mapping of the "terrestrial microhierarchy"), we must recognize that the welfare of our national society tends to be measured by the welfare of the "poorest," or least educated, or most victimized or deprived, persons. The analogy to a chain may have a lesson for us. We know that a chain is only as strong as the weakest link. The way to increase the strength of a chain is to identify the weaker links and work to strengthen them.

Now that thousands of people have refused to remain "hidden away from society" in ghettos, we may as a nation have to learn to listen to them. Doing so may require some basic changes in the social fabric or system. The point is that the breakdown of our democratic way of life seems to be coming unless we do make some basic changes one way or the other. The earlier this point is realized, the less drastic will likely be the change, and the greater will be the benefits to all of the resulting changes.

The same idea applies in the international realm, where the welfare of individual nations is the focus of attention.

At the local, state, national, international, and global social-system levels, the first need is to work to develop attitudes which are more favorable to the types of changes that are needed. One of the best ways to change attitudes is to work meaningfully with basic models in systems philosophy, and to provide mental exercises which open and stretch the mind. Such exercises are among the most powerful, most appealing, and most effective vehicles for attitudinal change.

Operation Man to Mankind[7] is conceived as a series of publications, conferences, and other projects designed to accelerate the transformation of provincial, self-centered social systems into the united organic world system discussed in this volume. Laszlo quotes the words of Glenn T. Seaborg, which inspired this project.[8] I shall set forth the quotation since it describes the need for the project beautifully.

> This [mankind as an organic whole] is more than a utopian dream. *It is the new imperative*—the goal toward which we must all move. Though they may each phrase it somewhat differently, this seems to be the consensus of most of the thoughtful philosophers, scientists and educators speaking, writing and thinking today. And there seem to be few if any viable alternatives to moving in this direction.[9]

In order to translate our claims for systems philosophy into the embryo of a global plan for educating mankind generally, we must, as this commentator pointed out elsewhere, "provide simply, inexpensively, imaginatively, and operationally the conceptual capability of converting our fragmented complex of domestic, defense, and space problems into an integrated pattern of global opportunities."[10]

The term, *eco-cybernetics,* which permeates the statement of strategy and procedural objectives, might be viewed here as a special case of the general *system-cybernetics* outlined by Laszlo. Briefly stated, eco-cybernetics is a socio-cybernetic process whereby ecosystems and other self-directed systems or organizations adapt to their environment and "solve their problems."

---

[7]Jere W. Clark, "*Operation Man to Mankind: A Proposal for a Long-Range, Global, Trans-disciplinary Futurology Program . . . ,*" unpublished manuscript, September 4, 1970.

[8]Chapter 5 (iii), "Social Systems," above.

[9]Glenn T. Seaborg, "Mankind as a Global Civilization," United Church of Christ *Journal* (May, 1968) (italics in original).

[10]Jere W. Clark, "Designing a Global Network of Centers for Universal Systems Education," for publication in a forthcoming issue of *International Associations.*

"Eco-cybernetics" consists of a common-sense, metaphorical blending of the following generalized elements of general *eco*logy, *eco*nomics, and *cybernetics*:

a. The broad scope and species-balance function of general ecology. (This might be called the "bird's eye view" of the "apple of the universe.")

b. The decision-making, priority-setting, allocation framework of economics generalized to embrace the use of means generally, to achieve designated ends generally, as effectively as possible. (This might be considered to be the "worm's eye view" of the same "apple.")

c. The information-handling operations highlighted in general cybernetics and related disciplines. (This might be viewed as the figurative "sap" flowing through the apple tree which makes both the apple, the worm, and the bird possible!)

The general approach to implementing this mission profile is centered around what Kenneth Boulding has called the "indispensable minimum of knowledge" required to enable a student to program his own mind in such a way that he will be able to re-program it whenever he needs and for whatever purposes might emerge during his entire lifetime. Three basic concepts guide our efforts to identify and apply this indispensable minimum of knowledge. One is the *economy-of-thought* function of science generally. Another is the concept of *heuristics* as advocated for decades by George Polya. The third is the idea dramatized by Buckminster Fuller for several decades, namely, *synergistics*. Let us now consider what each of these terms means in turn.

Economy

In his book, *The Science of Mechanics,* Ernst Mach stated near the turn of the century, "the purpose of science is economy of thought." In a similar vein, Laszlo states in Chapter 8 ("System: Framework for an Ontology"), "economy of thought is an ever present motivation of systematic theories, and systems philosophy is no exception."

Hence economy, as used here, is much more than the "bargain basement" approach to getting shopworn or otherwise inferior merchandise for a low price. In common-sense terms it means optimizing the fruits of a given amount of resources. Alternatively, it means to achieve a given result (goal) with a minimum of resources. Hence, *economizing* is the process of maximizing the benefits (economic, political, ethical, aesthetic, etc.) of resources or means of whatever variety (economic, social, psychic, aesthetic, etc.).

Alfred North Whitehead's description of a "sense of style" expresses the general idea quite elegantly. In *The Aims of Education*[11] he says:

> It is an aesthetic sense, based on admiration for the direct attainment of a foreseen end, simply and without waste. Style in art, style in literature, style in science, style in logic, style in practical execution have fundamentally the same aesthetic qualities, namely, attainment and restraint. . . .
>
> Style, in its finest sense, is the last acquirement of the educated mind; it is also the most useful. It pervades the whole being. The administrator with a sense for style hates waste, the engineer with a sense for style economizes his material, the artisan with a sense for style prefers good work. Style is the ultimate morality of the mind.

Heuristics

Heuristics are essentially principles for the intuitive exploration of problem situations with a view toward formulating simple, common-sense approaches to them. Heuristic probing is usually a prelude to, rather than a substitute for, rigorous, precise analysis.

Synergistics

Synergistics, as used here, is the process of combining and utilizing the component parts of anything in ingenious ways so as to produce an overall result or dividend that exceeds the sum of the parts. This excess or surplus might be called a bonus. In the economic realm, it is analogous to the entrepreneur's profits. The term, *synergistics,* has been stressed for decades by the architect-philosopher, Buckminster Fuller. By linking the simplest and most basic organizational components (common denominators) of the universe heuristically and cybernetically with a universal metalanguage within the context of a systems theoretical meta-model, we have a basis for identifying, developing, and utilizing what Boulding called the "indispensable minimum of knowledge." This approach seems to be one of the most effective ways to translate Ernst Mach's idea of economy of thought into reality.[12]

---

[11]New York, 1929.

[12]Strategies and procedures for attaining the long-range goal of global-system education are being developed at the Center for Interdisciplinary Creativity at Southern Connecticut State College with the assistance of several other organiza-

In closing this commentary we should like to draw on the words of Sir Julian Huxley: "we need a science of human possibilities, with professorships in the exploration of the future . . . [to integrate] science with all other branches of learning into a single comprehensive and open-ended system of knowledge, ideas and values relevant to man's destiny."[13] Although we obviously have a long way to go, we feel that in proportion to our ability to get developed and to utilize pioneering volumes such as the present, success should be forthcoming on a global level—hopefully before the "crisis of crises" develops.

---

tions and individuals. The more active organizations in the venture are the Task Force on General Systems Education (of the Society for General Systems Research), the Consortium on Systems Education (in New Haven), The Council for Unified Research and Education (in New York City), Total Education in the Total Environment (in Wilton, Connecticut), and Perception Systems, Inc. (in New York City). Also cooperating are The World Future Society, The Creative Education Foundation, The Center for Integrative Education, and The Institute for General Semantics.

[13]"The Crisis in Man's Destiny," *Contemporary Religious Issues,* Donald E. Hartsock, ed., pp. 178–179.

BIBLIOGRAPHY

RECENT WORKS ARTICULATING ASPECTS AND SOURCES OF THE SYSTEMS PHILOSOPHY PARADIGM

A listing of books and articles which, in the writer's opinion, have made special contribution to the emergence of the systems paradigm in contemporary science and philosophy.

Works marked with * indicate symposium volumes containing a number of significant papers by writers in addition to the editor(s).

Allport, Floyd H., *Theories of Perception and the Concept of Structure,* New York, 1955.

Allport, G. W., *Pattern and Growth in Personality,* New York, 1961.

——, *Becoming: Basic Considerations for a Psychology of Personality,* New Haven, 1955.

——, "The Open System in Personality Theory," *Journal of Abnormal and Social Psychology,* 1960.

Ando, Albert; Fisher, Franklin M., and Simon, Herbert A., *Essays on the Structure of Social Science Models,* Cambridge, Mass., 1963.

Angyal, Andras, "The Structure of Wholes," *Philosophy of Science* (1939).

Ashby, W. Ross, *An Introduction to Cybernetics,* New York, 1956.

——, *Design for a Brain,* New York, 1960.

——, "Principles of the Self-Organizing System," *Principles of Self-Organization,* Foerster and Zopf, eds., New York, 1962.

Bayliss, L. E., *Living Control Systems,* San Francisco, 1966.

Beckner, Morton, *The Biological Way of Thought,* New York, 1959.

Beer, Stafford, *Cybernetics and Management,* London, 1959.

*Bennis, Warren, *et al.,* eds., *The Planning of Change,* New York, 1962.

von Bertalanffy, Ludwig, *General System Theory,* New York, 1968.

——, "General Theory of Systems: Application to Psychology," *Social Science Information,* **VII,** 6 (1967).

——, *Problems of Life: An Evaluation of Modern Biological Thought,* New York, 1952.

——, *Robots, Men, and Minds,* New York, 1967.

——, "Theoretical Models in Biology and Psychology," *Theoretical Models and Personality Theory,* David Krech and George S. Klein, eds., Durham, North-Carolina, 1952.

*von Bertalanffy, Ludwig; Rapoport, Anatol, eds., *General Systems: Yearbook of the Society for General Systems Research,* 1956 and seq.

Bitterman, M. E., "The Evolution of Intelligence," *Scientific American,* **212** (1965).

Blauberg, I. V., Sadovsky, V. N. and Yudin, E. G., "Some Problems of General Systems Development," *Unity Through Diversity,* W. Gray and N. Rizzo, eds., New York (in press).

Blum, H. F., *Time's Arrow and Evolution,* Princeton, 1968 (3rd ed.).
Bohm, David, "Further Remarks on Order," *Towards a Theoretical Biology,* C. H. Waddington, ed., Chicago, 1970.
Boulding, Kenneth E., "General Systems Theory—The Skeleton of a Science," *Management Science* (1956).
——, "Some Questions on the Measurement and Evaluation of Organization," *Ethics and Bigness,* H. Cleveland and H. D. Laswell, eds., New York, 1962.
——, *The Image,* Ann Arbor, 1956.
——, *The Organizational Revolution,* New York, 1953.
Brams, Steven J., "Transaction Flows in the International System," *American Political Science Review* (1966).
Brillouin, Leon, "Life, Thermodynamics, and Cybernetics," *American Scientist,* 1949.
——, *Science and Information Theory,* New York, 1956.
——, "Thermodynamics and Information Theory," *American Scientist* (1950).
Bronson, Gordon, "The Hierarchical Organization of the Central Nervous System: Implications for Learning Processes and Critical Periods in Early Development," *Behavioral Science* (1965).
Bruner, Jerome, *Toward a Theory of Instruction,* Cambridge, Mass., 1966.
*Buckley, Walter, ed., *Modern Systems Research for the Behavioral Scientist,* Chicago, 1968.
——, *Sociology and Modern Systems Theory,* Englewood Cliffs, N. J., 1967.
Burbidge, Margaret, and Burbidge, Geoffrey, "Formation of Elements in the Stars," *Science,* **128**, No. 3321 (1958).
Burhoe, Ralph W., "Values Via Science," *Zygon,* **4**, 1 (March, 1969).
Burns, W.; Brachs, G. D., "Designing Curriculum in a Changing Society", *Educational Technology,* April, 1970.
Cadwallader, Mervin, "The Cybernetic Analysis of Change in Complex Social Organizations," *American Journal of Sociology,* 1959.
Campbell, Donald T., "Adaptive Behavior from Random Response," *Behavioral Science* (1956).
——, "Blind Variation and Selective Retention in Creative Thought as in Other Knowledge Processes," *Psychological Review* (1960).
——, "Common Fate, Similarity and Other Indices of the Status of Aggregates of Persons as Social Entities," *Behavioral Science* (1958).
——, "Variation and Selective Retention in Socio-Cultural Evolution," *Social Change in Developing Areas,* Herbert R. Barringer, George I. Blanksten, and Raymond Mack, eds., Cambridge, 1965.
Carroll, John, ed., *Language, Thought and Reality: Selected Writings of B. L. Whorf,* New York, 1956.
*Cartwright, D., ed., *Field Theory in Social Science,* New York, 1951.
*Cherry, Colin, ed., *Information Theory: Third London Symposium,* New York, 1956.
Chomsky, Noam, *Cartesian Linguistics,* New York, 1966.
——, *Syntactic Structures,* The Hague, 1957.
Clark, Jere W., "Designing a Global Network of Centers for Universal Systems Education," *International Associations* (in press).
——, "Operation Man to Mankind: A Proposal for a Long-Range, Global Transdisciplinary Futurology Program . . ." (Sept. 4, 1970). (MS).
——, "Quantum Leaps in Education," *Connecticut Industry* (May, 1969).

——, "The Crisis of Fragmentation," *International Associations* (Feb., 1970).
*Cleveland, H.; Laswell, H. D., ed., *Ethics and Bigness: Scientific, Academic, Religious, Political and Military,* New York, 1962.
Demerath, N. J., and Peterson, R. A., eds., *System, Change and Conflict,* New York/ London, 1967.
Deutsch, Karl W., *The Nerves of Government,* New York, 1963.
——, "On Social Communication and the Metropolis," *Daedalus,* **90.**
——, "Some Notes on Research on the Role of Models in the Natural and Social Sciences," *Synthèse,* **7** (1955).
Dobzhansky, Th., *The Biology of Ultimate Concern,* New York, 1967.
Easton, David, *A Systems Analysis of Political Life,* New York, 1965.
*——, ed., *Varieties of Political Theory,* Englewood Cliffs, N. J., 1966.
Eckland, B. K., "Genetics and Sociology: a Reconsideration," *American Sociological Review* (1967).
Eckman, Donald P., ed., *Systems: Research and Design,* New York, 1961.
Einstein, Albert, *Sidelights on Relativity,* New York, 1922.
——, *The World As I See It,* New York, 1934.
Eisenstadt, S. N., "Social Change, Differentiation and Evolution," *American Sociological Review* (1964).
Elsasser, W. M., *Atom and Organism, A New Approach to Theoretical Biology,* Princeton, 1966.
Errington, Paul L., *Of Predation and Life,* Ames, Iowa, 1967.
Etzioni, Amitai, *The Active Society: A Theory of Societal and Political Processes,* New York, 1968.
Feibleman, James, "Theory of Integrative Levels," *British Journal for the Philosophy of Science* (1954).
Firth, Raymond, *Elements of Social Organization,* London, 1956.
*Foerster, H. von, and Zopf, G. W., Jr., eds., *Principles of Self-Organization,* New York, 1962.
Frankl, Victor E., "Reductionism and Nihilism," *Beyond Reductionism: New Perspectives in the Life Sciences,* Arthur Koestler and J. R. Smythies, eds., London and New York, 1969.
Fuller, Buckminster, *Operating Manual for Spaceship Earth,* Carbondale, Ill., 1970.
Garner, W. R., *Uncertainty and Structure as Psychological Concepts,* New York, 1962.
George, F. H., "Cybernetics and Automation," *Cybernation and Society,* London, 1960.
——, *Cybernetics and Biology,* Edinburgh, 1965.
Gerard, R. W.; Kluckhohn, C., and Rapoport, A., "Biological and Cultural Evolution," *Behavioral Science,* **1** (1956).
——, "Hierarchy, Entitation, and Levels," *Hierarchical Structures,* Whyte, Wilson, and Wilson, eds., New York, 1969.
——, "Units and Concepts of Biology," *Science,* **125** (1957).
*Gorman, T., ed., *Freedom and the Human Sciences,* New York (in press).
Gouldner, Alvin W., "Reciprocity and Autonomy in Functional Theory," *Symposium on Sociological Theory,* Llewelyn Gross, ed., New York, 1962.
*Gotesky, Rubin; Laszlo, Ervin, eds., *Evolution and Revolution: Patterns of Development in Nature, Society, Man and Knowledge,* New York, 1971.
*Gray, W.; Duhl, F. D.; Rizzo, N., eds., *General Systems Theory and Psychiatry,* Boston, 1969.

*Gray, W.; Rizzo, N., eds., *Unity Through Diversity: Festschrift in Honor of Ludwig von Bertalanffy,* New York (in press) (2 Volumes.)
Grene, Marjorie, "Biology and the Problem of Levels of Reality," *New Scholasticism* (1967).
*Grinker, Roy R., ed., *Toward a Unified Theory of Human Behavior,* New York, 1956.
Grodin, F. S., *Control Theory and Biological Systems,* New York, 1963.
Hall, Arthur D., *A Methodology for Systems Engineering,* Princeton, New Jersey, 1962.
Hall, E. T., *The Hidden Dimension,* New York, 1966.
Hanson, N. R., *Patterns of Discovery,* Cambridge, 1958.
Hardin, Garrett, "The Cybernetics of Competition: A Biologist's View of Society," *Perspectives in Biology and Medicine,* 1963.
*Harris, Dale, ed., *The Concept of Development,* Minneapolis, 1957.
Harris, Errol E., *The Foundations of Metaphysics in Science,* London, 1965.
Harrison, E. R., "The Mystery of Structure in the Universe," *Hierarchical Structures,* Whyte, Wilson, and Wilson, eds., New York, 1969.
Hebb, D. O., *The Organization of Behavior,* New York, 1949.
Henry, Jules, "Homeostasis, Society, and Evolution: A Critique," *The Scientific Monthly* (1955).
Herbert, P. G., "Situation Dynamics and the Theory of Behavior Systems," *Behavioral Science* (1957).
Huxley, Julian, *Evolution in Action,* New York, 1953.
*Jeffries, L. A., ed., *Cerebral Mechanisms in Behavior,* The Hixon Symposium, New York, 1951.
*Jones, Marshall R., ed., *Nebraska Symposium on Motivation,* Lincoln, 1960.
*Kalmus, H., ed., *Regulation and Control in Living Systems,* New York, 1966.
Kaplan, David, "The Superorganic: Science or Metaphysics?" *American Anthropologist* (1965).
Katchalsky, A., and Curran, P. F., *Nonequilibrium Thermodynamics in Biophysics,* Cambridge, Massachusetts, 1965.
Khailov, K. M., "The Problem of Systematic Organization in Theoretical Biology," *General Systems,* 1964.
*Kilpatrick, F. P. ed., *Explorations in Transactional Psychology,* New York, 1961.
Kluckhohn, Clyde, "Values and Value-Orientations in the Theory of Action," *Toward a General Theory of Action,* T. Parsons, ed., Cambridge, 1951.
——, "Universal Categories of Culture," *Anthropology Today,* A. L. Kroeber, ed., Chicago, 1953.
Knorr, K.; Verba, S., eds., *The International System: Theoretical Essays,* Princeton, 1961.
*Koch, Sigmund, ed., *Psychology: A Study of a Science,* Vol. 2, New York, 1959.
Koestler, Arthur, *The Act of Creation,* New York, 1964.
——, *The Ghost in the Machine,* New York, 1967.
*Koestler, Arthur; Smythies, J. R., *Beyond Reductionism: New Perspectives in the Life Sciences,* London and New York, 1969.
Kohlberg, L., "Early Education: A Cognitive-Developmental View," *Child Development,* 1968.
*Krech, David, and Klein, George S., eds., *Theoretical Models and Personality Theory,* Durham, North Carolina, 1952.
——, "Dynamic Systems as Open Neurological Systems," *Psychological Review* (1950).

Kremyanskiy, V. I., "Certain Peculiarities of Organisms as a 'System' from the Point of View of Physics, Cybernetics, and Biology," *General Systems,* **5** (1960).
Kroeber, A., *Configurations of Culture Growth,* Berkeley and Los Angeles, 1944.
Kuhn, Alfred, *The Study of Society: A Unified Approach,* Homewood, Illinois, 1963.
——, "Toward a Uniform Language of Information and Knowledge," *Synthèse,* **13** (1961).
Kuhn, T. S., *The Structure of Scientific Revolutions,* Chicago, 1962 (new ed. 1970).
Lange, Oskar, *Wholes and Parts: A General Theory of System Behavior,* New York, 1965.
Lashley, K. S., "Coalescence of Neurology and Psychology," *Proc. American Phil. Soc.,* 1941.
Laszlo, Ervin, *Beyond Scepticism and Realism,* The Hague, 1966.
——, "Cognition, Communication and Value," *Communication: The Ethical and Moral Issues,* Lee Thayer, ed., New York (in press).
——, *Essential Society,* The Hague, 1963.
——, "Integrative Principles of Art and Science," *Integrative Principles of Modern Thought,* H. Margenau, ed., New York, 1971.
——, "Range of Functions of an Unstable Subsystem," *Scientia,* **LXIV** (1970).
——, *System, Structure, and Experience,* New York, 1969.
——, *The Systems View of the World,* New York (in press).
*——; Stulman, J., eds., *Emergent Man,* New York (in press).
*——; Wilbur, J. B., eds., *Human Values and Natural Science,* New York, 1970.
*——; Wilbur, J. B., eds., *Human Values and the Mind of Man,* New York, 1971.
Le Chatelier, H., "Recherches experimentales et théoriques sur les equilibres chimiques," *Annales des Mines,* 8eme Serie, Paris, 1888.
Lee, Dorothy, *Freedom and Culture,* New York, 1959.
*Lerner, Daniel S., ed., *Parts and Wholes,* New York, 1963.
Lévi-Strauss, Claude, *The Savage Mind,* Chicago, 1966.
——, *Structural Anthropology,* New York, 1963.
Levins, Richard, "The Limits of Complexity," *Biological Hierarchies,* H. H. Pattie, ed. (in press).
Lewin, Kurt, *Field Theory in Social Science: Selected Theoretical Papers,* New York, 1951.
*Litterer, J. A., ed., *Organizations: Structure and Behavior* (Vol. I), New York, 1963.
Lotka, A., *Elements of Mathematical Biology* (1925), New York, 1957.
Maccia, Elizabeth Steiner, and Maccia, George S., *Development of Educational Theory Derived from Three Educational Theory Models,* Columbus, 1966.
Mace, C. A., "Homeostasis, Needs, and Values," *British Journal of Psychology,* **44** (1953).
MacKay, Donald M., *Freedom of Action in a Mechanistic Universe,* Cambridge, 1967.
——, "Mindlike Behaviour in Artefacts," *British Journal of Philosophy of Science,* **2** (1951).
——, "Towards an Information-Flow Model of Human Behaviour," *British Journal of Psychology,* **47** (1956).
*Margenau, Henry, ed., *Integrative Principles of Modern Thought,* New York, 1971.
——, *The Nature of Physical Reality,* New York, 1950.
Maruyama, M., "Metaorganization of Information," *Cybernetics* **4** (Reprinted in *General Systems Yearbook,* 1965).

——, "The Second Cybernetics: Deviation-Amplifying Mutual Causal Processes," *American Scientist,* (1963.)
*Maslow, A. H., ed., *Motivation and Personality,* New York, 1954 (new ed., 1970).
*——, ed., *New Knowledge in Human Values,* New York, 1959.
——, "A Theory of Metamotivation: The Biological Rooting of the Value-Life," *Journal of Humanistic Psychology* (1967).
——, *The Psychology of Science,* New York, 1966.
——, "The Fusion of Facts and Values," *American Journal of Psychoanalysis,* **23** (1963).
Mather, Kirtley F., "The Emergence of Values in Geologic Life Development," *Zygon,* **4**, 1 (1969).
——, *The Permissive Universe,* New York, 1971 (in press).
Matson, Floyd, *The Broken Image,* New York, 1964.
*Maziarz, E. A., ed., *Evolution of Man's Values,* New York (in press).
McCulloch, Warren S., *Embodiments of Mind,* Cambridge, 1965.
McHale, John, *The Future of the Future,* New York, 1969.
Menninger, K.; Mayman, M., and Pruyser, P., *The Vital Balance: The Life Process in Mental Health and Illness,* New York, 1963.
*Mesarovic, Mihajlo D., ed., *Views on General Systems Theory: Proceedings of the Second Systems Symposium at Case Institute of Technology,* New York, 1964.
Miller, George A.; Galanter, E.; Pribram, K. H., *Plans and the Structure of Behavior,* New York, 1960.
Miller, James G., "Living Systems: Basic Concepts," *General Systems Theory and Psychiatry,* William Gray, D. F. Duhl, and N. Rizzo, eds., Boston, 1969.
——, "Toward a General Theory for the Behavioral Sciences," *American Psychologist* (1955).
Modelski, George, "Agraria and Industria: Two Models of the International System," *The International System,* Klaus Knorr and Sidney Verba, eds., Princeton, 1961.
Monod, Jacques, *Le hasard et la nécessité. Essai sur la philosophie naturelle de la biologie moderne,* Paris, 1970.
Morgan, Lloyd, *Emergent Evolution,* London, 1923.
Mowrer, O. H., "Ego Psychology, Cybernetics and Learning Theory," Adams, *et al., Learning Theory and Clinical Research,* New York, 1954.
——, *Learning Theory and the Symbolic Processes,* New York, 1960.
Murphy, Gardner, "Toward a Field Theory of Communication," *Journal of Communication* (1961).
von Neumann, John, *The Computer and the Brain,* New Haven, 1958.
——, *Theory of Self-Reproducing Automata,* Urbana, 1966.
——; Morgenstern, O., *Theory of Games and Economic Behavior,* Princeton, 1947.
Northrop, F. S. C., "Introduction," in Werner Heisenberg, *Physics and Philosophy,* New York, 1958.
——, *The Logic of the Sciences and the Humanities,* New York, 1947.
——, "The Problem of Integrating Knowledge and the Method for its Solution," *The Nature of Concepts, Their Interrelation and Role in Social Structure,* Stillwater, Oklahoma, 1950.
——, "Toward a General Theory of the Arts," *The Journal of Value Inquiry,* **I** (1967).
Oparin, A. I., *The Origin of Life,* London, 1957.
Ostow, Mortimer, "The Entropy Concept and Psychic Function," *American Scientist* (1951).

Parsons, T., *Structure and Process in Modern Societies,* Glencoe, Ill., 1960.
——, *Essays in Sociological Theory Pure and Applied,* Glencoe, Ill., 1949.
——, *The Social System,* New York, 1957.
*——, and E. A. Shils, eds., *Toward a General Theory of Action,* Cambridge, Mass., 1951.
*——, Shils, E. A.; Naegele, K. D.; Pitts, T. R., eds., *Theories of Society,* New York, 1961.
Penfield, W., "The Permanent Record of the Stream of Consciousness," *Proceedings Inter. Congr. Psychology,* Montreal (June, 1954).
Pepper, Stephen C., *Sources of Value,* Berkeley, 1958.
——, "Survival Value," *Human Values and Natural Science,* Ervin Laszlo and J. B. Wilbur, eds., New York, 1970.
Phillips, J. C., Jr., *The Origins of Intellect: Piaget's Theory,* San Francisco, 1969.
Piaget, J., *Biologie et connaissance,* Paris, 1967. In English, *Biology and Knowledge,* Chicago and London, 1971.
——, *Genetic Epistemology,* New York, 1970.
——, *Les mécanismes perceptifs,* P.U.F., Paris.
——, *Le Structuralisme,* Paris, 1968.
——, *The Psychology of Intelligence,* London, 1950.
Pitts, W., and McCulloch, W. S., "How We Know Universals, The Perception of Auditory and Visual Forms," *Bulletin of Mathematical Biophysics* (1947).
Planck, Max, *Where is Science Going?* London, 1933.
Polanyi, M., "Life's Irreducible Structure," *Science* **160** (1968).
Popper, Karl R., *The Logic of Scientific Discovery,* New York, 1959.
Portmann, A., *Animal Forms and Patterns,* New York, 1967.
——, *Animals As Social Beings,* New York, 1961.
Powers, W. J.; Clark, R. K., and McFarland, R. I., "A General Feedback Theory of Human Behavior," *Perceptual and Motor Skills* (1960).
Prigogine, T., *Etude Thermodynamique des Phénomenes Irreversibles,* Paris, 1947.
Pringle, J. W. S., "On the Parallel Between Learning and Evolution," *Behavior,* **3** (1951).
Puligandla, R., "The Concepts of Evolution and Revolution," *Evolution and Revolution: Patterns of Development in Nature, Society, Man and Knowledge,* R. Gotesky and E. Laszlo, eds., New York, 1971.
*Quastler, H., ed., *Information Theory in Biology,* Urbana, 1953.
*——, ed., *Information Theory in Psychology,* Glencoe, Ill., 1955.
Rapoport, Anatol, "Mathematical Aspects of General Systems Analysis," *General Systems Yearbook,* 1966.
Raymond, Richard C., "Communication, Entropy, and Life," *American Scientist,* (1950).
Rice, S. A., *Quantitative Methods in Politics,* New York, 1928.
Rosen, Robert, *Optimality Principles in Biology,* New York, 1967.
——, "A Survey of Dynamical Descriptions of System Activity," *Unity Through Diversity: Festschrift in Honor of Ludwig von Bertalanffy,* Gray and Rizzo, eds., New York, 1971.
*Rosenau, James N., ed., *International Politics and Foreign Policy: A Reader in Research and Theory,* rev. ed., New York, 1969.
*——, ed., *Introduction to Linkage Politics,* New York, 1969.
Rosenblueth, Arturo, *Mind and Brain: a Philosophy of Science,* Cambridge, Mass. 1970.
——, and Wiener, N., "Purposeful and Non-Purposeful Behavior," *Philosophy of Science* (1950).

*Royce, J. R., ed., *Psychology and The Symbol: An Interdisciplinary Symposium,* New York, 1965.
Sachsee, Hans, *Die Erkenntnis des Lebendigen,* Braunschweig, 1968.
*Scher, Jordan, ed., *Theories of the Mind,* New York, 1962.
Schrödinger, Erwin, *What Is Life?,* Cambridge, 1945.
Seaborg, Glenn T., "Mankind as a Global Civilization," United Church of Christ *Journal* (May, 1968).
Sears, Paul B., "Utopia and the Living Landscape," *Daedelus,* **94,** 2 (1965).
Selye, Hans, *The Stress of Life,* Toronto, 1956.
Shands, Harley C., *The War With Words: Structure and Transcendence,* New York, 1971.
Shannon, Claude E., and Weaver, Warren, *The Mathematical Theory of Communication,* Urbana, 1949.
Shapley, Harlow, *Of Stars and Men,* Boston, 1958.
——, *The View from a Distant Star: Man's Future in the Universe,* New York, 1963.
Sherrington, C. S., *Man On His Nature,* New York, 1949.
Simon, Herbert A., "The Architecture of Complexity," *Proceedings of the American Philosophical Society,* **106** (1962).
——, *Models of Man,* New York, 1957.
——, *The Sciences of the Artificial,* Cambridge, Mass., 1969.
Simpson, G. G., *Principles of Animal Taxonomy,* New York, 1961.
Sinnott, Edmund W., *The Biology of the Spirit,* New York, 1955.
——, *The Problem of Organic Form,* New Haven, 1963.
Slobodkin, Lawrence B., *Growth and Regulation of Animal Populations,* New York, 1961.
Smith, Cyril Stanley, "Structure, Substructure, Superstructure," *Reviews of Modern Physics* (1964).
Sommerhoff, G., *Analytical Biology,* London, 1950.
Sorokin, P. A., *Social and Cultural Dynamics,* New York, 1937.
——, *Sociological Theories of Today,* New York, 1966.
Spencer, H., *Principles of Sociology,* New York, 1897.
Stanley-Jones, D., and Stanley-Jones, K., *The Kybernetics of Natural Systems: A Study in Patterns of Control,* New York, 1960.
Taylor, Alastair M., "Evolution-Revolution and General Systems Theory," *Evolution and Revolution: Patterns of Development in Nature, Society, Man and Knowledge,* R. Gotesky and E. Laszlo, eds., New York, 1971.
Taylor, James G., *The Behavioral Basis of Perception,* New Haven, 1962.
Teilhard de Chardin, P., *The Phenomenon of Man,* New York, 1959.
*Thayer, Lee, ed., *Communication: General Semantics Perspectives,* New York, 1970.
——, "Communication—*Sine qua Non* of the Behavioral Sciences," *Vistas in Science,* D. L. Arm, ed., Albuquerque, 1968.
——, ed., *Communication: The Ethical and Moral Issues,* New York, (in press).
Thorpe, W. H., *Learning and Instinct in Animals,* London, 1963.
Tinbergen, N., *Social Behavior in Animals,* London, 1953.
Toch, Hans H., and Hastorf, Albert H., "Homeostasis in Psychology," *Psychiatry* **18** (1955).
*Tou, Julius T., and Wilcox, R. H., eds., *Computer and Information Sciences: Collected Papers on Learning, Adaptation, and Control in Information Systems,* Washington, D.C., 1964.
Toulmin, Stephen, and Goodfield, June, *The Architecture of Matter,* New York, 1962.

Vogt, Evon Z., "On the Concepts of Structure and Process in Cultural Anthropology," *American Anthropologist,* **62** (1960).
Waddington, C. H., ed., *Towards a Theoretical Biology,* Chicago, 1970.
Walter, W. Grey, *The Living Brain,* London, 1953.
*Washburne, Norman, ed., *Decisions, Values, and Groups,* Vol. II, New York, 1962.
Watt, Kenneth E. T., *Systems Analysis in Ecology,* New York, 1966.
Weiss, Paul, *Dynamics of Development: Experiments and Inferences,* New York, 1968.
——, "Experience and Experiment in Biology," *Science,* **136** (1962).
——, "From Cell to Molecule," *The Molecular Control of Cellular Activity,* J. M. Allen, ed., New York, 1962.
——, ed., *Hierarchically Organized Systems in Theory and Practice,* New York, 1971.
——, "One Plus One Does Not Equal Two," *The Neurosciences,* Querton, Melnechuk, Schmitt, eds., New York, 1967.
Werner, Heinz, *Comparative Psychology of Mental Development,* New York, 1957.
——, "The Concept of Development from a Comparative and Organistic Point of View," *The Concept of Development,* D. Harris, ed., Minneapolis, 1957.
Weyl, Hermann, *Symmetry,* Princeton, 1952.
Whitehead, A. N., *Process and Reality,* New York, 1929.
——, *Science and the Modern World,* New York, 1925.
——, *Symbolism, Its Meaning and Effect,* New York, 1927.
——, *The Aims of Education,* New York, 1929.
——, *The Concept of Nature,* Cambridge, 1920.
*Whyte, L. L.; Wilson, A. G.; Wilson, D., ed., *Hierarchical Structures,* New York, 1969.
Whyte, Lancelot Law, *Internal Factors in Evolution,* New York, 1965.
——, *The Next Development in Man,* New York, 1950.
——, *Unitary Principles in Physics and Biology,* New York, 1949.
Wiener, Norbert, *Cybernetics,* 2nd ed., Cambridge, Mass., 1961.
——, *The Human Use of Human Beings: Cybernetics and Society,* Garden City, New York, 1954.
*——, and Schadé, J. P., eds., *Cybernetics of the Nervous System,* New York, 1965.
Woodger, J. H., *Biology and Language,* Cambridge, 1952.
——, *Biological Principles,* New York, 1966.
*Yovits, M. C.; Jacobi, G. T., and Goldstein, C. D., eds., *Self-Organizing Systems,* Washington D.C., 1962.

INDEX OF NAMES

INDEX OF SUBJECTS